SPACE POLICY FOR THE TWENTY-FIRST CENTURY

SPACE POLICY
FOR THE
TWENTY-FIRST CENTURY

Wendy N. Whitman Cobb
and Derrick V. Frazier

UNIVERSITY OF FLORIDA PRESS

Gainesville

Copyright 2024 by Wendy N. Whitman Cobb and Derrick V. Frazier
All rights reserved
Published in the United States of America

29 28 27 26 25 24   6 5 4 3 2 1

A record of cataloging-in-publication data is available from the Library of Congress.
ISBN 978-1-68340-406-4 (cloth)
ISBN 978-1-68340-412-5 (pdf)
ISBN 978-1-68340-427-9 (ebook)

University of Florida Press
2046 NE Waldo Road
Suite 2100
Gainesville, FL 32609
http://upress.ufl.edu

For Gus (Wendy)

For Sara (Derrick)

CONTENTS

List of Figures   ix

List of Tables   xi

Preface   xiii

List of Acronyms   xvii

1. Introduction   1
2. The Space Race during the Cold War: The Quest for Prestige and Security   23
3. Post–Cold War Space History   48
4. Presidents and Space Policy   70
5. Congress and Space Policy   93
6. NASA   116
7. Military Space and the Intelligence Community   140
8. The "Other" Actors   160
9. Nongovernmental Forces   179
10. Civilian Space Policy   201
11. International Politics and National Security Space Policy   222
12. Commercial Space Policy   246
13. Major Issues in US Space Policy   271

Notes   291

Bibliography   299

Index   313

FIGURES

1.1. NASA's budget in millions, 1959–2022 (constant 2021 dollars)   9
4.1. Presidential statements about space, 1957–2020   79
4.2. Presidential statements by type of space policy, 1957–2020   81
5.1. Congressional hearings by chamber on space, 1970–2020   103
5.2. Frequency of military/national security space hearings, 1970–2020   104
8.1. Space economy output in millions of current US dollars, 2012–2019   165
8.2. NASA cost and schedule overruns, 2013–2022   175
9.1. General Social Survey attitudes on space exploration spending, 1973–2018   182
9.2. General Social Survey interest in space, 2008–2018   182
9.3. Technology spectrum   194

TABLES

1.1. Types of space policy   13
2.1. Major space race milestones   25
11.1. Theoretical perspectives: actors, preferences, outcomes, and schools of thought   236

PREFACE

An appreciation for the importance of space has been a long time coming. While the launch of Sputnik in 1957 opened the space age and inaugurated a new dimension of the simmering Cold War, for the most part, it became apparent over the ensuing decades as something kind of interesting but not necessarily as important to someone's everyday life. Space wasn't as important to the average American as getting kids to school on time, making the rent payment for the month, or dealing with expensive health care. And for a while, that was true—space didn't help wake the kids up (or assist in their education!), space systems didn't underlie the entire global financial system enabling rent payments to be made, and its spin-offs didn't help make health care better.

The world has changed—slowly, quietly—but it has nonetheless changed. Space has become such an indispensable part of our everyday lives that it's probably easier to identify the things that space isn't involved in. But the irony of this importance is the subtle nature of its ubiquity. Space for most people is unnoticed, underappreciated, and little understood. Most of us take for granted the precise timing and location services that GPS satellites provide that wake us up on time and get us to the right place when we need to be there. We take for granted how communications satellites allow us to conduct business almost instantaneously around the globe. We take for granted how these same systems enabled millions of children to go to school or their parents to work from home during the COVID-19 pandemic. We take for granted the technologies that we now have because of investments in space technology. And yet, if we suddenly lost those things, we would *literally* (not to mention figuratively) be lost.

While this importance largely has escaped the notice of the public because of how seamlessly space has been integrated into our everyday lives, policymakers, elected officials, and others are now beginning to take seriously the importance of space and the systems we have placed there. The result has been greater interest in protecting those assets, finding cheaper and easier ways of putting them there, and trying to better understand the dynamics, political and otherwise, that underlie all of these activities. Because there is no doubt that space exploration and exploitation is a *political* activity. Sputnik and the space age it started were motivated by politics, a situation that continues today. It is only because the importance of space has grown so slowly and so quietly over time that it has escaped larger notice until now.

This greater attention has led to a marked increase in interest from academics. For a long time, not many people studied space in the social sciences and humanities. There were books recounting space history and political scientists here and there who understood space's importance and tried to study it more systematically. It was rare, however, for a course such as space policy to be offered in colleges and universities. One of this book's authors, despite having the opportunity to take a space policy course as an undergraduate, was later actively discouraged from pursuing the study further because it had little career potential. Thankfully, this advice was not taken and years of trying to better understand the politics of space are now resulting in the book you're holding.

As in the political world, things also have changed in academia. More people are studying space including the development of communications technology, orbital mechanics and engineering, international space law, space security and strategy, and, yes, space policy. This increased interest in educating students on space policy has led to an increase in space policy courses across the country, which this book aims to support. Our hope is that this book is useful not only for those already teaching the course but encourages others to develop a class knowing that new resources are available for them to use. But for the students, our hope is that this book enables you to think critically about how space policy is made, who's making it, and more importantly, what is motivating them to craft it. Space policy courses should not just be a recitation of the different approaches the United States and other states have taken, but a critical analysis of why, which leads to a better understanding of how policy is made today and what may come tomorrow.

This book would not have been possible without the support of several groups and individuals. First, the idea for the book arose from a presentation one of us gave at the 2021 Summer Workshop on Teaching Space (SWOTS) on teaching US space policy. In grappling with what that presentation should look like, Wendy presented the outline of the book that you see here. The enthusiastic support of the attendees was vital to encouraging this project to move forward. Second, the Department of Political Science at the US Air Force Academy graciously allowed Wendy to test run the book in fall 2022, teaching a space policy course to a group of enthusiastic cadets. A hearty thank-you goes to them, the future decision-makers and leaders, who were game to read something that no one else had before and provide their honest feedback.

The book is better because of you. Finally, we appreciate the thoughtful and comprehensive comments this book's reviewers provided, which helped to sharpen and improve this book tremendously.

Personally, Wendy would like to thank several people who have unknowingly aided this book's creation: Roger Handberg at the University of Central Florida, an early space scholar and longtime mentor who has unfailing continued to support my work; Larry Dodd who provided the foundational education in political institutions that has contributed so much to my study of space; my husband, Josh, who continues to support me in all that I do; and the rest of my family who have watched me on TV, read my articles and books, and gone with me to watch launches. A massive debt of gratitude is also owed to Derrick, without whom this book could not have been done. On Wendy's part, the book is dedicated to her dog, Gus (named after the astronaut Gus Grissom), who passed as this book was coming to fruition.

For Derrick, the better part of the last decade has been remarkable. I have been a part of so many space endeavors, from teaching space to the next generation of Air and Space Force strategists, to being a part of important discussions on the futures of space defense, space law, and commercial space. All of these experiences have undoubtedly made the book better. Experiences alone are not enough, however. I would like to think all of my space students at SAASS over the years, particularly my dissertation-level folks like Brad Townsend, Allen Rotter, and Amber Dawson who forced me to think more deeply about some areas of space that I might not otherwise have. Others, like Raj Agrawal were always good for insightful discussions and opening doors for me to peek behind. Kristi Lowenthal, my space teaching partner for several years did similarly. And, while we sometimes take them for granted, my thanks to Wendy who has been a great coauthor. Finally, to my family, Biljana, Sarah, Patrick, and even our dog Charlie, my deepest thanks. Without your love and support none of this would be possible. Thanks for letting me work! Of course, to many more I owe thanks. No one achieves the kind of success I've been lucky to experience without a long list. If I left you out, I promise to say thanks when our paths cross again!

Regardless of the many people who have helped shape and influence this book along the way, the remaining errors are our own. Additionally, the views presented here are not representative of the US Air Force, Department of Defense, or the US government.

ACRONYMS

ABM – Antiballistic missile
ABMA – Army Ballistic Missile Agency
APSCO – Asia Pacific Space Cooperation Organization
ARM – Asteroid redirect mission
ARPA – Advanced Research Projects Agency
ASAT – Anti-satellite weapon
ASD/SP – Assistant secretary of defense for space policy
AST – Office of Commercial Space Transportation
AVC – Bureau of Arms Control, Verification, and Compliance
BEA – Bureau of Economic Analysis
BIS – Bureau of Industry and Security
CASIC – China Aerospace Science and Industry Corporation
CBO – Congressional Budget Office
CCL – Commerce Control List
CIA – Central Intelligence Agency
Comsat – Communications Satellite Corporation
COTS – Commercial Orbital Transportation Services
DARPA – Defense Advanced Research Projects Agency
DDTC – Directorate of Defense Trade Controls
DIA – Defense Intelligence Agency
DNI – Director of national intelligence
DOD – Department of Defense
DOE – Department of Energy
DOS – Department of State
DOT – Department of Transportation
DSCOVR – Deep Space Climate Observatory
DSP – Defense Support Program
EAR – Export Administration Regulations

EMP – Electromagnetic pulse
EO – Executive order
EOSAT – Earth Observation Satellite Company
ERTS – Earth Resources Technology Satellite
ESA – European Space Agency
ESSA – Environmental Science Services Administration
EU – European Union
EUCOM – European Command
FAA – Federal Aviation Administration
FBC – Faster, better, cheaper
FCC – Federal Communications Commission
FOBS – Fractional orbital bombardment system
GAO – Government Accountability Office
GEO – Geostationary orbit
GEOINT – Geospatial intelligence
GOES – Geostationary Operational Environmental Satellite
GPS – Global Positioning System
GSS – General Social Survey
HLS – Human landing system
IC – Intelligence community
ICBM – Intercontinental ballistic missile
IGO – Intergovernmental organization
IGY – International Geophysical Year
INDOPACOM – Indo-Pacific Command
Intelsat – International Telecommunications Satellite Consortium
ISAM – In-space servicing, assembly, and manufacturing
ISR – Intelligence, surveillance, and reconnaissance
ISRO – Indian Space Research Organization
ISS – International Space Station
ITAR – International Traffic in Arms Regulations
ITU – International Telecommunications Union
JAXA – Japanese Aerospace Exploration Agency
JCS – Joint Chiefs of Staff
JPL – Jet Propulsion Laboratory
JPSS – Joint Polar Satellite System

LEO – Low Earth orbit
MEO – Medium Earth orbit
MIDAS – Missile Defense Alarm System
MILAMOS – Manual on International Law Applicable to Military Activities in Outer Space
MODS – Military Orbital Development System
MOL – Military Orbital Laboratory
NACA – National Advisory Committee for Aeronautics
NASA – National Aeronautics and Space Administration
NASC – National Aeronautics and Space Council
NAVSTAR – Navigation System with Timing and Ranging
NDAA – National Defense Authorization Act
NDS – National Defense Strategy
NEAR – Near Earth Asteroid Rendezvous
NEO – Near Earth object
NGA – National Geospatial-Intelligence Agency
NIST – National Institute of Standards and Technology
NMS – National Military Strategy
NNSA – National Nuclear Security Administration
NOAA – National Oceanographic and Aeronautical Administration
NPR – Nuclear Posture Review
NTIA – National Telecommunications and Information Administration
NRO – National Reconnaissance Office
NSA – National Security Agency
NSC – National Security Council
NSpC – National Space Council
NSS – National Security Strategy
OES – Bureau of Oceans and International Environment
OIRA – Office of Information and Regulatory Affairs
OMB – Office of Management and Budget
OSD – Office of the Secretary of Defense
OST – Outer Space Treaty
OSTP – Office of Science and Technology Policy
OTA – Office of Technology Assessment
OTE – Organize, train, and equip

PE – Punctuated equilibrium
PGM – Precision-guided munition
PNT – Position, navigation, and timing
POES – Polar Operational Environmental Satellites
PPD – Presidential policy directive
PPWT – Treaty on the Prevention of Placement of Weapons in Outer Space
SALT – Strategic Arms Limitation Talks
SBIRS – Space-Based Infrared System
SCIF – Sensitive compartmented information facility
SDI – Strategic Defense Initiative
SEI – Space Exploration Initiative
SIG-Space – Senior Interagency Group on Space
SLS – Space Launch System
SMDC – Space and Missile Defense Command
SOE – State-owned enterprise
SOPs – Standard operating procedures
SPD – Space policy directive
SSA – Space situational awareness
SSTO – Single-stage-to-orbit
STEM – Science, technology, engineering, and math
STM – Space traffic management
STS – Space transportation system
TIROS – Television Infrared Observation Satellite
TTP – Tactics, techniques, and procedures
ULA – United Launch Alliance
UN – United Nations
UNCOPUOS – United Nations Committee on the Peaceful Uses of Outer Space
USGS – United States Geological Survey
USML – United States Munitions List
USSF – United States Space Force
USSPACECOM – United States Space Command
USSR – Union of Soviet Socialist Republics
USSTRATCOM – US United States Strategic Command
VSE – Vision for Space Exploration
WMD – Weapon of mass destruction

1

Introduction

One can imagine the sense of excitement and adventure felt by many children watching Neil Armstrong's Moon landing in 1969. Their imaginations, undoubtedly, ran high. Today it was the Moon; by the time they were adults, the human race would have colonized Mars and would likely be exploring beyond the inner solar system. At the time, it's likely many adults shared similar thoughts with their children. When their efforts were put to good use, what could the Americans and maybe even the Russians do when it came to exploring the heavens? The possibilities were endless!

More than five decades later, we are far from such lofty images of human space exploration; yet space is integral to most aspects of our daily lives in ways that would have been difficult to foresee in 1969. The path to space has not been a linear one, however, nor has it been evenly disbursed across civil, commercial, and military sectors. Nonetheless, the path has been rapidly moving forward in ways that require critical attention for anyone interested in understanding the future of the space domain.

In this book we contribute to this understanding by examining the components of US space policy. We aim to provide a sense for what is important about space for the United States as well as why. As you'll see, the answers to these questions primarily depend on the outcomes of interactions among political actors, including those within and outside the US government. In studying these interactions, we will also observe how space policy is formed. Knowing what is important about space, why it is important, and how this is determined provides students of space policy a better idea of what to expect from the United States when it comes to its behavior in the space domain. Ultimately, an understanding of these behaviors and resultant outcomes will help inform how the space community engages with some of the more difficult problems associated with space, a domain that has been typically thought of as a part of the global commons, where no country alone can decide its future fate, yet every country depends on its effective management.

The Importance of Space

Since the 1950s and the opening of the space domain, space has played a vitally important role in everything from international relations and military activities to how we communicate and move around the world. Space-based assets are absolutely vital in many everyday activities from getting gas or money from an ATM, to monitoring hurricanes and wildfires and tracking economic activity. Space systems underpin entire segments of the global economy and yet they are incredibly fragile, vulnerable to the smallest piece of debris. And as our public and military dependence grows, space systems become increasingly attractive targets to potential adversaries who would seek to disrupt not just military capabilities and actions but everyday civilian activities.

While most people have generally taken the US dependence on space for granted, the space domain is changing rapidly. In the latter half of the 20th century, space was generally seen as a sanctuary, while used for military purposes in many cases, it was not a place where active combat would take place. As such, the notion of space as a sanctuary allowed countries and companies to feel relatively secure in placing satellites in orbit and sending humans into space. However, several developments over the past 20 years have disrupted the notion of space as a sanctuary. In the words of the 2011 National Security Space Strategy, space has become congested, contested, and competitive.[1] In the first instance, the difficulties and cost of getting to space have historically limited how many people have been able to utilize it. In the past 15 years, however, technological advancements in terms of reusable rocket technology and increasing use of off-the-shelf technology in cubesats and nanosats have greatly increased the ability of people, companies, and countries to use space. This increased usage has resulted in a concomitant increase in the number of satellites in Earth's orbit along with dangerous space debris. While the idea of "space" might make it seem as if it is an unlimited resource, near-Earth space is rather scarce. The growing number of actors and space-based systems is leading to a situation where space is significantly congested. This congestion in turn creates greater difficulty in operating safely in the domain.

Because a growing number of actors can access space, it is also becoming more competitive. Since the Cold War, the United States and the Soviet Union (now Russia) have competed in the space domain to not only benefit their military activities, but to demonstrate technological superiority to the rest of the world. Today, other countries have entered the domain to do precisely the same. Additionally, companies are competing in the space domain to provide commercial services such as remote sensing, mapping, and monitoring. As the cost of reaching space has come down, more and more companies are seeking out economic opportunities, including the potential for space-based solar energy and mineral mining on the Moon or nearby asteroids. Therefore, space is competitive not just at the state level, but on a very real commercial level.

Finally, given how space is becoming more congested and competitive, along with the inherent vulnerability of satellites, space is more contested than it has ever been. With very limited international law and custom to guide activities in outer space, states are stepping up their efforts to not only protect and defend their assets, but to demonstrate superiority if not outright control over the space domain. These efforts are demonstrated by moves in the past several years by the United States to stand up an independent military branch dedicated to the mission of protecting and defending US assets, the US Space Force (USSF). Other countries have similarly followed suit including France, Germany, and Australia. Countries including China, the US, and Russia have all stepped up their military activities in space, with some declaring that space is now weaponized. This heightened state of threat in space and the consequences it could have for all other space activities has in turn garnered greater political and public attention.

If all we knew was how dependent on space the public is, space policy would be worthy of study in its own right. However, the significant dependence on space coupled with the increasing threats presented in the space domain make space policy, and the study of it, even more important. How does the United States decide what to do in space? How are different activities prioritized? How does space policy relate to military activities, the economy, and the pursuit of scientific exploration? How does the nature of the space environment and the unique difficulties that it presents impact space policy? What agencies and organizations are involved in carrying out space policy? And perhaps most importantly, *who* is making these decisions and *why*? These questions, and others, are all intrinsically important in understanding the state of space today, how we got here, and where we're going.

This book is an attempt at not just answering these questions but understanding and exploring them. Although there are discussions of space history, this book is not about regurgitating major space milestones or recounting specific policy decisions. Instead, this book draws from the disciplines of political science and public policy to guide us in an understanding of space policy dynamics, recognizing that multiple institutions and actors are involved and that many decisions are interrelated and have consequences for other policy areas. Importantly as well for understanding what this book is and is not, we are most directly interested in American space policy. While the actions of other countries greatly influence US decisions (and vice versa), how other countries determine their own space policy is not in our purview.

Because this book draws on the literature of political science to provide a framework for understanding US space policy, we now turn to examine various approaches and definitions that you will encounter in later chapters before laying out some of the space-specific terminology and ideas that you will also find in the remainder of the text.

Approaches to Public Policy

Political scientists (and others) study public policy in many ways. Some look for general patterns in how policies are created, adopted, and enacted. This includes approaches that identify stages of the policymaking process and some basic ideas about policy change, including incrementalism and punctuated equilibrium (PE). Others, seeking out the unique characteristics of individual policy areas, try to understand the more specific nuances that might lead to disruptions in these normal patterns; for instance, John Kingdon's policy streams model proposes that policy change comes about when three "streams" converge: the recognition of a policy problem, the existence of a solution, and the political opportunity to bring about change.[2] The study of public policy is about acknowledging both of these ideas—that there are some general patterns to how policy is made and carried out but that the uniqueness of some policy areas, like space, present particular challenges that result in differences from what we might typically expect. While the rest of this book will focus specifically on space policy, it is also important to discuss some of the more general concepts of public policy upon which this later examination is predicated. To be sure, this is not an exhaustive summary or exploration of the ways in which scholars analyze public policy, something that is out of this project's scope. However, understanding concepts such as the stages of the policymaking process, patterns of policy change, and policy actors is helpful in understanding the nature of American space policy.

Stages of the Policy Process

As anyone with a familiarity of American politics probably recognizes, policymaking in the US is complicated, difficult, and often confusing. While policymaking often resists a strict framework, in studying public policies like space, it is helpful to try to identify a rough set of stages through which policy is made even if these stages may overlap or remain incomplete. One framework that is often used to assess policy has five stages: agenda setting, policy formulation, adoption, implementation, and evaluation. Agenda setting is the process through which political leaders and policymakers become aware of policy problems that need solving. This happens through crisis or the work of individual policy entrepreneurs advocating on behalf of a policy. When trying to understand different policies, understanding how policymakers become aware of a problem can be important in understanding how they might react to it.

At the same time, policy entrepreneurs or others may be advocating for a particular approach that can be used to correct the perceived policy problem. What policies are formulated and how is influenced by the people involved in their creation, the institutions in which they are being considered, and how the individuals involved define the problem or problems associated with them.[3] In examining how space policy is made, we can look at who is coming up with any given policy—is it the president, is it Congress, is it the National Aeronautics and Space Administration (NASA), or is it someone else? All these actors have different motivations (discussed in later chapters) that are likely to influence their approach to space. Therefore, having a sense of how policies are formulated and by whom can help explain policy choice.

One variable that may impact the writing of policy, however, is its chances of being adopted. Thinking about policy adoption makes us think about who is doing the adopting (for example, is it Congress via legislation or NASA through different approaches or regulations) and the various policies involved in making those decisions. Many policies may be floated, written, and considered, but not all of them become official policy or law. This fact emphasizes just how difficult it can be to adopt policy in the US for various reasons: political, partisan, cost, time, etc. Further, choices often must be made between different types of policy—even though some members of Congress may wish to spend more on space exploration, when space is considered against other policy priorities like defense or social spending, its importance often wanes. Policy adoption then is about analyzing these choices and the ways in which they are made.

Adopting a particular policy is only half the battle—once political leaders agree to adopt an approach, it must then be implemented. While this seems like a relatively simple thing to do, it is anything but easy. Legislation and other policy statements are often phrased in generalities, leaving specific meanings and tactics to be worked out later. It falls to the implementing agencies to make hard decisions about what those generalities mean and to put them into practice as best they can. This is a particular problem for space where members of Congress or others involved in policy formulation may not have much expertise in the technology of space exploration or exploitation. NASA, the Department of Defense (DOD), and others must make choices to put policies into action, and those choices can be just as difficult as deciding what policy approach to adopt in the first place.

The last stage in the policy process is one that is often ignored—evaluation. Once a policy has been adopted and implemented, we can assess how well it is performing in addition to how well it addresses the original policy problem. Ideally, evaluations lead to a new cycle of policy development wherein the problems with the policy as written are identified and changed in a cycle of continuous improvement. In reality, evaluations are often hampered, not taken seriously, or do not lead to significant policy change. Problems that were once thought of as serious enough to warrant a place on the political agenda are seen as fixed until the next crisis arises, and leaders are forced to consider the problem anew.

As we examine space policy throughout this book, you will find that looking at space through the lens of policy stages is helpful in explaining outcomes. Different policy actors have motivations that can lead to them supporting and advocating for particular policy approaches. The ways in which institutions like NASA view themselves also influence what they believe to be the "correct" approach to space policy. The difficulty of adopting policy in general and space policy in particular often leads to compromises and shortchanged budgets that significantly impact how policies are adopted and implemented. Finally, evaluation is often ignored, with a series of policy crises and challenges that largely drive space policy, forcing the issue back onto the political agenda.

Policy Change

Another lens through which we can understand and explain public policy is through larger patterns of policy change. Given that policy is difficult to create, adopt, and implement, it's not surprising that it may not change often. This approach to policy is called incrementalism: policies tend to change only slightly when looked at year to year, particularly in a budgetary sense.[4] Because political leaders must consider so many different policy problems on a regular basis, if a given area is not experiencing significant difficulties, it makes sense to make small changes annually. It also takes time to implement new policies so if they change in a major way year after year, these changes can lead to major problems in implementing policy and assessing its performance.

Incrementalism helps to explain space exploration in the United States. Figure 1.1 shows NASA's budget in constant dollars. Aside from the large increase it experienced in the 1960s, the amount of money NASA has received has been fairly stable with only slight changes in most years. However, there are also other smaller jumps in NASA's budget, including the mid-1980s and early 2000s. When we consider these comparatively large increases, using incrementalism to understand policy change no longer suffices. An alternative means of thinking about these changes, then, includes one concept known as PE. First introduced by Frank Baumgartner and Bryan Jones, the theory proposes that, in general, policies are relatively stable and establish what they term a policy monopoly, or a shared understanding among political actors of both the policy problem and the policy solution. When these monopolies are in place, it is natural to see little variation year over year. However, when events lead to the policy monopoly being challenged or upset, significant change can occur.[5] Importantly, though change *can* occur, it does not mean that change *will* occur. In looking at NASA's budget again, the budget bump that comes in 1986 and 1987 is a result of the space shuttle *Challenger* accident. However, while NASA's budget received an increase to support the construction of a replacement orbiter, the major approach to human spaceflight and space exploration did not change.

Thinking about what conditions lead to policy change in space is important given how unique space is. The technology required to get to space is expensive and difficult. It often takes many years to implement policies and observe changes. Because we do not necessarily "see" what's happening in space, these types of issues may be ignored or taken for granted. Further, when compared to other policy areas, neither the public nor many political leaders rank space as a leading problem that must be dealt with immediately. This can contribute to incremental approaches that may lead to a major crisis if a policy fails in the future.

Types of Policies

Another way of understanding public policy is by breaking the broad term "policy" into more specific types. Depending on the typology imposed, scholars may see more specific differences in how policies are considered or the ways in which they change over time.[6] Two policy distinctions are important for our consideration of space policy: foreign vs. domestic and military vs. civilian.

Foreign policy is a type of policy that deals with how a country interacts with states, international organizations, and other entities located outside of its own sovereign territory. Foreign policy, while certainly a more specific categorization, still involves a variety of topic areas and can include such things as aid provided to other governments or nongovernmental organizations, climate change, international laws, treaties, and agreements, the deployment of military forces, arms control, tax policy, and imports and exports. Domestic policy, on the other hand, includes policy that is meant to apply and affect only those within a state's sovereign territory. It is similarly broad and many issues from social spending and welfare to education to roads and bridges and voting rights can fall under its umbrella. While developments and approaches to domestic policy may impact a country's foreign policy or vice versa, the major difference between these two categories is the target population, within a country or outside of it.

Political scientists who have studied differences in domestic and foreign policy recognize that while the two interact, there are considerable differences in how politicians and the public approach them. Foreign policy can often be more difficult for the public to understand and form opinions on, leading to a greater disengagement with these issues in comparison to domestic policies. This disengagement is further reinforced by the fact that domestic policies do indeed have a greater impact on a citizen than many foreign policies do. While the public may have certain general views on foreign policy, they often are influenced by elite opinion on the matter.[7] Because domestic issues are likely at the top of voters' minds, domestic policy tends to be more susceptible to changes in public attitudes as their elected representatives may be unwilling to go against public opinion. However, with the increase in media and social media, some scholars suggest that the public is indeed becoming more informed on foreign policy. Even when they're not, however, foreign and domestic policies often interact. For instance, the negotiation of free trade agreements such as the USMCA or its predecessor NAFTA can have consequences for domestic employment and the presence of business and job opportunities.

Space policy, while appearing to be largely a domestic policy issue, is one that straddles these lines. The formation and execution of US space policy is strongly influenced by foreign countries, the US's current approach to foreign policy in general, and the larger geopolitical context. The 1960s space race is an excellent example of this combination of influences. The Apollo program was formulated and adopted as a response to Soviet actions in space. Similarly, while the United States works with several states including Russia on the International Space Station today, its decision to not cooperate with China in space is based significantly on concerns about China's international behavior and more aggressive actions in space. This relationship also works in the reverse—American actions in space (domestic space policy) has the capability of influencing foreign policy and international relations. A major by-product of the Apollo program in the 1960s was increased foreign support for the United States along with a recognition of the US as a technological leader in space by the end of the 1960s. Today, US military actions regard-

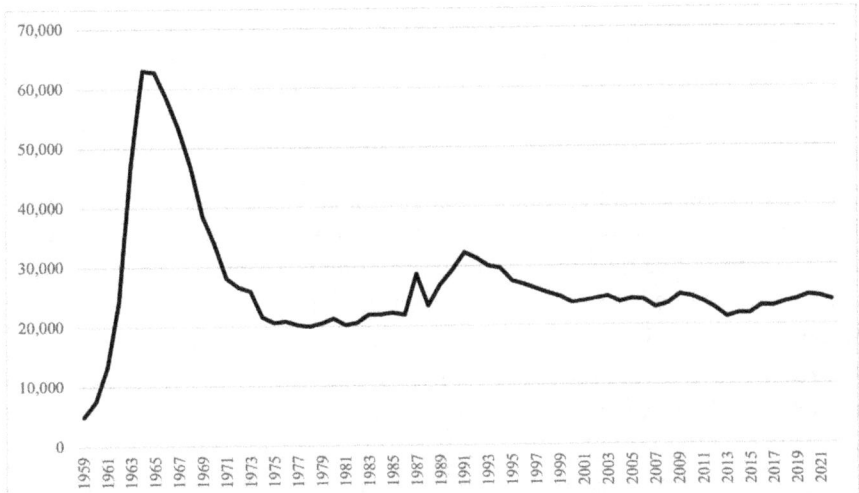

Figure 1.1. NASA's budget in millions, 1959–2022 (constant 2021 dollars). Figure created by the authors with data from The Planetary Society.

ing space can cause changes in how other countries, specifically Russia and China, may respond.

Another distinction in policy types is military versus civilian policy. While both may be considered domestic policy (like space, military policy also straddles the foreign/domestic divide) but in this case, civilian refers to nonmilitary matters. Because military policy relates so closely to national security, there is often greater support for a strong military in terms of its people and presence. Similarly, military policy is seen as a mandatory budget issue meaning that it must be funded, often at a particular level, each year. On the other hand, civilian policies are relatively more negotiable or discretionary and are more likely to change year to year.

Much like the distinction between domestic and foreign policy, the dividing line between military and civilian is similarly fuzzy. Aspects of military policy such as basing decisions have impacts on civilians in the surrounding areas and their prospects for economic success. Health care policy and approaches to wellness may in turn influence how many people may be able to serve in the military in a time of crisis. Choices to fund major military projects or weapons systems have inherently civilian and domestic impacts on the places where they will be carried out. Given this, it is natural to wonder what the use or purpose of such a distinction is. Obviously, the rhetorical argument of whether something is military or civilian may influence who is likely to support a given policy, but it also says something about the acceptability of the policy or action depending on who is carrying it out. This aspect in particular has become quite important for space policy.

As will be discussed further below, much of the technology used in space and for space exploration is considered dual use—it can be used for both military and civilian purposes. If the military is developing a certain rocket technology with plans to deploy it (military policy), then other actors may see that as a threat or as representing a particular danger. On the other hand, if a civilian agency like NASA does the same thing but for the purposes of space exploration, the activity is likely to be seen as having a greater sense of legitimacy. Similarly, the public may not support space exploration for the sake of research or scientific achievement, but if those same activities may be tied to military or national security, the public may be more supportive. Even though the same fundamental technology is being used in each case, its purpose, to serve either military or civilian policy, is important in assessing reaction to it and its possible consequences at both the domestic and foreign levels.

To be sure, the types of policies and models discussed here are meant to serve as a very brief introduction to some of the ideas and frameworks you will encounter in this book. While space is a unique and nuanced policy area, these concepts are helpful in trying to broadly understand and even predict the course of space policy in the United States. While there is no absolute right way to think about or approach space policy, these concepts are meant to serve as a helpful guide as you consider the topic.

Space-Specific Terminology and Ideas

Just as understanding more general approaches to public policy can assist us in understanding US space policy, there are ideas and concepts unique to space that are equally important. While understanding space policy may seem daunting given how scientific and technical it can be, this section is meant to explain in plain language some of the major and more difficult aspects such as types of orbits, launches, and the types of rocket technologies that we use. Before delving into that, however, this section begins with a brief discussion of the types of *space* policies.

Types of Space Policies

While "space policy" seems more specific and targeted than a larger category such as domestic or civilian policy, it is still fairly generic. Before examining space policy specifically, it is necessary to define what policy is. Politics is often defined as "who gets what, when, and how"—it is those decisions, the what, when, and how, that constitutes public policy. Sometimes policies can be quite broad (for example, advance American interests in space) with several different specific activities constituting *how* the policy is implemented (human spaceflight, planetary exploration programs, etc.). In this book, we will use the terms "space policy" and "space activities" interchangeably—even if a specific action like the Artemis program to return the US to the Moon is part of a larger policy, decisions on what to do and how to do it still constitute space policy in the United States. Space policy encompasses several different activities and policies including military, civilian, scientific, and commercial. While they are all related (and not just in the sense that they all have to do with space), there are unique aspects to each that must be considered. The ways in which we can organize these types of policies is by the actor undertaking them as well as by the ultimate purpose as summarized in table 1.1.

Military space concerns those space activities that the military undertakes for its purposes. This category can and does contain several activities that militaries find useful: navigation, remote sensing and intelligence gathering, communications, and missile warning are a few examples. There is an important distinction to make here, however, in terms of military space activities and the presence of weapons—because space has been utilized for military purposes, it is accurate to say that space has been *militarized*. On the other hand, it may not necessarily be *weaponized* in the sense that weapons are in orbit or have been used from orbit. While some countries, including the United States, have claimed that countries like Russia and China have weaponized space, there is no open-source data to indicate that—at present—weapons are in space. Given the secrecy often required in such activities and prohibitions on weapons of mass destruction in space, it is not surprising that if weapons were indeed placed in space, countries would like to keep those actions secret.

Civilian space policy, while utilizing similar space systems as the military, can be defined as space activities that a government undertakes for nonmilitary purposes. Examples of this can include remote sensing systems that use cameras, radars, and other types of sensing equipment to look back on Earth from orbit. In the military sphere, these systems can be used for locating targets or intelligence purposes, but the civilian realm also finds them helpful for Earth monitoring, weather forecasting, environmental tracking, and emergency management among other purposes.

Commercial space policy is characterized as space activities undertaken by private actors for private economic gain. While some of the earliest commercial space activities included various forms of communication and Earth monitoring, the commercial space sector has ballooned in recent decades with private companies now involved in everything from providing launch services, space tourism, low-latency internet services, satellite radio and television, and even to economic and environmental monitoring. The size of the space economy has grown significantly in recent years reaching an estimated worth of $447 billion in 2020 with commercial space representing 80 percent of that figure.[8]

Commercial space policy poses a difficult question for countries like the United States because of who is held accountable for actions in space. The 1967 Outer Space Treaty (OST), one of the only pieces of international law to deal with the space domain, holds states responsible for their own behavior in space along with the behavior of private entities in space who fall under their control. In other words, the US government is held responsible for activities that companies like Planet, SpaceX, and Blue Origin carry out in space. Because of this responsibility, governments have an intrinsic interest in regulating the behavior of commercial space companies both to ensure the public's safety as well as adherence to international norms of behavior. These needs must also be balanced with the economic benefits that these companies accrue and the services that they in turn provide to government customers. It cannot be forgotten that in the United States, private companies, not the government, provide launch services. Similarly, the US government may have an incentive to allow remote sensing companies to develop and deploy highly sensitive instruments so that they can in turn purchase the service.

As the importance and value of the commercial space sector grows, it is important to consider this a separate area of space activity but one that has significant overlap with both military and civilian space policy. Because of this interrelationship, the US government has often had to balance the needs of one area against the others. To take but one example, data from remote sensing satellites can be quite sensitive if being deployed for military purposes—any given country might not want their adversary to have the same intelligence that they do. Thus, a country like the US may move to restrict what private companies with remote sensing systems can launch or who they can provide their services to. At the same time, if too many restrictions are put in place, they may hamper civilian or commercial space policy goals. Companies may not be able to compete in the global market or civilian authorities may lack information needed to carry out their own jobs. Similarly, the US may have an interest in which countries are buying space technologies like rockets and satellites and restrict potential customer bases for private companies. The need to balance competing priorities has often marked US space policy with no easy answer as to how to accomplish it.

Table 1.1. Types of space policy

Category	Actor	Purpose
Military	Government	Military
Civilian	Government	Nonmilitary
Commercial	Private	Economic gain
Scientific	Private or public	Scientific exploration

A final area of space policy is scientific, which either private or government actors carry out with the ultimate purpose of scientific exploration. Activities that fall under this heading include planetary exploration like rovers, space telescopes, and physical experiments in space. With costs for space launch and satellites coming down over time, more private actors can use space for scientific exploration and experiments. However, the bulk of scientific efforts in space have taken place under the guidance of civilian government agencies like NASA or the European Space Agency (ESA) because of the significant cost usually associated with them. This has often hampered space science because, when weighed against other space activities, science often has a lower priority. This lower prioritization is often the result of competing government needs in more expensive space areas such as human spaceflight and military space.

Like the other areas of space policy, there are places in which space science policy interacts with the other areas of space policy. One such example is human spaceflight. When we consider it based on our classification scheme for space policy, it has been typically carried out by civilian government agencies for different purposes including both civilian and scientific. One of the most significant debates about human spaceflight is whether it serves any scientific purpose at all. As early as the 1950s, scientists argued that space science can be better carried out *without* the intervention and involvement of humans. Robotic or automated probes can be built at a lower cost and without the safety and environmental mechanisms needed to keep humans alive in space, and yet still carry out the same scientific functions. History, however, has shown that humans can and do perform valuable scientific experiments on orbit. At the same time, there is also overlap in terms of civilian and military policy as competition between countries in human spaceflight, which has often had geopolitical and military ramifications. During the Cold War space race, the building of rockets to launch humans to space was used as a proxy to demonstrate military power and might. Both countries also considered launching humans to orbit for a military purpose, with the Soviet Union actually doing so under its Almaz program. Even basic scientific programs have often been seen through a military lens, with some commentators claiming that China's recent probe sent to the far side of the Moon demonstrates a significant military capability that comes under the guise of peaceful scientific activity.[9]

Given the commonalities among civilian space activities, some may argue that the distinction between policy areas is not worth making. While this is true to an extent, as noted previously, different kinds of activities may be more or less acceptable depending on their ultimate purpose. Building ever more powerful rockets to send humans deeper into space appears to be more acceptable an activity than using the same rockets to launch advanced weaponry. Building advanced optical technology for use as a spy satellite may seem like a better use of government money than building it for a space-based telescope. Allowing commercial companies to sell their services to foreign countries may only be acceptable depending on who it is being sold to. Thus, in the United States, the government has often had to balance the needs of different types of space policy in ways that advanced the national interest, whatever that may be.

Getting to Space

No matter the type of space activity we discuss, they all share a common need to actually get *to* outer space. Access to space has historically been the most complicated and most expensive part of space exploration because of the energy required to lift mass out of Earth's gravity well and into orbit. Although technically the weakest physical force, Earth's gravity is so strong that, in order to break it, launch vehicles must reach a stunning 25,039 miles per hour to break free and leave Earth's orbit entirely. If a vehicle only wants to reach orbit and not leave Earth completely, the orbital velocity needed is still over 17,000 miles per hour.

These speeds might not be so difficult to reach if the mass you wanted to send to space was small, but typically, this has not been the case. As the mass of a payload increases, the amount of fuel needed also increases. The fuel adds to the mass of the payload, requiring even more fuel. This is especially true as a rocket first lifts off and is experiencing both its heaviest weight and its lowest altitude. As a rocket travels higher, the force of gravity is reduced as well as its weight. To take advantage of the more favorable conditions as a rocket climbs, many modern launch systems come in stages—as one part of the rocket exhausts its fuel, its engines shut off and are ejected from the rest of the craft allowing a new set of engines and fuel to take over, reducing weight and increasing its ability to get to orbit.

Theoretically, any power source might do to build a rocket as long as it fulfills Newton's third law of conservation of momentum: for every action there is an equal and opposite reaction. If an engine creates a force downward, it can push its mass upward. To create this downward thrust, rocket engines combine a fuel with an oxidizer. When the two are combined and the engine is "lit," they produce a force that is harnessed through engine nozzles downward to launch the rocket. The earliest rocket engines utilized solid fuels, which have the fuel and oxidizer packed together into a cylinder. This solid fuel is easy to transport and work with but once lit, the controlled explosion must continue until it has burned itself out. Newer rockets utilize liquid fuels, which feature the chemicals in liquid form. Far harder to work with, liquid fuels are more powerful and the force of the reaction can be controlled. Some types of launch vehicles like the space shuttle utilize both: the shuttle featured two solid rocket boosters and three main engines on the shuttle, which utilized liquid fuel provided by the external tank on liftoff.

The "tyranny of the rocket equation" helps to explain why launch systems must be so large, but it only partially explains why they are so expensive. Because rockets have had to travel so high and so fast, the ability to recover and reuse rockets has been limited. In the 1960s, all US rockets were expendable meaning that they were used once and then either burned up as they reentered Earth's atmosphere or were left to sink to the ocean floor. The space shuttle was partially reusable—the solid rocket boosters could be recovered and reused, but the external tank was not. The orbiter was largely reusable, although it did have to go through a significant recovery process after each launch, often involving the replacement of hundreds of intricate tiles on its belly to protect it from the heat of reentry. This long turnaround time and cost significantly limited how often the shuttle could be launched.

Only recently have private companies developed rockets that can be reused. In 2015, SpaceX successfully recovered the first stage of its Falcon 9 rocket by landing it on a drone ship at sea following a launch. Since then, they have demonstrated that the recovered booster can be reused successfully with one booster having been flown 10 times. Blue Origin has also developed similar technology allowing its New Shepard suborbital rocket to land itself remotely after expending its rocket fuel. (Suborbital launches are those that do not go high enough and fast enough to reach orbit or break Earth's gravity but do go high enough to technically reach space.) These reusable technologies mean that rockets no longer have to be thrown away after each launch, significantly reducing the cost of getting to orbit.

In Space

Getting to orbit is only half the battle—once there, you must go to the place that can best serve your purpose. It might be strange to think of different places in space, as it inherently has few landmarks and is endlessly open. However, in the space immediately surrounding Earth, we can identify particular areas that are more or less useful depending on your purposes. To begin, low Earth orbit (LEO) is 180 to 2,000 kilometers in altitude. Here, you will find Earth monitoring satellites, scientific satellites, weather satellites, and the International Space Station along with most human space missions. These satellites move quickly, orbiting the Earth roughly every 90 minutes. Medium Earth orbit (MEO), is 2,000 to 35,780 kilometers. These satellites move slower and thus tend to linger over certain locations meaning that they can provide longer coverage over Earth. The American GPS system is placed in this altitude. Finally, geostationary orbit (GEO) begins at 35,780 and goes out to the Moon. Satellites traveling at this altitude actually orbit the Earth every 24 hours making their location appear stable over a point on Earth. This type of orbit was first suggested by the science fiction author Arthur C. Clarke in the 1940s to provide consistent global communications.

The choice of an orbit for a satellite depends on its purpose and the state of the technology being used. The farther away the satellite, the better and more sensitive the technology needs to be. For example, a remote sensing satellite at GEO needs a stronger camera to monitor a location on Earth compared to LEO. However, at LEO, the satellite might be moving too fast to get a good picture. In GEO, fewer satellites are needed to provide global coverage (at a minimum, three), but at MEO and LEO, more satellites are needed. Space operators must also consider the presence of the Van Allen radiation belts, which are areas of high radiation around Earth ranging from 640 to 58,000 kilometers in altitude. While some satellites may be able to operate there, it is often undesirable because of the need for additional shielding from radiation.

The vast majority of satellites in orbit currently are in LEO with this number growing. As of summer 2022, there are more than 5,400 satellites with 4,700 in LEO.[10] In the coming years, various companies are planning on expanding their space presence with mega-constellations in this orbit. This includes SpaceX's Starlink and OneWeb, which both intend to provide low-latency internet satellite around the world. Lower launch costs and technology improvements for the satellites have enabled huge increases in the use of satellites over the past two decades. Small, standard-shaped satellites variously called cubesats or smallsats have enabled further miniaturization of technology and decreased payload mass, reducing costs. Additionally, smallsats can be mass-produced since their low orbit and smaller technological package can be balanced out by placing more of them in orbit.

As mentioned previously, there is a growing problem in terms of congestion in near-Earth space. Satellites cannot simply be bunched together closely in orbit—because they need to communicate with Earth (and other satellites), they must be far enough away from each other so that ground stations can hear their radio communications. Further, each satellite is moving at such a high rate of speed that were satellites any closer together, the chances of collisions between them would greatly increase. Added to this problem is the fact that most satellites have only a limited amount of maneuvering fuel on board and satellite operators are reluctant to move a satellite unless absolutely required.

Internationally, orbital slots are regulated by the International Telecommunications Union (ITU). First established in the 19th century to connect telegraphs around the world, this intergovernmental organization (IGO) has evolved over time to regulate the global use of the radio spectrum leading to its relationship with space. Under the OST, no one country can claim sovereignty or control of space, including orbits. That being said, orbital slots and radio frequencies are limited so under a previous radio treaty, countries have agreed to allocate radio frequencies via the ITU. The ITU reserves particular frequencies and orbits for all countries, which then must license those spectrums internally.[11]

In addition to the physical congestion of satellites and radio crowding, the increasing number of satellites must also contend with increasing amounts of space debris. Space debris includes old rocket bodies (or pieces of them), old spacecraft that have ceased to function, lost tools or parts from human missions, or pieces of satellites that have been destroyed in space. Space debris is an increasing hazard for several reasons. Not only does it complicate operations in an already complicated place, but a large amount of it cannot be tracked because it is too small. And while it might be too small to be tracked, it is not too small to cause major problems. Recall that to be in Earth's orbit, an object must be going at least 17,000 miles per hour. A piece of debris less than 10 centimeters in size but traveling at that speed can destroy a satellite or spacecraft depending on where and how it hits. This risk increases the costs of building satellites if they need to be given extra protection from debris as well as increases the hazard for humans in Earth's orbit. Astronauts on the International Space Station are often asked to take "shelter" because a piece of debris is approaching too close to the station.

Because of the increase in space debris, a situation called the Kessler syndrome becomes more likely and more dangerous. The term describes the dangers of a cascading collision: if one piece of debris strikes an object, that object in turn can become many more pieces of debris, all of which have chances of striking other objects causing more debris. This is not unlike the situation depicted in the movie *Gravity*. While the movie was, for obvious reasons, quite dramatized, the problem increases slowly over time. Anything that increases the amount of debris in space, including the use of weapons that kinetically destroy satellites, contributes to the debris problem and thus increases the chance of a Kessler syndrome–type event.

While the problem of space debris is quite clear, how to solve it is not. Countries like the United States have adopted various mitigation measures including requirements that satellites de-orbit themselves after a certain period of time or be moved to "parking" orbits, but that does not solve the problem of the debris that is already there. Some companies are developing technology that can remediate the problem by removing debris, but this is complicated technologically, politically, and legally. Technologically, it is difficult to maneuver in close enough to a piece of debris, latch on to it somehow, and move or de-orbit it. Legally, objects launched into space remain the property of the country that launched them so any efforts at removing debris must have the permission of the country that generated the debris in the first place. If ownership is unclear or unable to be tracked, it cannot be removed legally. Finally, politically, this type of technology is dual use—if a piece of debris can be latched on to and destroyed, the same technology can be used to destroy a working satellite. Thus, if countries were to directly invest in and build debris remediation systems, they may be accused of trying to weaponize space.

There are many other considerations when it comes to getting to and operating in space, but these are some of the ones that have significantly influenced American space policy. In many ways, what is possible in terms of policymaking is significantly limited in terms of the technology available to make it happen and the constraints of operating in a harsh domain like space. The high costs required limit what the US and other countries can either afford or are willing to pay for. In addition to these materialistic limits, there are some international legal and political regimes that also greatly impact space policy.

The International Space Regime

Science fiction franchises like *Star Trek* often portray a future free of conflict on Earth and many different types of people (and aliens!) working together to explore and settle the final frontier. Unfortunately, the space domain today is not free from conflict over its use and exploitation. Indeed, from the beginning of the space race, it has been treated as an extension of Earth-based politics. Because of the possibility of conflict extending into space and the very limited nature of near-Earth space, a critical question has been and continues to be what rules of the road can all countries agree on to regulate the use of space? While we will take this question up in greater depth in later chapters, this section briefly introduces the major treaties that currently guide behaviors in space.

The first and most important international treaty regarding space is the OST, signed in 1967. While in the early years of the space race the United States and the Soviet Union resisted any formal negotiations limiting their behavior in space, events in the early 1960s such as the Cuban Missile Crisis, atmospheric nuclear testing, and the increasing human presence in space forced the two superpowers to reconsider. The result was the OST, the main principles of which are as follows:

> the exploration and use of outer space should be carried out for the benefit and in the interest of all countries;
> space shall be the "province of all mankind";
> outer space is not subject to any claims of sovereignty;
> states shall not place weapons of mass destruction or nuclear weapons in space or on any celestial bodies;
> the Moon and other celestial bodies should be used for peaceful purposes;
> astronauts are the envoys of mankind;
> launching states are responsible for all national space activities;
> states are liable for damage caused by their space objects; and
> states shall avoid harmful contamination of space and celestial bodies.[12]

While many, if not most, of these principles appear noncontroversial, like anything, the devil is in the details. One of the more difficult provisions is the prohibition on the placement of nuclear weapons and/or weapons of mass destruction in space. The OST does not define what is meant by "weapon" or "weapon of mass destruction." Dual-use technology makes distinguishing between what is a legitimate activity and what is illegitimate incredibly difficult in space. Additionally, the definition of what is a "weapon of mass destruction" may also be different in space. A rocket that deliberately hits a satellite and creates a cloud of debris that in turn destroys other satellites or renders an orbit unusable might be said to have created "mass destruction." While it was not the destruction of life or Earthly property, given the extent to which the world and the global economy depends on space-based assets today, the destruction of satellites might indeed have massive implications.

Just as the notion of a space weapon was left undefined in the OST, many of these other provisions also were left unclear. As a result, other treaties were drafted to put these principles into action. These include the 1968 Rescue Agreement, which requires signatory states to provide any and all assistance to astronauts who have landed within their territory whether accidentally or not; the 1972 Liability Convention, which reaffirms the idea that states are responsible for launched objects and puts into place rules under which states can claim damages; and the 1974 Registration Convention, which requires states to register all objects launched into space with the United Nations.

While these treaties were relatively uncontroversial since they were elaborating on broader principles agreed to in the OST, one final treaty proved more difficult, the Moon Treaty. While the OST stated that no country could claim sovereignty over the Moon or any other celestial bodies, the Moon Treaty was more ambitious in scope, aiming to put into place a framework of rules and regulations under which celestial bodies might be managed and governed. It asserts that the Moon represents the "common heritage of mankind" and, as such, would create an international body representing all countries to govern activities there. It also requires the equitable use and distribution of any and all lunar resources and banned all military activities in space. While the Moon Treaty represents the most fully detailed attempt to institute a comprehensive governing regime for space, many of its provisions proved controversial for major powers. To date, only 18 states have ratified the treaty, none of which are major spacefaring states.

Given the increased use of outer space for various purposes by various actors, provisions for behavior in space remain problematic. Definitions remain hard to come by and states in general resist efforts that might limit their options in the future. Thus, while various international agreements have been proposed and considered, no further treaties have been forthcoming. Different organizations including the United Nations Committee on the Peaceful Uses of Outer Space (UNCOPUOS) and the United Nations (UN) Disarmament Commission have worked to formulate informal norms of behavior that, while not official, can help guide what actors do and do not do in space. Like formal treaties, however, these informal agreements also suffer from the problem of enforcement, specifically, who will do so and how.

Plan of the Book

Much about space remains unknown. The uncertainty and lack of knowledge about space, what is possible to do there, and even what *should* be done there contributes to the state of American space policy today. The enormous cost and high technology places further limits on the actions of governments with limited budgets and competing priorities. But the reality is that space continues to hold a certain fascination for a good chunk of the public and provides another arena for state competition and power seeking. The challenge for space policy has not only been to balance all of these factors but to do so in a democratic society with its requisite checks and balances and difficulties in governing.

Thus, this book is not just about space and what it takes to make policy for such a domain, but it is a book about American government, its institutions and the people who make them up. The rules of the road that these institutions impose greatly condition how space policy is made and to what extent it may be successful. Democracy, as we well know, is imperfect. Governing is messy. The same critiques we make of policymaking in other areas are also present for space, sometimes to a greater extent. The result is that patterns in US space policymaking, while sharing some similarities with other areas, are unique in their own right. It is these patterns that this book tries to illuminate.

To do so, the book is divided into three parts. The first part sets the stage. In addition to this first chapter, chapters 2 and 3 focus on space history, Cold War and post–Cold War, respectively. While this is not a space history book, a brief familiarity with some of the major events, actors, and policies sets a good foundation for understanding the more political dynamics presented in later chapters. Part two then focuses on the different actors involved with space policy. Chapter 4 explores the role of presidents in setting the agenda and the extent to which they may (or may not) be successful at influencing space policy. Chapter 5 takes up the role of Congress, including the tools through which they can influence space policy and the competing influences on legislators. Any study of space policy in the US would be incomplete without examining NASA's role, which is chapter 6's task. NASA, as the implementor of policy and a center of space expertise, has significant power over how policy is implemented and even conditions what might be possible to begin with. NASA is only part of the story, however. Chapter 7 explores military space policy, along with the role of the intelligence community. This chapter includes not just what the US has done militarily in space, but how it is organized for space, particularly with the development of the USSF. Chapter 8 details the other government agencies also involved in space policy including the DOD, the Federal Aviation Administration, the Department of State, and others. Finally, chapter 9 explores the nongovernment forces at play including technology, economics, and public opinion.

The final section of the book looks at specific areas of space policy in further detail. Chapter 10 explores civilian space policy and activities including the debate over robotic and human exploration, space traffic management, and problems of debris. Chapter 11 details the role that international politics plays in US space policy given that global actors and the global environment have often driven American behavior in space. With the recent rise of commercial actors in space, chapter 12 examines these more fully including their increasing power and ability to influence norms of behavior in space. Finally, chapter 13 takes up the major unanswered questions that will likely need to be confronted in space policy in the near future.

Summary Points

- The importance of the space domain has increased as the world has become more dependent on space-based capabilities for everything from communication to economic transactions.
- Space policy isn't just a recitation of different actions or programs the United States has undertaken over time. It's understanding who has influence and power, the institutions and mechanisms through which that is exercised, and the motivations of the various actors.
- Political scientists try to understand policy in a variety of ways including the stages of the policymaking process and concepts such as path dependency, incrementalism, PE, and policy typologies.
- There are four different types of space policies discussed in this book: civilian, military, commercial, and scientific. They are divided based on the actors involved and their ultimate aim and purpose.
- Though this book focuses on space policy, some understanding of the space environment is required including the problem of launch, types or orbits, problems in the space domain including debris, and the various international agreements and institutions that are involved in space issues.

Discussion Questions

1. In what ways do you use space-based services in your everyday life?
2. In what category of space policy do you believe human spaceflight best fits? Why?
3. Based on what you know so far, why do you think creating rules of the road for space has been so difficult for major states?

For Further Reading

Mendenhall, Elizabeth, "Treating Outer Space Like a Place: A Case for Rejecting Other Domain Analogies," *Astropolitics*, vol. 16, no. 2 (2018). https://www.tandfonline.com/action/showCitFormats?doi=10.1080/14777622.2018.1484650.

Pettit, Don, "The Tyranny of the Rocket Equation," *NASA*, May 1, 2012, https://www.nasa.gov/mission_pages/station/expeditions/expedition30/tryanny.html.

Riebeek, Holli. "Catalog of Earth Satellite Orbits," *NASA Earth Observatory*, September 4, 2009, https://earthobservatory.nasa.gov/features/OrbitsCatalog.

2

The Space Race during the Cold War

The Quest for Prestige and Security

Several key moments such as the successful launch of Sputnik on October 4, 1957; the Apollo Moon landings in 1969 and the early 1970s; and the Apollo-Soyuz mission in 1975 have defined Cold War space history. Rather than solely focusing on great moments, however, we examine the Cold War space era as part of a long, historical narrative that, at its core, was a struggle for global leadership between the United States and the Soviet Union.

The Cold War dates to a much earlier period, perhaps as early as the 1917 Russian Revolution, when the United States under President Woodrow Wilson, decided to intervene in the Russian Civil War against the Bolshevik Communists in 1920. In refusing to recognize the Soviets' rule at the end of the country's civil war despite their clear victory, many scholars suggest that this American disdain for the Communists forced Soviet leaders like Vladimir Lenin and his successor Joseph Stalin into a distrustful, antagonistic relationship. This antagonism existed despite both sides allying with each other to defeat Nazi Germany during World War II.

Throughout the period of communist rule, the fundamental incompatibility between capitalism and communism made it impossible for either the United States or the Soviet Union to fully reconcile its way of life with the other. Instead, the two systems and their peoples pushed toward constant competition in all facets of life, to demonstrate which system (and perhaps which people) were inherently superior. A continuous struggle for power mixed with this clash of ideologies, as the two states sought to gain allies to become the global leader of the new world order emerging from the ashes of World War II.

While the outcome of the race for global leadership remained uncertain for some time to come, the post–World War II period found the United States and the Soviet Union seeking prestige and security advantages over the other. The space domain was one important area in which this took place, given that space represented opportunities to gain in both areas. In this chapter we explore the history of the space race during the Cold War from the perspective of these two themes. The prestige to be gained in space came through the need to master the complex scientific and technological challenges presented by achieving spaceflight, whether by technology or people. With the mastery of these scientific and technological challenges came their application to national security; space advancements served both as an acknowledgment of moral superiority and military dominance.

With technology, while both sides shared similar goals, how they sought to achieve them differed. The USSR focused on utility and simplicity, allowing them to achieve success at an initially rapid pace. This approach facilitated the Soviet imperative to place the first artificial satellite into orbit faster than the United States. As detailed in table 2.1, it also allowed the Soviets to achieve a range of other firsts, such as the first animal to observe and report on the cosmos near and far from Earth. The United States, however, chose a more technical route; it was one that took more development time but incurred less risk. In the end, these differences generally meant quick gains in prestige for the Soviets but slow and steady gains for the Americans.

The one area where this was most disconcerting was in rocket technology and its resultant impacts on security. Ballistic missile technology and its mastery paved the way for the development of the nuclear arms race between the two sides that epitomized their existential rivalry. The Soviet mastery in rocket technology early on, along with the country's massive military industrialization, led the United States to fear a sizable missile gap between the two countries such that the Soviet Union was believed to be the more capable. The development of spy satellites to accurately verify this information would become quite important for the United States in its assessment of both sides' strategic readiness and force postures.

There were differences in terms of people as well. For the Soviet Union, individualism always remained obscured by duty to the state; people were important to the extent that they served the state, because it was the state that mattered. What did this mean in terms of prestige? Certainly, the focus on the Soviet Union illuminated its power and magnified its significance. Service to the state could lead to the Soviet Union doing great things, which in turn would reward all Soviet people versus a select few. This was the sort of narrative that the Soviets wanted to share with the world, with their cosmonauts and scientists serving as their preferred messengers.

Table 2.1. Major space race milestones, 1957–1975

Date	Event	Country
October 4, 1957	First satellite (Sputnik)	Soviet Union
November 3, 1957	First animal (Laika)	Soviet Union
August 18, 1960	First spy photography from space (Discoverer 14)	United States
August 19, 1960	First live animals returned from space (Belka and Strelka)	Soviet Union
April 12, 1961	First human in orbit (Yuri Gagarin)	Soviet Union
May 19, 1961	First planetary flyby (Venera 1)	Soviet Union
June 16, 1963	First woman in orbit (Valentina Tereshkova)	Soviet Union
June 19, 1963	First Mars flyby (Mars 1)	Soviet Union
March 18, 1965	First spacewalk (Voskhod 2)	Soviet Union
March 16, 1966	First spacecraft docking (Gemini 8)	United States
December 21, 1968	First humans to orbit the Moon (Apollo 8)	United States
July 20, 1969	First humans on the Moon (Apollo 11)	United States
April 19, 1971	First space station (Salyut I)	Soviet Union
November 14, 1971	First spacecraft to orbit another planet (Mariner 9)	United States
July 15, 1975	First multinational human-crewed mission (Apollo-Soyuz Test Project)	Soviet Union and United States

In contrast, individuals were center stage for the United States. As Kennedy remarked, it was about what individuals could do for the country and not the other way around. This focus on individuals was not to say that the state did not share in the glory—the Apollo missions getting the United States to the Moon did so on behalf of the country—but it was the individual and his or her American spirit that did great things. These great things reflected the unique nature of an American society that was structured to allow individuals the room to succeed and benefit from the success of their own labor.

Early on then, the stage for the Cold War was set. The Soviet Union and the United States competed for prestige and security, with two different systems and two different groups of people seeking similar goals and faced with dire prospects for losing. The competition was palpable after the end of World War II, and for both sides it was existential. As you will see in this chapter, the space race provided an outlet for the competition that was less confrontational than other aspects of the rivalry but competitive, nonetheless. For US space policy in particular, the space race during the Cold War would leave behind a long-lasting legacy that remains to the current day.

Prestige and the Legacy of World War II

While many conceive of the space race as a period within the Cold War that begins with Sputnik and ends with the Apollo Moon landings, we take a broader, more comprehensive approach in this chapter. Given that we are interested in US space policy as the subject of study, this type of approach allows us to use history as a way from which we can observe the evolution of US space policy through the challenges presented by both external and internal influences. Throughout the Cold War, the Soviet threat along with domestic politics interacted and worked together to influence the direction of American space policy.

Conceived in this way, the story of the space race between the US and USSR probably begins with what Walter McDougall acknowledges was the first endorsement and support by a government of spaceflight.[1] Under the leadership of Vladimir Lenin, the Soviet Union began the serious study of rocket research in the early 1920s. Rocket pioneers like Konstantin Tsiolkovsky examined the use of propellants for interplanetary flight, while others focused on the military applications of rockets. While the work remained heavily theoretical, it was not long before experiments began testing the utility of chemicals such as liquid oxygen for long-range propulsion. The work of Tsiolkovsky and other Russians, however, had little use to the Soviet government if their work could not be put toward the defense of the nation. Thus, rocket research in the USSR became a function of the state, with efforts geared toward the research and development of military applications versus the more abstract considerations of spaceflight. These efforts stood in contrast to the interests of those scientists and researchers who developed the technology and in no small part, these efforts limited the Soviets' ability to quickly move toward space. Nonetheless, such progress remained important and would help shape Soviet preparedness for space after World War II.

Prior to the war, German scientists also had begun working on rocketry for the purposes of spaceflight, but the Nazi war machine redirected efforts toward the use of rockets in conflict. The Vergeltungswaffe 2, or V-2, rocket represented the apex of these efforts, unleashing terror upon its victims at the end of World War II. As its name translates, the V-2 was used for revenge by Nazi Germany on its adversaries, particularly the United Kingdom, to punish the country for its war efforts against Germany. While the V-2 was just one of several last-minute efforts by Germany to win the war, or at least negotiate a better peace, the proven technology behind the V-2 made it appealing to both the United States and Soviet Union as both sides began preparing for a future postwar world. As we discuss next, there was an immediate recognition of the V-2's derivatives as future weapons of war.

Prewar theorists like Emilio Douhet envisioned airpower as the future weapon of choice for coercion. Policymakers were enamored with the idea of being able to reach beyond ground troops on the front lines, and of shortening wars by being able to threaten leaders and their people directly. Such an approach stood in stark contrast to the recent world wars with their long, drawn-out conflicts on well-defined battlefields. Depending on one's perspective, World War II either validated such thoughts or made it clear that it would be the next generation of airpower weapons who determined the validity of such ideas. Many military leaders and policymakers, for example, viewed the Combined Bomber Offensive against Germany, in which Allied bombers attacked the country's industrial centers and cities, as a good illustration of what massive amounts of airpower could bring to the conduct of war. When combined with the technology of the V-2 rocket, the United States and the Soviet Union recognized the need to quickly develop how such weaponry could credibly defend against threats from its rival.

Many of the advances in rocketry were made possible by the technology encapsulated in the V-2. As such, it was important for both the US and USSR to capture as much of this technology as possible toward the end of World War II; this included bringing in the scientists responsible for the program. For the United States, these efforts manifested in Operation Paperclip, the US Army and Special Operations mission to capture both the scientists, such as Chief Scientist Werner von Braun and his associates, in addition to the remains of any usable V-2 rocketry. For the USSR, these efforts led to similar attempts to capture German rocket scientists but also to gain control over the rocket factories and means of production so that they could be duplicated back in the USSR.

While the Soviet Union possessed geographic advantages in securing technology and scientists, given their rapid advancement and physical control over much of Germany, the Americans had won the war of perception, with scientists wanting to defect and/or secure capture by the Americans, for whom they viewed as civil, humane, and even friendly toward prisoners. Most Germans viewed the Soviet Union, in contrast, as cruel and barbaric, with many believing (rightly so in many cases), that they would be sent to work in prison camps for the duration of their lives. In the case of von Braun and his closest colleagues, they took great efforts to escape the Soviet threat, locating an American battalion to which they could surrender. In the end, the race for V-2 technology and the scientists and engineers attached to the project ended up favoring the United States, particularly with the apprehension of von Braun. Nonetheless, the Soviet Union was able to capture its own share of people and technology, paving the way for the next phase of competition in the development of intercontinental ballistic missiles and the ability to send things into orbit.

We should note here that while the efforts to acquire the V-2 were important to both sides, the US and USSR had familiarity with and previous research on rocket technology as well. In the United States, it was the work of Robert Goddard, who created the first liquid-fueled rocket in 1926, that helped von Braun and the Germans fully develop the V-2. Unfortunately for Goddard and the United States at the time, the US government was not particularly interested in rocket development for military use. For the USSR, we return to the legacy of rocket research by Konstantin Tsiolkovsky. His work would serve as the foundation for Russian aerodynamics research and rocket design. Tsiolkovsky's efforts, which even included a 1903 design of a future spacecraft, also served to motivate two new Soviet space pioneers, Sergei Korolev and Valentin Glushko, as they ushered in a new era of Soviet space in the postwar period.

In many ways, the competition for the technology of the V-2 encapsulated the perceived character of the two sides in the burgeoning Cold War. The Americans tended to view themselves as having achieved the early moral high ground, a position they perceived as helping them carry the day in achieving greatness in the face of others' moral inferiority. Beyond US borders, many took this attitude as arrogance, and for the Soviets in particular, it was critical that the Americans demonstrate that these perceptions were inaccurate. In other words, it was their view that the Soviet Union represented morality and progress, in contrast to the Americans' vice and morally bankrupt opulence.

The question for both sides was how best to develop the technology. Which side could harness its human knowledge to build the machines necessary for a progressive future? Which country could lead the world into a new, advanced industrial age and serve as the example for other states to follow? The country that could do this would gain a level of prestige that would make it the unquestioned leader of the international system. The issue of prestige then becomes a central theme in the space race, and as we explore next, it very much sat at the forefront of concerns for the US and Soviet Union.

Prestige through Civilian Space

Why did prestige matter for the two superpowers and how did this manifest in the early stages of the space race? First, prestige has always been relevant for states. In part, this is tied to the idea that there is a direct correlation between power and prestige. For some international relations scholars, the question is whether power determines prestige or whether prestige is a conduit to power. Although we will return to this question in a later chapter, here, the direction of causation is somewhat irrelevant. For the Soviet Union, prestige was necessary to demonstrate to the rest of the world that a new approach to governance was a legitimate way of improving the human condition. For the United States, prestige was necessary to validate society's trust in the role of the individual and the invisible hand of the free market to spur ingenuity, innovation, and overall improvement in humanity. Both sides attempted to create a narrative that offered alternatives to the fascist ideology that had risen to preeminence in Europe throughout the interwar period.

While Joseph Stalin did not appear overly concerned about prestige throughout his rule, focusing more on the development of Soviet industrial and technological strength in terms of raw capabilities, Soviet Premier Nikita Khrushchev understood the importance of prestige in competition with the United States. Having risen through the ranks as a true believer in the great Marxist-Leninist experiment, Khrushchev internalized the ideas of scientific-technological progress as a foundational pillar of Marxist-Leninist superiority over capitalism. In short, Soviet scientists were motivated by efforts to produce for the greater good and were supported by other Soviet citizens. These citizens, through their own efforts, provided the resources necessary for the state to provide all the raw materials, education, and collective spirit required to advance knowledge for all humankind. This knowledge could come through both scientific and military prowess.

While this sounded like propaganda to Western audiences, such a firm belief made it imperative to demonstrate the power of this philosophy to the wider world. The space domain provided the stage in which to do it, with the International Geophysical Year in 1957 providing the push both the Soviets and Americans needed to identify the goal of achieving space orbit by an artificial satellite. Success would mean an early victory in the race for the heavens.

The International Geophysical Year (IGY) was an international program of scientific research sponsored by the International Council of Scientific Unions. The goal was to spur international efforts in conducting geophysical research from 1957 to 1958, most of which focused on topics related to solar activity, cosmic radiation, and ionospheric physics. What was important in these research efforts as they related to space was the use of rocketry and related equipment for high-altitude data collection. Both the Soviet Union and the United States viewed this as an opportunity to send an artificial satellite into space, gaining the prestige that would come along with such an accomplishment.

Putting a satellite into orbit, however, required rocket capabilities that allowed for long-range ballistic trajectories, the same sorts of trajectories that would allow rockets to be used as long-range weapons of destruction. This was important as the newly minted rivals began exploring the ideas of marrying intercontinental ballistic missiles (ICBMs) with nuclear armaments, ushering in the nuclear missile age just prior to the space age. The dual utility of missile technology would and continues to maintain the close relationship between nuclear weapons and space. The close relationship also ties together notions of prestige and power for those states seeking to demonstrate an elevated place in the hierarchy of power among nations. This elevation of place between the US and USSR during the Cold War thus served as a precursor to other state efforts at seeking great power status and the associated prestige that comes with it.

While the struggles to achieve victory in the IGY are narrated in detail elsewhere, the Soviet Union's success with Sputnik in 1957 officially placed the Soviets in the lead when it came to the competition for prestige. This accomplishment was important as it demonstrated technological superiority, at least from the perspective of a global audience. When combined with the 1949 Soviet detonation of the country's first atomic bomb, four years after the Americans' use of them on Japan, and the decree of the 1947 Truman Doctrine calling for the containment of communism, the Soviets made it clear to the world that the United States would not necessarily be the ones dictating the direction of the global order.

For American policymakers, the "Sputnik moment" as it was called, created mixed feelings. Public perceptions pushed President Eisenhower to speed up the US efforts to send a satellite into orbit; but important in this regard was that the United States do this through civilian means. Unfortunately, this meant that the best rocket available for the US, the Jupiter-C, would not be used as a first choice since it was used by the US Army for military purposes. This also effectively sidelined von Braun's team working for the Army in Alabama. Instead, the US pinned its hopes on the Navy Research Lab's Vanguard rocket for an attempt in 1957 to match Sputnik. The explosion of the rocket on the launch pad in late 1957, unfortunately, reinforced public perception of the US lagging behind the Soviets in space. Of course, this was problematic for the space race itself, but it was also problematic as it related to the wider Cold War competition and the ability of the United States to outperform the Soviet Union. The United States did manage to launch Explorer 1 into orbit in January 1958, signaling at least technological know-how in the race to keep pace with the USSR. Despite the success of Explorer 1, as mentioned in the introduction, the Soviet Union would go on to a longer list of space "firsts," taking an early and clear lead in the space race.

From a US space policy perspective, the Soviet Union's "firsts" created both challenges and opportunities with respect to the problem of prestige. In terms of challenges, US presidents Eisenhower and Kennedy needed to find ways to gain ground on the seemingly widening gap between the two rivals. Regarding opportunities, both administrations were uniquely positioned to shape the direction of US space policy for decades to come. President Eisenhower, for example, was able to take advantage of the Sputnik launch to set the precedent for territorial overflight. Prior to the launch, there were no clear distinctions made regarding the boundaries of a state's sovereign airspace. Once Sputnik completed an orbit and overflew the territories of other countries, the USSR was not able to contest future overflight from US satellites. This precedent presented a long-term gain for the US as the DOD and Central Intelligence Agency (CIA) considered the use of satellites for data collection on Soviet and other potential adversary targets. The first efforts to use this technology successfully took place in 1960 with the launch of the MIDAS early warning satellite, and the Discoverer 14 mission.[2] The latter was a part of the Corona program, which, at the time, was a top-secret group of intelligence gathering satellites used to photograph restricted areas in the Soviet Union. The effectiveness of these missions helped pave the way for the establishment of the National Reconnaissance Office (NRO) in 1960 and its mission continues to remain a key part of US data collection from space.

Clandestine data collection, however, would take the US down a path that focused more on security; the prestige component in terms of US space policy would come from two key sets of decisions. First was the creation of the National Aeronautics and Space Administration (NASA) in 1958, from the National Aeronautics and Space Administration Act, signed on July 29. Second was NASA's subsequent initiation of the manned missioned projects of Mercury, Gemini, and Apollo.[3] For the Americans, and certainly part of the appeal of the manned mission projects, space was the new frontier from which their culture and spirit were not just compatible but something for which they practically were made. This narrative at least was a vision that drove the early motivations behind NASA's exploits.

As mentioned previously, President Eisenhower wanted a civilian face to its space endeavors. NASA would be the civil component of US space and while initially closely tied to the DOD's own agenda, NASA has been viewed by most US policymakers since Eisenhower as needing to be independent in this regard. The United States wasn't simply trying to show military superiority in its quest for prestige, something that would become more important later, America sought to demonstrate to the free world that the civil sector was just as important. Independence, then, would mean concentrating on exploration and scientific discovery; this focus would compete with the Soviets head-on in terms of the challenge brought about by claims of Marxist-Leninist scientific and technological superiority.

What then was the overarching policy approach to NASA during the Cold War? In short, it was to fund missions that would extend the basic science and research associated with space exploration; NASA would operate according to principles that focused on the sanctuary approach to space. Arguably, this was not done benignly; a singular civil institution that could serve as the face of US space allowed the DOD to concern itself with other aspects of the domain, ones that provided for security and ones that were less concerned with sanctuary. NASA would become the prestige arm; the DOD could focus on keeping the country safe, and projecting power. Ironically, however, it would be the DOD and the development of its capabilities that gained the prestige for the United States in the long term, a point to which we later return.

Despite a straightforward policy approach in its establishment, the story of NASA during the Cold War was one of competition and struggle both from external and domestic challenges. As we examine in more detail in the NASA chapter, the tension between Congress and the president over control of NASA's agenda was and is a challenge for the organization and the creation of long-term, stable policy. During the Cold War, especially, this challenge was problematic as politicians sought to use NASA for their own political ends. These ends were most acutely illustrated in budgetary decisions regarding space. As the space race continued to heat up, President Eisenhower wanted to rein in expenditures, but the era of manned spaceflights, beginning with Alan Shepard as the first American in space in 1961 and culminating with Neil Armstrong's first steps on the Moon in 1969, received significant domestic support that Eisenhower was powerless to stop.

NASA spending was justified to catch up and surpass the Soviets in space. Again, the broader Cold War context mattered. During this period of the Cold War, the Soviet Union had already placed a person (Yuri Gagarin) in orbit in 1961 and had created what became known as the Iron Curtain in Eastern Europe. The most visual representation of the curtain was the building of the Berlin Wall in 1961. President Kennedy's failed Bay of Pigs invasion in Cuba in April 1961, just days after Gagarin's flight also magnified the need for victory in space, as the march of communism and Soviet leadership seemed determined to spread across the globe. This march included the Western Hemisphere, which, since the era of President James Monroe, had been viewed as inhospitable to other great powers.

The United States needed something remarkable to turn the tide against the successes of communism. For President Kennedy, a successful landing of an American on the Moon represented a monumental goal that could decisively demonstrate American leadership and technological prowess in space. Of course, such a goal was fraught with challenges. However, as President Kennedy offered in his special speech to the US Congress in May 1961, there was no option but to try:

> Recognizing the head start obtained by the Soviets with their large rocket engines, which gives them many months of lead-time, and recognizing the likelihood that they will exploit this lead for some time to come in still more impressive successes, we nevertheless are required to make new efforts on our own. For while we cannot guarantee that we shall one day be first, we can guarantee that any failure to make this effort will make us last. We take an additional risk by making it in full view of the world, but as shown by the feat of astronaut Shepard, this very risk enhances our stature when we are successful.[4]

Kennedy's announcement also meant seeking financial support from Congress for NASA since this goal required the type of resources that would comprise a sizable portion of the country's national budget. While other options to catch the Soviets may have been possible, correspondence with then-NASA administrator James Webb and Secretary of Defense Robert McNamara made it clear that the US needed a mission that would focus on national prestige, it was this more than any real distance in capabilities that was lacking in the competition with the Soviets.[5] Prestige, would come at a cost, however, with NASA taking about $3.7 billion to operate per year by 1963.[6]

Domestic politics regarding NASA had additional influences on the race to the Moon, with decisions made during this period remaining relevant to the present day. In particular, the Apollo Project represented numerous opportunities to bring in federal money to contractors in districts involved in the research and development areas of the space industry. Thus, began the expansion of space infrastructure in the United States, with southern states like Texas, Alabama, Florida, and Virginia in particular attracting significant amounts of federal investments that would cement their places and roles in US space policy for the long term. Thus, having created the domestic incentive for support in going to the Moon, President Kennedy and his successors bolstered the rationale for an endeavor of this scope.

At the same time, NASA's requirements for astronauts were so stringent and drew largely from the ranks of military test pilots that women and Black Americans were largely excluded from the most visible space participants, the astronaut corps. Particularly after the Soviets launched the first woman to space, Valentina Tereshkova, there was some debate over the role of women in space. A group of female pilots that came to be known as the Mercury 13 were put through the same medical tests as the astronauts had undergone by the same doctors with great success suggesting arguments that women could not tolerate the pressures of space were largely irrelevant.[7] Still, social norms served to keep women out of the astronaut corps, perhaps best exemplified by a statement made at a congressional hearing on the topic by John Glenn, the first American to orbit the Earth, that "the men go off and fight the wars and fly the airplanes and come back and help design and build and test them. The fact that women are not in this field is a fact of our social order."[8] Regardless, there were women employed at NASA who proved essential to the space including Katherine Johnson, Dorothy Vaughan, and Mary Jackson, the Black female calculators featured in *Hidden Figures* and Margaret Hamilton, a software engineer responsible for the development of the software needed to run the Apollo command and lunar modules.[9]

Black Americans were also largely excluded from the ranks of NASA during the space race. Following President Kennedy's suggestion that a Black man be chosen to be part of the astronaut corps, Air Force (USAF) pilot Ed Dwight applied for the third astronaut class but was not ultimately chosen for reasons that have remained unclear.[10] Black Americans faced other disadvantages as well in participating in the space race as many NASA facilities were located in the segregated South. These NASA centers played an important role in desegregating their local communities as NASA threatened to withhold funding from them in the late 1960s.[11]

After successfully pulling back from the brink of World War III during the Cuban Missile Crisis in 1962, the Kennedy and later administrations began to gain confidence in their ability to manage the Soviet threat. As we discuss in the next section on security, questions regarding a missile gap were resolved and American policymakers began to feel on more solid ground regarding US-Soviet competition. With respect to the Apollo Project, six successful Moon landings, beginning with Neil Armstrong and Apollo 11 in 1969 and ending with Apollo 17 in 1972, secured the US position as a leader among spacefaring nations, redressing the imbalance left by the string of Soviet firsts in the 1950s and early 1960s. The success of the Apollo missions stood in stark contrast to a string of Soviet failures. Due to errors with lunar landers, such as that of *Luna 15,* and the inability to successfully launch their N1 heavy rockets, the Soviet Union was unable to duplicate Moon landings for cosmonauts. Finally, the United States had overtaken Soviet prowess in the race for the heavens.

The Apollo missions, unfortunately, represented the height of American prestige in space for the foreseeable future. The 1970s brought a more peaceful period between the US and USSR, cooling the competition between the two nations while increasing cooperation. As noted at the outset of this chapter, space served as a backdrop to the global competition that the Cold War had come to represent. During the détente of the 1970s, leaders of the two superpowers tried to lower the temperature on what had quickly become a heated power struggle between them. Conflicts over Cuba, the Middle East, and Latin America, along with widespread political discontent, led the superpowers toward constant confrontation. With both sides not wanting to trigger nuclear war, the two countries began to find ways to accept the other's sphere of influence and open communications such that mistakes and misperceptions could be avoided. In short, the two countries started learning how to coexist despite their differences. The most symbolic image of this cooperation was the Apollo-Soyuz mission that ended with the "handshake in space" between Apollo and Soyuz crews in 1975.

Apollo-Soyuz offered hope, at least symbolically, that the two states could cooperate for the betterment of humankind. While the coordination and cooperation proved challenging, again reflecting efforts to jockey for overall prestige, the mission represented a bright spot in Cold War relations and space cooperation more broadly. Simultaneously, however, it did offer signs that space would be a more competitive environment, leaving open the question of whether space would remain a sanctuary or be open for international competition, military or otherwise.

Enduring Competition for Prestige

The space race for prestige certainly reached its apex in 1969, but competition continued throughout the remainder of the Cold War. What one observes, however, is a shift away from the civilian emphasis toward one directed toward security. We examine this in more detail next, but first, it's worth discussing two additional legacies on the civilian side that brought prestige to both countries in their quest for space leadership: space station development and space transport.

Placing a person in orbit or even sending one to the Moon were remarkable feats in the 1960s, but true mastery of space implied a long-term, more permanent status. Building on their Soyuz capsule concept, the Soviet Union reached another first in launching their Salyut 1 space station in 1971. Unfortunately, this space station was associated with disaster, as the three-person mission of Soyuz 11, the first crew to reside in the space station, died upon reentry back to Earth. Despite the severe conditions of space, the three cosmonauts are the only human travelers to have perished in this environment. Overall, the Salyut program helped achieve a long-term goal of maintaining a permanent Russian presence in LEO, with the Salyut capsules being used to build the modular Mir space station in 1985 and the Russian part of the ISS in the 1990s.

The United States would not be far behind the Soviet Union in building and launching its own space station, Skylab, in 1973. Occupied through 1974, Skylab hosted three crews, with the crew of Skylab 4 setting a record (albeit temporarily) of 84 days in space. The success of Skylab and the continuation of space station building by the Soviets in the 1970s led to considerations of a larger and more permanent US station, but policymakers scrapped the idea due to cost and the opportunity to create the ISS at the end of the Cold War.

Maintaining a space station also meant having a reliable way to supply it. As a follow-on to the Apollo missions, President Nixon agreed to support the development of a Space Transportation System (STS) that would create the first reusable spacecraft. The STS, or space shuttle, could carry people and equipment into orbit and return to Earth, landing like a glider. The program represented a step down from the ambitious Moon efforts, but also kept the United States at the leading edge of space, maintaining its advantage over the Soviet Union. With its first launch in April 1981, *Columbia* would usher in a new era for spaceflight.

Among the firsts of this time was the inclusion of women and Black Americans in the astronaut corps. While no law stated that women (or Black Americans for that matter) could not be astronauts, the passage of the Equal Employment Opportunity Act in 1972 put legal pressure on NASA to allow women to apply to the astronaut program.[12] Further, the creation of the new position of mission specialist in the space shuttle program meant that individuals no longer had to meet stringent requirements such as having military test pilot experience or a minimum number of hours flying a jet to be admitted. The result was that in 1978, the latest class of astronauts featured three Black men, six women, and one Japanese American man. From this class, Sally Ride would go on to be the first American woman in space, and Guion Bluford became the first Black American in space.

While the shuttle program proved a tremendous success in demonstrating the leadership of the United States in the Cold War space race, particularly with the Soviets having tried and failed to launch their own shuttle, 1986 would prove a turning point in space exploration. The space shuttle *Chal-

lenger, having completed nine missions, including carrying the first American female and African American astronauts, met disaster on January 28, 1986, exploding just 73 seconds into its flight. The disaster shook the nation to its core, leading many to question the importance of space exploration and the expenditure of resources on space overall. Although the STS program continued, a subsequent accident with *Columbia* in 2003 brought significant pressure to end the program and replace it with a new platform.

As you will see in the next chapter, the US would be unable to provide a timely replacement for the shuttle, leaving transport to the ISS dependent on Russia, again striking a blow to US civilian space in the post–Cold War period. As it turns out, the USSR had invented tried-and-true technologies to achieve orbit, even if they were perceived as rustic and crude in nature. For the US, this period resulted in stagnation that would pave the way for what has been labeled "New Space." It is these new efforts at commercialization and hybrid approaches that perhaps may be responsible for the salvaging of US civilian space as the leader in the domain in the 21st century. Before turning our attention, however, a discussion of the security dimension of space remains.

Cold War Space History: Seeking Security

For some, the idea that the bomber would "always get through" seemed to be less realistic in the Cold War period, leading to the idea that perhaps manned aircraft was not always the best approach to stable deterrence. As mentioned earlier, both the US and USSR recognized the potential for ballistic missile capabilities after witnessing the technological success of the V-2 rocket. The combination of ballistic missile capability with that of nuclear weapons led to the consideration of space for security objectives beyond espionage. As the Cold War began to heat up in the 1960s, for example, it was clear that military superiority and thus space superiority would rival the prestige gained by demonstrations of technological know-how. For policymakers in the US then, space became a military domain far earlier than the more recent claims brought about by the clamor surrounding and the subsequent commissioning of the USSF.

The relationship between space and security during the Cold War (and beyond), was multidimensional in nature. First, as we observed earlier, space provided the opportunity for information collection, beyond missile early warning. This opportunity led to innovative efforts seeking to maximize efficiencies in data collection to directly influence strategy and policymaking. Second, ultimate security through nuclear deterrence required working *through* the space domain in terms of weapons and *in* the domain with respect to early warning systems. Third, space technology became an enabler, providing communication, timing, and navigation for military units and thus for the United States, revolutionizing its ability to project power. While the effectiveness of space for power projection would not be recognized until the end of the Cold War (during the Persian Gulf conflict with Iraq), it is important to know that it was during the Cold War period that the stage was set for the United States to achieve primacy after the fall of the Soviet Union.

The use of satellites for observation and intelligence gathering became an early priority for the United States at the outset of the Cold War. The closed and secretive Soviet Union made it difficult for the United States to gather information on the state of its security programs. In the nuclear age, it was paramount to acquire information on the status of the USSR's nuclear capabilities and overall force posture. By the mid-1950s, US policy under Eisenhower focused on the development of satellites and a high-altitude plane (the U-2) to accomplish this task. In comparison to the U-2, satellites provided a less risky solution than manned aircraft. Even though the U-2 could fly at altitudes well out of range of any Soviet defenses, the capabilities required to do so left no fail-safes or redundant systems on the aircraft, making any problem in flight a critical one that jeopardized the pilot's life.

Even though the launch of Sputnik dealt a blow to the US reputation, it was a positive in that it established a precedent for overflight by satellites. The National Security Council recognized the importance of this principle as early as 1955 in top-secret discussions on the future of US space policy; thus, once precedent was set, the US could engage with surveillance satellites as the primary means of gathering intelligence on the Soviet Union's nuclear and military capabilities. In the meantime, the U-2 would serve as the preferred and only means of reliable data collection, particularly with respect to the deep interior of the Soviet Union where many secret facilities and testing sites were located. The CIA and USAF would be the agencies in charge of these programs.

At a more strategic level, the New Look policy of the Eisenhower administration focused on maximizing effectiveness in meeting strategic priorities at lower costs. This approach placed a premium on spy satellites and nuclear weapons. While bombers remained the delivery vehicle of choice, there were significant budget increases for intermediate range ballistic missile (IRBM) and ICBM development, particularly given fears of being outperformed by the USSR where Premier Khrushchev boasted they could be turned out in their factories as easy as sausages. Between 1955 and 1957, funding for the

Atlas, Thor, Titan, Minuteman, and Jupiter missiles had increased tenfold in the overall DOD budget.[13] With most of the missile development taking place within the USAF, its primary role in the space race became focused on the fielding of these weapons.

Building off the establishment of the Western Development Division in California in 1954, run by Lieutenant General Bernard Schriever, the USAF worked on making the Atlas ICBM a more reliable component of nuclear deterrence. Like the problems faced by the R-7, Russia's first ICBM, early versions of the Atlas were vulnerable to attack (at least the perceived possibility of attack) from the USSR and not credible for second strike purposes. Again, both sides faced the dilemma of being unprepared for possible advances in ICBM technology by the other side; the result led to the speeding up of programs by the two countries, which in turn would lead the states closer to greater deterrence stability, assuming they both progressed at the same rate.

While fitting in with larger, national priorities, this focus was not the preferred choice institutionally. The USAF and the DOD more broadly sought to drive the space race discussion, bringing to the fore the wide range of military applications that could ensure peace and keep the Soviets at bay. Unfortunately, the DOD could not agree on the division of labor among the services, effectively weakening it and the services' overall position. Given President Eisenhower's desire to keep space as a domain of peace, the inability of the DOD to present a unified front made it easy for the Eisenhower administration to prioritize civilian space over that of the military. It was in this context that NASA was created, which would focus attention on civilian efforts in space for the next two administrations.

So, what then for military space missions and the security of the nation? The Eisenhower administration was responsive to this problem, just not in the way the DOD would have preferred. The creation of the Advanced Research Projects Agency (ARPA) provided a way in which military space could be coordinated but done without policymaking being subjected to the whims of interservice rivalry. ARPA control over military space also maintained the appropriate civilian-military balance to keep the services in line with national priorities. This control allowed policymakers to credibly address defense concerns without turning space into a warfighting domain, something that resonates in the more recent debates on the USSF.

ARPA control did not last long, however, and eventually the USAF would find itself as the main service for space. What then for the USAF in terms of space? The fallout of Sputnik led to a focus in three areas for the USAF and its space program. First, was the continued development of IRBMs and ICBMs, which we have addressed. Second, was the continuation of its reconnaissance satellite program, WS-117L, a program that was meant to take advantage of the precedent of overflight and allow US intelligence gathering in a safer and more thorough way than the U-2 program. Third, was an effort to participate in manned spaceflight, directly competing with NASA projects.

The WS-117L program would evolve into the development of a wide range of capabilities that included the Corona, Samos, and Missile Defense Alarm System (MIDAS) programs. While Corona and Samos focused on photo surveillance, MIDAS was unique in that it was an early warning system and used radio frequencies to transmit information. This was vital in an age of ICBMs where there would be very little time to respond to an incoming launch. While the success of MIDAS was limited, it paved the way for other programs like Defense Support Programs (DSPs) and Space-Based Infrared System (SBIRS) programs that would serve as the standard for missile warning from the middle of the Cold War through the current day.

The competition between the USAF and NASA for manned spaceflight did not end well for the USAF. Despite claims about space being a logical extension of the air domain (after all, one cannot get to space without first going through the air), civilian policymakers were not keen on duplicating efforts. Putting people into orbit was meant to win prestige and support the idea that space would be used for peaceful purposes—something that was clearly a NASA-run mission. Undiscouraged, the USAF sought to develop a space plane (Dyna-Soar) in lieu of being denied the man in space mission. The problem remained the same. There was no interest at the policy level for creating a weapons system that could be used from LEO or when suborbital. The USAF did make efforts to present the system in a different light, claiming, for example, that it could be used to test hypersonic flight possibilities and downplaying its offensive nature. Despite the efforts, the program was eventually canceled, with the only "space plane" being the shuttle program run by NASA. Even the quest for space stations was abandoned due to the decisions to focus on civilian-crewed projects. This meant the start-up and quick departure from plans like the Military Test Space Station in the late 1950s and early 1960s, the Military Orbital Development System (MODS) shortly thereafter, and its follow-on, the Military Orbital Laboratory (MOL) in 1963.[14]

The USAF efforts should be placed in greater context, however. The Soviet Union had clearly positioned itself to utilize space in a warfighting capacity. As early as 1960, the Soviets were engaging in research for a space station and the practical application of space-to-space and space-to-ground weapons.[15] This research included the development of a fractional orbital bombardment system (FOBS) to counter any efforts by the US to develop antiballistic missile defense.

The civilian-military space divide during this period also continued to occupy policymakers' thinking. While there was clarity about the importance of space for security, the best way to achieve it remained difficult to determine. Nuclear weapons and their role in space would begin to drive the discussion in ways that pushed for greater caution in an arms race that was already heated given the ICBM competition. Both the US and USSR started testing the effects of nuclear weapons in space in 1958, ratcheting up their tests in terms of total yields. The Starfish Prime test in July 1962, for example, yielding about 1.45 megatons, caused extensive damage to the first commercial relay communication satellite, Telstar, in addition to generating a massive electromagnetic pulse (EMP) that disrupted electronics back on Earth. The explosive yield was enough to temporarily distort the Van Allen radiation belt in the magnetosphere.

Collectively, these nuclear tests made it evident to both sides that nuclear weapons in space would likely degrade the use of the domain in the long term. With mutually assured destruction already guaranteed through ICBM and later submarine-based launch capabilities, both the US and USSR had a shared interest in keeping the environment pristine enough for other uses such as intelligence gathering that would prove invaluable for maintaining security at home. It was this mutual interest that led to early discussions about how to regulate behavior in space.

The Outer Space Treaty

At the United Nations, the launch of Sputnik generated the establishment of an ad hoc committee, UNCOPUOS, in 1958. While little progress was made there early on, the danger illustrated by successive nuclear tests led to renewed efforts to negotiate a treaty to codify acceptable and unacceptable behavior in space. Perhaps the most important event during the Cold War as it relates to space policy was the development and entry into force of the OST in 1967. To date, the OST represents humanity's best effort to codify behavior and institutionalize norms associated with the protection of the space domain and the regulation of space operations. While there are many critics of the OST, especially today, it is important to place into context the treaty's development.

For the USSR and the US, the newfound ability to destroy the world in a nuclear cataclysm made it imperative to find ways to cooperate on the most sensitive issues. Despite this, the rivalry between the two sides, however, made it difficult to find a basis for trust. Thus, while both sides wanted to control arms and establish predictable patterns of behavior to avoid misperceptions, writing agreements both sides could adhere to and that could be verified were challenging at best. The OST was written in this light—it was a way to agree on principles if not specifics.

The UN General Assembly began work on an agreement as early as 1962, with the final provisions agreed on in 1966 and opened for signature in 1967. The basic principles include: space exploration and use for the benefit of all mankind, free use of space for exploration, non-sovereign status over space, the restriction of nuclear weapons and weapons of mass destruction in space (on orbit), the use of celestial bodies including the Moon for peaceful purposes, the responsibility of states for space activities, the envoy status of astronauts, liability of states for damage from their space equipment, and agreement on avoiding harmful contamination of space and celestial objects. Achieving such an agreement was difficult, particularly because, prior to the signing of the Limited Test Ban Treaty between the two countries in 1963, the Soviet Union wanted to negotiate space "non-armament" as a part of a broader disarmament agenda. Even with the agreement in place, as we discuss in later chapters, the ambiguity of the treaty has left a legacy that many claim is very unhelpful moving forward. The OST, for example, does not ban weapons in space, only those considered nuclear or weapons of mass destruction. Nor is the OST clear on the militarization of space, which as we have noted was very important for both the US and USSR. Thus, while the OST provides a framework for space behavior and norms, it is far short of a comprehensive agreement on interactions in the space domain.

Security after the OST

The creation of the OST in the context of the Cold War generated a lot of opportunities for competition between the US and USSR to cool down. Having come to the brink of war during 1962, the space race shifted away from security for the US and back to prestige with the Apollo Project. Security-related space efforts would be directed toward defensive goals: protecting the country from incoming missiles with early warning systems, collecting intelligence with spy satellites and utilizing space to better provide for stable deterrence. After the Moon landings in 1969, the Nixon administration ushered in a new period of détente with the Soviet Union, leading to improved relations with the communist bloc and an agreement on antiballistic missile defenses (ABM) and strategic arms limitations (SALT), all focused on stabilizing deterrence and made possible by both sides having the ability to spy on the other from space. The space race thus helped facilitate cooperation due to mutual interest, while also allowing trust to develop between the two states in agreements on nuclear weapons. The 1970s, then, represented payoffs of hard lessons learned throughout the early days of the Cold War.

The end of the 1970s through the end of the Cold War would question these payoffs. For the United States, the Soviet invasion of Afghanistan represented the failure of détente. In space, the Soviets renewed testing of previously developed anti-satellite (ASAT) capabilities beginning in 1976, prompting concerns in the US over its vulnerable systems. In the first national space policy since the Eisenhower administration, President Carter highlighted the need to protect against this vulnerability, acknowledging the increasing importance of space across the range of US activities in the civil, military, and commercial sectors. The US also further developed on its own ASAT capabilities to match Soviet efforts. Another important development that illustrated this increasing importance across all the sectors was the launch of the Navigation System with Timing and Ranging (NAVSTAR) satellite in 1978.

NAVSTAR utilized the concepts behind an earlier naval satellite navigation system that helped guide submarines in the 1960s. Using ground receivers, differences in the timing of signals received from multiple satellites allows for the calculation of location anywhere on Earth's surface. This capability provides numerous benefits to military forces including the ability to track others and to accurately deliver weapons. Additionally, precise timing could be maintained globally, allowing computer systems to remain synchronized, an important requirement for the advanced computing required in all space sectors. Eventually fielding 24 satellites, NAVSTAR would constitute the Global Positioning System (GPS) that we are all familiar with today.

While the Carter administration signaled a reignition of the space competition amid the re-intensification of the Cold War, the Reagan administration left no doubt that US space policy was no longer focused solely on a civilian approach to peace to maintain security. In 1985, the DOD stood up the United States Space Command (USSPACECOM) to better coordinate space operations across all the services. This followed an internal realignment in the USAF in 1982, which established Air Force Space Command to centralize USAF space efforts. Rhetorically, however, national policy continued to point toward the use of space for peaceful purposes and space as sanctuary.

What differed, however, was what qualified as peaceful and how aggressive the US would be in seeking to ensure security. The most indicative example of this shift in US policy was the establishment of the Strategic Defense Initiative (SDI) in 1983. SDI was branded as a way for the United States to unilaterally provide for its security, breaking away from the stability of mutually assured destruction that had been the basis for security interactions between the US and USSR for the previous three decades.

SDI involved an elaborate network of sensors, space-based kinetic energy weapons, communications, computers, and command and control systems all working together to destroy incoming ballistic missiles. At this point in the space age, there was growing enthusiasm in the technological prowess of the United States to develop and field the types of advanced capabilities needed to effectively operate such a complex system. SDI would need a lot of resources to develop; politically this meant making sure that the Soviet threat was clear. President Reagan's narrative of the USSR as the "evil empire" helped clarify the threat the Soviets presented. While SDI never came together as a fully fledged system, research on the program continued throughout the remainder of the Cold War and into the post–Cold War period.

Throughout the 1980s, technological development continued, rapidly evolving to create capabilities that could support efforts in civil, military, and the fledgling commercial space sector. The use of satellite phones and ship-to-ship communications via satellites, as well as the relay of information through computer systems linked to satellites, revolutionized the speed at which military institutions could make decisions and conduct operations. Precision navigation and timing also allowed improvements with precision weapons and computer targeting. All these advances, from a US perspective, facilitated the development of a much leaner, stronger, and qualitatively superior military force able to deploy and project power anywhere in the world in a relatively short period of time. The culmination of these advances during the Cold War would come at its end, during the Persian Gulf War in 1990. It was during the conflict that the US military was able to employ an array of space capabilities (e.g., GPS, command and control, military communications, and reconnaissance) that allowed it to project power in a manner never achieved by another country to date. With the implosion of the Soviet Union in 1990 and its dissolution in 1991, the United States was left unrivaled in its space supremacy.

Conclusion

The Cold War space race was one focused on prestige and security. Though they were emphasized differently at different times, they both were important in creating a host of legacies that remain essential to the space sector today. In terms of US space policy, the salience of prestige and security created a struggle between prioritizing civilian and military space initiatives. The civilian side painted a picture of space as sanctuary, focusing on the utilization of space for the betterment of all. The military side produced an image of interservice rivalry and aggressive behavior that while possibly gaining security, was also just as likely to lead to war.

Policy decisions were not driven solely by domestic considerations, however. The actions and rhetoric of the USSR also helped shape US behaviors. The early days of prestige seeking provided an outlet for the intense competition in other areas, but the specter of weaponization and utilization of space for war always lingered over policymakers' decisions about how best to engage the adversary in space.

James Moltz, in his work on the politics of space security, opens with a definition of security as "the ability to place and operate assets outside the Earth's atmosphere without external interference, damage, or destruction."[16] This definition allows us to think about what happens in the domain and is rather helpful in this regard. In the context of the Cold War, however, the concept of space security was tied directly to both the US and USSR's national security, given the association between space and ICBMs. While the Cold War ended with the implosion of the Soviet Union and the unrivaled primacy of the United States both globally and in the space domain, the legacy of the Cold War for US space policy endures, and the tensions between prestige and security remain.

As we discovered in this chapter, the ability to think about prestige and security evolved from a history of interactions between the US and USSR over what was and was not acceptable behavior in space, in addition to what were and were not important goals to achieve. At the outset of the Cold War and the beginning of the space race, both countries were uncertain of those answers. This uncertainty and the heightened security competition between the two rivals led to two areas of concern with respect to gaining advantage over the other. First, was the use of space as a showcase for technological and scientific prowess. Which state and which system represented the best approach for others to follow? Second, was the notion of space as a potential battleground, to include the use of space-based weapons, reconnaissance, space stations, communications, and the use of space to facilitate or deter nuclear war? While both areas overlap and are dependent on the other, each area led to policies of prestige and security that remain relevant for the United States.

Settling in on a pattern of interactions was not an easy and quick transition. The evolution of cooperation and competition in space, and how the US and USSR managed to deal with some of the more complicated issues arising out of operating in this domain, took decades. Even then, the distrust between the two states led to repeated efforts to outmatch the other in terms of capabilities, sometimes leading to crisis and instability. That stated, much of the interactions and norms that are pointed to today as being relevant for holding actors accountable for their behavior in space originated during this period. Worth consideration is whether the same dynamics that existed during the Cold War exist today in ways that make this period useful for understanding the future of space and the security concerns that come with its use. In the next chapter we begin to address this question in our examination of space in the post–Cold War period.

Summary Points

- The Cold War period was a time of intense competition of which space was a part. Given the initial focus on prestige, space provides an environment for competition that is not based on conflict. While the Soviets have a lot of initial success, the US claims leadership with its successful Apollo Project.
- Security concerns quickly become an important part of the space race, putting pressure on the US and USSR to explore weaponization of the space domain.
- Cold War dynamics, including threat of nuclear war, offer opportunities for the US and USSR to settle into a pattern of interactions in space that eventually become stable, despite new challenges as technology improved.
- The end of the Cold War places the US in a unique position to both utilize space and drive its development through the early 2000s.

Discussion Questions

1. Why are missiles and space so intimately related? What are the implications of that for the space domain then? Now?
2. What was the importance of prestige to the Cold War competition between the Soviet Union and the United States? Was that more important than the security dimension?
3. How have successive presidential administrations balanced civilian and military space policy? Which do you believe was more significant in the course of Cold War history?
4. Are there any parallels you see between space history in the Cold War and space policy today?
5. Was the decision to separate the military and civilian space programs a good one? Why or why not?

For Further Reading

Brinkley, Douglas, *American Moonshot: John F. Kennedy and the Great Space Race* (New York: HarperCollins, 2019).

Launius, Roger D., *Reaching for the Moon: A Short History of the Space Race* (New Haven, CT: Yale University Press, 2019).

Logsdon, John M., ed., *Exploring the Unknown: Selected Documents in the History of the US Civil Space Program, Volume I: Organizing for Exploration* (Washington, DC: NASA, 1995). https://history.nasa.gov/SP-4407/ETUv1.pdf.

McDougall, Walter A., . . . *the Heavens and the Earth: A Political History of the Space Age* (New York: Basic Books, 1985).

Sambaluk, Nicholas Michael, *The Other Space Race: Eisenhower and the Quest for Aerospace Security* (Annapolis, MD: Naval Institute Press, 2015).

3

Post–Cold War Space History

As the previous chapter showed, the Cold War was pivotal to the course of US space policy. Competition with the Soviet Union not only provided a compelling rationale for pursuit of space exploration and exploitation but helped to keep the Cold War from getting hot by providing a nonmilitary competitive option. While the Cold War may no longer provide such a direct and compelling driver of US space policy, its legacy is important in two ways. First, space policy, like other policies, is path dependent. Choices about what policy directions to go in and how to implement them at one point in time influence the available policy options at a later point in time. What this means for space policy is that choices made during and driven by the Cold War influence later policy by making some policy options more appealing than others. One example of this is the idea that space should be seen and treated as a "sanctuary," safe from open kinetic conflict. While both the US and the Soviet Union began developing space weapons in the 1960s, both realized that their deployment and use would litter the space environment with debris and make future operations quite difficult. As a result, space was treated in such a way as to protect it from such incidents thereby allowing both countries' use of the domain. The idea of space as a sanctuary has persisted well into the post–Cold War era and has directly influenced the design of American space systems.

The second legacy of the Cold War is that US space policy lost much of the driving rationale that it acquired because of superpower competition. As long as the Soviet Union was continuing its advances in all areas of space, supporters of wider US efforts in space could point to Soviet actions to justify claims for additional resources. Reagan's decision to support the development of a space station can be seen in this light as the Soviet Salyut station was used to help justify such an expensive proposal. In his 1984 State of the Union address announcing the decision to build a space station, Reagan explicitly mentioned the need to demonstrate US leadership in technology and space.[1]

With the Cold War over and no other state coming near the US in terms of space capability, American space policy lost perhaps its strongest foundation of support. The result of these two factors is that US space policy, at least in the early post–Cold War period, continued along much the same path as it had previously but without a strong reason to do so.

This chapter explores US space policy from the end of the Cold War to today to understand how these dynamics set the stage for current space policymaking. We note here that the dividing line between the Cold War and the post–Cold War periods is somewhat hazy. While the fall of the Berlin Wall came in 1989 (itself a metaphor for the crumbling of the Soviet Union), the Soviet Union did not dissolve until late 1991. For this reason, we pick up several elements of space policy in this 1989–1991 period, particularly since they carried over into the 1990s. While space's path dependency certainly influenced the course of events for the US in space, this chapter also shows how changing geopolitical situations and increasing capabilities have combined to slowly shift US space policy from one that takes space and its advantages for granted to one that increasingly recognizes the importance and value of space for all dimensions of life.

Military Space Policy

With satellite and rocket technology still in development during the Cold War, there were not many opportunities for direct military utilization of space. As such, most national security space activities centered on satellites that could provide intelligence, surveillance, and reconnaissance (ISR), missile warning and tracking, and monitoring for nuclear explosions. That did not mean, however, that more extensive military use of space was not considered or developed. As discussed in the previous chapter, Reagan's SDI proposed far more extensive use of space weaponry primarily as a means of missile defense. In the more immediate term, however, perhaps one of the most significant military assets under development was the GPS.

First conceived in the early 1970s, GPS was developed as a constellation of approximately 24 satellites that could provide precise location and timing services to the US military and its allies. GPS as conceptualized would assist the military in several ways, from allowing troops on the ground to know their location and destination to enabling easier navigation for ships. In the mid-1980s, spurred on by bombings that had not hit their targets, the military proceeded with development of precision-guided munitions (PGMs) or bombs that could utilize GPS to hit a precise target. Developed in secret, by the time Saddam Hussein ordered his Iraqi troops to invade Kuwait in August 1990, the new PGMs were ready for use even if the GPS system was not fully deployed.

Many authors have called the 1991 Persian Gulf War the "first space war" for its use of PGM, ISR assets, and GPS. Space did play a significant role in the conflict, but not without several caveats. First, several problems arose with GPS. The full constellation had yet to be launched, leaving some gaps in coverage depending on where the satellites were in their orbits. Additionally, the military did not have enough GPS receivers on hand. Because GPS had been designed for civilian use in addition to military use, satellites gave off two sets of signals, one highly accurate for military and government use and a second that allowed for some error in calculations for civilian use. To mitigate the lack of military receivers, the military bought off-the-shelf civilian receivers and allowed them to pick up the more accurate military signal. Next, to counter the advantage that GPS gave coalition forces, Iraq deployed signal jammers capable of disrupting signals in particular areas. These deployments demonstrated that the GPS signal was highly susceptible to such actions. Finally, though PGMs were used during the conflict, only several dozen PGMs were available at the time and not all of them hit their intended target. Richard D. Easton and Eric F. Frazier thus summarize space's effects on the Persian Gulf War writing that it "served as a sort of beta test for GPS-guided munitions but offered more of a full-dress rehearsal for the system as a navigational aid."[2]

The Persian Gulf War helped to solidify for the United States the importance of space as a means of supporting terrestrial conflict. Even though Iraq's efforts at jamming were problematic, it did not stop the US from completing the GPS constellation and planning upgrades to it. The war also encouraged military planners to think more about how space assets might be used in terms of missile warning and tracking, communications, and data relays, all things that might clear the fog of war during intense conflict. The impact of space systems on military conflict did not go unnoticed by other countries either. Some scholars have argued that for China in particular, the Persian Gulf War not only demonstrated the potential military uses for satellites, but it also demonstrated how dependent on space the United States might become in the near future. If the US were to become so dependent on military space systems, that might offer potential adversaries an avenue to attack; in other words, by disrupting systems on orbit that the US had come to depend on, China might increase its ability to win in a future conflict.

With the end of the Cold War, the United States became the last standing superpower not just on Earth but in space. While the Russians maintained their Soviet space architecture (including the orbiting space station Mir), the fear that they might use it against the US was suddenly gone. While the lack of an adversary did not stop the US military from continuing to develop new space technologies, it also did not make them stop to consider how fragile such assets might be. Though the US intelligence community had been calling for some time to make American space assets more resilient to potential attacks, lacking an adversary in the 1990s, there was no immediate pressure to do so. This was especially true given the continuing belief that space might be treated as a sanctuary in terms of active conflict. Thus, at the moment that the US was further enhancing its dependence on space militarily, it also had no reason to consider or plan for potential threats.

Civilian Space Policy

Following the *Challenger* accident in 1986, most of the US efforts in the civilian space realm focused on returning the shuttle to flight as well as securing a replacement orbiter. While it was put on the back burner for a while, plans for a space station also continued in development. These efforts, however, were disrupted rather suddenly in early 1989 when word came from the White House that the new George H. W. Bush administration was looking to introduce a new approach to space policy on the upcoming 20th anniversary of the Apollo 11 Moon landing. What came to be known as the Space Exploration Initiative (SEI) was announced by then-president Bush on the steps of the Smithsonian National Air and Space Museum with the Apollo 11 crew behind him. It proposed continued development and construction of the space station, a return to the Moon, and flights to Mars.

Short on details, the SEI was subject to pushback both from members of Congress and NASA which, as an institution, preferred to focus on shuttle and space station operations rather than a new lunar effort. NASA's ensuing 90-day study put the price tag of such a plan at $500 billion over 20–30 years—an enormous cost considering larger budget cuts and tax raises that the administration was planning on. Further, as it became clear that the Cold War was quickly coming to an end, there was no larger concern about the Soviet Union that could drive such an expensive proposal. Lacking support from within NASA and Congress, Bush's SEI was quietly ended with the incoming Clinton administration.

The budget pressures and lack of a Cold War motivation that helped doomed SEI also threatened NASA's burgeoning space station efforts. Though NASA had embarked on planning efforts for a space station in the early 1980s and received the Reagan administration's support for it in 1984, the program had multiple fits and starts in terms of design. With development costs already rising in the mid-1980s, NASA brought in several international partners including ESA and Japan to build modules for what was then being called Space Station Freedom. However, by 1993, the program came under increasing budgetary pressure and within several votes of being denied funding in the House of Representatives. As it became clear that the space station lacked significant political support, then-NASA administrator Dan Goldin attempted a new strategy for civilian space policy, one that worked *with* Russia rather than against the Soviet Union.

While all of Russia suffered economically from the breakup of the Soviet Union, its space and missile industry was in a particularly precarious situation. Lacking a stable source of income, specialists in missile and rocket design—let alone weapons of mass destruction—were vulnerable to efforts from countries like North Korea, Pakistan, China, and Iran to hire them away. Recognizing the potential for this sensitive information to proliferate to adversarial states, Goldin lobbied the Clinton administration to support a cooperative space station effort between the US and Russia. Russia would get a much-needed financial lifeline that kept sensitive knowledge within the country and the United States would benefit from Russian knowledge and support in the building of the space station. In September 1993, the US and Russia agreed to the joint venture that became, with the continued participation of the ESA, Japan, Canada, and others, the International Space Station (ISS).

Even with the infusion of political support and Russian involvement, development of the ISS still took several years. The initial Russian modules were launched in 1998 with the first crew arriving in 2000. ISS construction continued to keep the space shuttle program busy with a sustained mission. Though the space shuttle delivered the "final" ISS component in 2011, new modules continue to be designed and launched, most recently the Russian research module Nauka in 2021.

Space Science

NASA's most significant issue in terms of science and exploration moving into the 1990s was the flawed Hubble telescope. In development since the late 1960s, the large space telescope suffered initial delays due to lack of funding. However, by the late 1970s, additional funding was secured through both NASA and the ESA. Nonetheless, the Hubble's launch was further delayed by the *Challenger* accident, leaving the shuttle *Discovery* to finally launch and deploy Hubble in April 1990. Unfortunately, officials discovered a major flaw in its primary mirror shortly thereafter; a misshapen lens resulted in blurry images, effectively making the Hubble inoperable. Given the expense and time it took to develop Hubble, there was significant political fallout for NASA. Coming as it did just a few years after *Challenger*, politicians continued to question NASA's competence and capability for carrying out complex missions. Despite the lack of confidence, NASA scientists did quickly develop a solution that called for the development and installation of a new system to counteract the minuscule mirror flaw, acting as glasses or contact lenses would to correct a person's flawed vision. This solution, along with other upgrades to Hubble, were installed by the crew of the shuttle *Endeavour* in late 1993.

With the successful Hubble repairs completed by a human crew, NASA officials made it clear that even with the end of the Cold War, the organization's focus would remain on human spaceflight. In terms of the larger scientific mission, however, Dan Goldin, as NASA administrator, brought a renewed focus and different approach to the scientific exploration of space. A former engineer, Goldin was nominated to the NASA post by George H. W. Bush after the previous administrator was pushed out by Vice President Dan Quayle in the debate over SEI. Goldin was retained by the Clinton administration and eventually became NASA's longest-serving administrator.

Building on his previous experience in private industry, Goldin introduced a new approach to NASA missions in general, what he called "faster, better, cheaper" (FBC). NASA's planetary and science missions, to this point, had tended to be expensive, complex, and few in number. Because NASA's science budget was perennially lower than that allocated to human spaceflight, this meant that NASA not only had fewer resources for science missions, but used them on a few, large exploration missions. Instead, Goldin proposed moving toward smaller missions that utilized more off-the-shelf technology, thereby bringing costs down. This would enable the faster development of more missions; should one or two missions fail, then the higher number of missions enabled under this paradigm would be able to make up for it.

As expected, the FBC approach enabled some stunning missions. Unfortunately, the approach also came with some heartbreaking failures. Successes included the Mars Pathfinder mission, which deployed the Sojourner rover on the red planet in 1997, the Clementine lunar probe, Mars Surveyor, and the Near Earth Asteroid Rendezvous (NEAR) mission to the asteroid Eros. The Mars missions in particular, with the Surveyor landing coming on July 4, 1997, captured the public's imagination with Mars as it became the first wheeled vehicle to drive on another planet. On the other hand, FBC contributed to several mission failures, including the Mars Polar Lander and the Mars Climate Orbiter. Though the successful missions provided evidence of FBC's potential, several factors limited its overall success, including lack of resources, low political acceptability of risk, and a clash with NASA's already engrained organizational culture.[3]

While elements of the FBC approach remain, NASA's space science approach continues to focus on large programs coming out of four different science directorates: astrophysics, Earth science, heliophysics, and planetary. Guided by decadal surveys about what NASA's priorities should be in each of these areas, mission teams put forward proposals that must compete for what continues to be a limited pot of money. These missions have included the Mars *Curiosity* (2011) and *Perseverance* (2020) rovers, the Chandra X-ray Observatory, and the Parker Solar Probe, which set the record for closest approach to the sun. Even when approved by NASA, however, funding for such large-scale programs is extremely contingent on other political and budgetary priorities. For instance, when NASA canceled a final Hubble servicing mission in the wake of the *Columbia* accident discussed later, congressional lobbying forced NASA to restore the mission even with increased risks. Additionally, massive cost overruns on the James Webb Space Telescope (finally launched in late 2021) meant cuts and pauses to other scientific missions throughout the 2010s.

While these details paint a rather bleak picture of limits to science and exploration missions in space, they remain an important part of NASA's activities, particularly on the ISS. While the space shuttle initially served as a place for the deployment of science missions and the performance of different experiments, following *Columbia,* its missions were devoted to finishing the construction of the ISS. As the ISS was growing, however, it continued to build capacity and crew with which to engage in and tend to scientific exploration. To ensure equitable access to the space station, the United States organizes its scientific mission on the ISS under a framework known as the ISS National Laboratory. Established in 2005, it oversees, approves, and plans all scientific missions and experiments taking place on the ISS. Between 2011 and 2021, more than 600 scientific investigations have flown aboard the ISS, with many of them coming from private industry.[4]

The Commercial Space Boom and Bust

As discussed in the previous chapter, the US began preparing as early as the 1980s for a growing commercial involvement in space. While this did not emerge as quickly as some US policymakers had hoped, by the late 1990s, commercial interest in space appeared to be picking up, bolstered in part by the rise of the internet and dot-com companies as well as renewed interest in and availability of expendable launchers.

In terms of launch vehicles, several factors were at play. First, with the dissolution of the Soviet Union, Russia became much more interested in selling its launch services as a means of income for its struggling rocket industry. Additionally, Chinese launchers also became more widely available on the international market following an agreement between the US and China that would prevent China from underselling its launch competition. Finally, with the *Challenger* accident in 1986 came the end of the US government's commitment to launch all satellite payloads via the space shuttle. The Commercial Space Act of 1998 made it federal law that the government procure launch services through the private sector wherever possible, and several additional pieces of legislation encouraged NASA to explore the use of commercial services for the ISS. With a new market for government contracts opening, companies like Boeing and Lockheed Martin renewed their interest in a new generation of expendable launch vehicles that could provide service to commercial industries as well as the US government.

In addition to the growing availability (and slightly decreasing cost) of launchers, more commercial companies became interested in providing space-based services. Primary among these was a system developed for Motorola called Iridium. As originally envisioned, it would include 66 satellites that could provide voice communications via handsets that originally weighed one pound. Using a mixture of American, Russian, and Chinese commercial rockets, Iridium launched 95 satellites by 2002. However, cost overruns and the widespread deployment of cellular telephone systems seriously eroded the prospects of the company, and it was soon forced into bankruptcy. Through a commitment by the US government to purchase Iridium's communications services and a reorganization, Iridium reemerged and continues to provide global communications capabilities.

In addition to voice and data communications, the rise of the internet in the late 1990s sent some entrepreneurs seeking means of providing internet access via satellites. One such scheme involved the backing of Microsoft billionaire Bill Gates. Taking advantage of a reduction in launch costs with the wider availability of launchers, the Teledesic constellation planned to provide space-based internet via an originally planned 840 satellites. Though it built and launched a demonstration satellite, the failure of Iridium and Globalstar (another communications satellite constellation) as well as the bursting of the dot-com bubble led to Teledesic's eventual demise.

While companies like Lockheed Martin and Boeing had anticipated strong demand for their services, the failures of these companies led to significant problems for the American launch market. The US military, which continued to be Lockheed and Boeing's primary customer, faced a problem: increasing launch costs as the commercial market dried up. Looking to save both companies and reduce launch costs, the US government forced the two companies to form a joint operation, the United Launch Alliance (ULA), to combine effort and reduce cost. Though some concern was expressed at the time about what would become a ULA monopoly on American launch services, the military's need to preserve capabilities and reduce cost were paramount.

By 2003, then, the once promising commercial space industry appeared to once again be in decline. With little market beyond the governments of major countries and the space shuttle still dominating the American scene, there was little incentive to develop new, lower-cost launch systems. All of this would change in January 2003.

The Shuttle *Columbia*

By early 2003, the space shuttle was making steady progress in launching and constructing portions of the ISS. On January 16, 2003, the shuttle *Columbia* launched on its 28th mission from the Kennedy Space Center. Reviewing footage from the launch, some NASA officials became worried about large chunks of foam that had fallen off the shuttle's external fuel tank during the launch. While this had been a relatively routine occurrence throughout the shuttle program, the concern with this launch was whether the foam had hit the shuttle orbiter. While NASA made some limited attempts to assess whether there was any damage once in orbit, more extensive efforts were not carried out as it was considered unlikely that the damage, if there was any, could be fixed on orbit. As one of the few shuttle missions that was not going to the ISS, there would also be no other place for the astronauts to go. Unfortunately, *Columbia* had indeed sustained fatal damage as the chunk of foam hit its left wing.

On the morning of February 1, *Columbia* began its reentry procedures exposing its fragile underbelly to temperatures in the thousands of degrees as the friction of the atmosphere continued to build. The hole in the left wing allowed super-heated gases surrounding the shuttle to penetrate the orbiter leading to a loss of the orbiter and all seven of its crew as the shuttle broke up over Texas and Louisiana on its way back to Florida.

Much like the *Challenger* accident, the events leading up to the loss of *Columbia* had been seen in previous launches and, since no damage had ever been found, considered to be a risk that was acceptable. Engineers did not consider foam, a rather lightweight material, to be dangerous. However, later tests showed that the foam, moving at hundreds of miles per hour as it did during a launch, can cause the catastrophic damage that *Columbia* suffered.

Columbia marked a turning point for NASA and the space shuttle program. After suffering the loss of two shuttles over roughly 22 years, it became clear that technology developed in the 1970s was quickly becoming out of date and potentially more dangerous. With shuttle flights temporarily paused as NASA investigated the accident, NASA also had to consider how to continue with construction of the ISS and whether the shuttle program should remain as it was. These issues were also being considered at the White House as the Bush administration considered larger policy consequences stemming from the accident.

Nearly a year later, in January 2004, then-president George W. Bush put forward a new proposal not that dissimilar from his father's 15 years prior. The Vision for Space Exploration (VSE) proposed to end the space shuttle in 2010 (allowing it time to finish construction of the ISS), and a return to the Moon and missions to Mars. Retiring the space shuttle would allow NASA in turn to focus on a new generation of crew vehicle and launch system that would enable future missions. Unlike the doomed SEI, NASA officials welcomed the new proposal, and a commission was quickly established to consider how to implement it. Further, members of Congress were also generally supportive of the endeavor providing initial funding later in 2004 and fully endorsing it in the 2005 NASA authorization act. The follow-on program, Project Constellation, proposed a new crew vehicle, the Orion, and rocket family, the Ares.

A New Commercial Boom

While the VSE might have simply marked a change in approach for NASA and civilian space policy, in reality, it opened a window for change in all elements of American space policy. Because of the high cost of space exploration, NASA agreed with Congress to pursue Constellation on a pay-as-you-go approach. In other words, NASA would only be able to develop those things that received funding. While this is good in theory, it soon became apparent that Constellation would never be funded at the level that was necessary to have it ready by the time the shuttle retired. Therefore, as a hedge against expensive and long development timelines for the Constellation hardware, then-NASA administrator Mike Griffin initiated the Commercial Orbital Transportation Services (COTS) program to provide small amounts of seed money to commercial launch services to help them develop the needed systems for cargo and crew delivery to the ISS.

Though NASA had been encouraged to consider commercial services to the fullest extent possible, officials had never been entirely supportive of the effort, fearing that it would take needed support away from the shuttle. With the reality of the shuttle's retirement looming, attitudes within NASA shifted slightly and new space companies saw a window of opportunity with the shuttle gone. Even still, most within NASA remained hesitant of allowing private companies to develop and operate space launch programs that had historically been expensive, difficult, and dangerous. The failure of private space endeavors combined with the recent bust of the commercial space sector made NASA's attitude understandable.

NASA's reticence did not stop a crop of newly formed and emerging space companies from engaging with the agency. Despite the wave of problems in the late 1990s and early 2000s, some in the technology field saw space as a new industry ready for disruption. Inspired by science fiction, particularly the work of Gerard O'Neill, Jeff Bezos, the Amazon founder, established Blue Origin in 2000. Though highly secretive at first, Bezos's goal with Blue Origin was to help establish O'Neill's dream of space colonization as described in *The High Frontier*. Somewhat similarly, PayPal cofounder Elon Musk established SpaceX in 2003, frustrated with NASA's lack of progress on planetary exploration. Both tech moguls not only invested their own funds in these efforts, but they also came to the same conclusion as to what needed to be done first to make such goals plausible: reduce the cost of space launch through the development of reusable systems.[5]

Reusable launch systems are both highly coveted but difficult to achieve. The primary advantage of reusable technology is that in doing so, costs are decreased significantly. Musk often refers to airplanes to explain this principle: if you had to throw an airplane away after each flight, the cost of a ticket would be much higher than it is. However, reusability is difficult to achieve when the body that is flying is going at supersonic speeds and being exposed to all the dynamics and pressures of spaceflight. NASA, understanding the benefits of reusability, had initially sought a system that was completely reusable with the space shuttle. However, it quickly became clear that the cost of doing so would be far higher than what the political system could absorb. The result was a space shuttle that was only partially reusable and required significant time between flights for service and repair. If the new generation of space companies could achieve something close to reusability, however, launch costs that had proved rather steady over the history of the shuttle, could be reduced dramatically.

SpaceX moved first in this effort, designing and building the Falcon 1 rocket. In addition to reusability, SpaceX also pursued several other efforts to reduce its cost including using as much in-house manufacturing and off-the-shelf technology as possible. After three failed attempts, Falcon 1 made its first successful launch in 2008, making it the first privately developed liquid-fueled rocket to reach orbit. Shortly after this successful flight, SpaceX was notified that it, along with Orbital Sciences, had won another round of funding in NASA's COTS program. With that award, SpaceX moved onto the next version of the Falcon, the Falcon 9 and its cargo vehicle, the Dragon.

By this point, it had become clear that NASA's Constellation was suffering badly and that there would be a significant gap between the retirement of the space shuttle and Constellation. The new Obama administration, sensing a problem, established the Augustine Commission to assess the state of human spaceflight at NASA in 2009. The commission concluded that Constellation was so far behind schedule that it would no longer be feasible. While the commission proposed additional goals for human space exploration, they added destinations other than the Moon and Mars including near-Earth objects (NEOs) or even the moons of Mars. The result was that then-president Barack Obama announced a major shift in how the US would approach human spaceflight. Building on the advances of SpaceX and others, Obama directed NASA to allow private companies to provide launch access to LEO and redirect their attention on more advanced missions. Additionally, instead of focusing on a return to the Moon, Obama proposed to shift attention to a rendezvous with a near-Earth asteroid.

Political opposition to Obama's proposal was swift. Members of Congress whose states and districts were contributing to Constellation were deeply opposed to its cancellation. Additionally, NASA and some members of Congress were also hesitant to depend on private companies for access to space. The result was something of a compromise: some elements of Constellation remained including the Orion crew vehicle paired with a new NASA-led launch system, but commercial companies would begin providing LEO services.

The Rise of New Space Actors

As the story of SpaceX demonstrates, the early 2000s were a period in which several new actors became involved in space. These included both states like China and India as well as non-state actors like SpaceX and Blue Origin. Both state and non-state entrants alike have had a significant impact on the space domain, the space industry, and American space policy in recent years. These consequences are important for considering where space policy is today and where it is going in the near future.

In terms of non-state actors, SpaceX is not the only company encountering success. Drawing on Burt Rutan's Ansari XPRIZE winning design, Richard Branson's Virgin Galactic offers suborbital flights to the edge of space. While the company has been slower to develop its air-launched system and suffered a serious setback with the crash of a test flight in 2014, in summer 2021, Branson was aboard the company's first official flight with passengers. Similarly, Blue Origin developed its New Shepard system for suborbital spaceflights. Designed to be fully reusable, the rocket can loft up to six passengers at a time before the capsule returns to land. Blue Origin's founder Jeff Bezos flew onboard its first flight, just a week after Branson's. Blue Origin is also developing a larger reusable rocket, the New Glenn, and has been developing plans to build a commercial space station called Orbital Reef.

While companies that have been building rockets have drawn the most attention, other companies have focused on satellites and space-based systems. There are, of course, the ones that might be most familiar including satellite television and radio providers (DirecTV, Dish, SiriusXM) but others include Planet and OneWeb. Planet (formerly known as Planet Labs) operates a constellation of small satellites called Doves, which can image the entire Earth at least once a day. This remote sensing data can then be purchased by other companies, scientists, and even governments. While small satellites might not be able to provide the best resolution, the low cost of the satellites decreases the cost to the consumer and allows even more Doves to be launched. In turn, the high revisit rate offered by the system can allow almost instantaneous tracking of things like traffic to stores, weather, farming data, and the like. OneWeb also plans to use a large constellation of small satellites to provide internet service across the world, something that SpaceX is also looking to achieve with its Starlink program.

Due in part to the rise of non-state space actors that can provide lower-cost satellites and launch access to space, several new state actors have also become interested in space. First among these is China. While China made some initial forays into space in the 1960s, Mao's Cultural Revolution all but put a stop to Chinese space development. In the 1980s, this was reconsidered, and China began development of their own rockets, heavily based on old Soviet designs. By the early 1990s, their launch services were available for international purchase, and they renewed a program of human spaceflight. In 2003, they launched their first *taikonaut* and have launched eight successful crewed missions to date. This also includes the development and launch of China's first space station, the Tianzhou.

In addition to the human spaceflight program, China has also stepped up the pace of the development and launch of satellites for civilian and military purposes. By 2018, China surpassed the United States in terms of total launches with 39 compared to 31 for the United States.[6] It is not a coincidence that China's increased pace of activities in space has come along with their more forceful positions internationally. International prestige has long been a driver of space policy among major countries as the previous chapter demonstrated. China is also increasingly relying on satellite systems for the same services that those in the US have used them: communications, remote imaging, data transfer, etc. Finally, China is increasingly using these space systems for military purposes to both counter American advantages and develop their own. This includes the testing of an anti-satellite (ASAT) weapon in 2007 that added thousands of pieces of space debris around Earth and the development of advanced satellites to rendezvous and potentially dock with adversary satellites.

As detailed previously, with the end of the Cold War, Russia's space industry found itself in trouble. Without the supporting rationale of the Cold War and without resources, Russia became somewhat dependent on its relationship with the US via the ISS as well as its sale of rockets for commercial launches. With Vladimir Putin's rise to power in the early 2000s and increased state oil revenue, Russia's space sector rebounded to both reestablish their own capabilities and develop and deploy new ones. This includes the development of new weapons systems including a ground-based ASAT, which was tested in late 2021. However, there are signs that this renewal might be rather shallow. As SpaceX has come to dominate the commercial launch market and reduce costs, Russia has had lower demand for its services, once again reducing its resources. While Russia launched a new research module to the ISS in 2021, there were problems almost immediately with it as its engines fired unexpectedly and caused the ISS to tumble briefly. Even still, Russia remains a dominant player in space with long experience and significant capabilities.

In terms of human spaceflight, the most difficult space activity undertaken to date, the only other state actor that has pursed such a program independently is India. Similar to China, India's rocket efforts date to the late 1960s where they focused primarily on the development of rockets. In 2007, the Indian Space Research Organization (ISRO) decided to undertake an initial program of development for human spaceflight and in 2017 decided to proceed to a full program. In the meantime, their fleet of rockets including one designed for polar launches, provide commercial services for other companies and states.

With the development and availability of commercially available rockets such as those from the US, Russia, China, the ESA, and India, many other countries can also be involved in space without having to develop their own homegrown space program. As such, since the 1960s, numerous countries have taken advantage of such services to launch satellites including the United Kingdom, France, Australia, Brazil, Ethiopia, South Africa, and Saudi Arabia. One country, the United Arab Emirates, even took advantage of commercial services to both build and launch a mission to Mars in 2021. Further, as costs of launch have decreased, more and more countries can afford to launch their own national satellites and benefit from new space technologies.

Since the early 2000s, space has become what the 2011 US National Space Strategy called congested, contested, and competitive. In other words, as more people, countries, and companies can access space, the physical space around Earth has become more congested. Where once only several hundred satellites operated, LEO is now home to thousands with that number projected to increase dramatically over the next several years. As such, space has become more competitive especially for states like the United States. As China has looked to increase its space presence and Russia has revitalized its own, countries are continuing the Cold War pattern of competition in space. Further, other countries, seeing the benefits of space operations and the prestige that can flow from them, have also entered this competitive domain. Finally, because states are operating in a high-pressure, competitive, dangerous place like space, the domain has become contested. Countries and companies are all seeking to operate in a place that is actually quite limited. While countries might contest orbital slots or capabilities, companies are also seeking out their own slice of the pie and rights in space. This situation has led to a renewed focus in the US and elsewhere on the importance of space.

Space as Sanctuary or Warfighting Domain?

As discussed in the previous chapter, while space could have easily become the destination for kinetic conflict between the US and the Soviet Union during the Cold War, both countries understood that to do so would threaten their own ability to operate safely. If the orbits around Earth became littered with space debris, in other words, neither country would be able to use space for valuable national security pursuits such as intelligence, surveillance, and missile warning. As a result, space was given a sort of sanctuary status, at least informally, and kinetic provocations were discouraged. The notion of space as a sanctuary was further reaffirmed with the end of the Cold War and active cooperation and collaboration in space between the US and Russia. As such, the US paid little attention to the vulnerability of its assets in space, continuing to build on the success it experienced during the Gulf War and enhancing military capability with space-based systems.

Throughout the 1990s, the fact that space was informally considered a sanctuary was taken for granted by most American policymakers. By the turn of the century, however, warning signs began to emerge that more consideration was needed given how dramatically the US military had come to rely on satellites in space. First, it became clear that China not only intended to step up the pace of its space development but was doing so in part through intellectual property theft and outright spying. In 1999, then-Representative Chris Cox headed a commission to investigate such matters. The Cox Report found that China had covertly stolen highly sensitive technical information from the United States on systems including nuclear warheads. Following closely on the heels of the revelations in the Cox Report, in early 2001, the Commission to Assess United States National Security Space Management and Organization chaired by Donald Rumsfeld warned of the potential for a "space Pearl Harbor." Recognizing the increasing pace of space activities from other countries and the relative vulnerability of the US to losses in space, the Rumsfeld commission urged policymakers to take space threats more seriously and make space security a priority for the United States. With the commission's chair, Rumsfeld, set to take over as secretary of defense in the new Bush administration, the prioritization of space might have happened if the attacks of September 11 had not pulled political and military attention elsewhere.

The ensuing war on terrorism drew attention away from the need to make space systems more resilient and less vulnerable. This was especially the case as adversaries in both Afghanistan and Iraq had limited capabilities for interfering directly with space systems beyond local jamming and electronic warfare activities. Regardless, after witnessing the US use of space systems during the Gulf War, China continued to rapidly develop its own space capabilities that not only provided them similar military and national security services but could counter US assets as well. While China and Russia had promoted some formal limits on weapons in space through the United Nations, the Bush administration regarded them as a nonstarter, not just fearing that such an agreement would limit the US ability to use space but also skeptical on how such an agreement might be enforced. In several Bush administration documents, leaders continued to assert the US right to operate in space, to respond to any potentially hostile acts, and to not engage in arms control negotiations that would fundamentally limit the US ability to operate in space.

In 2007, the space terrain shifted once again. Despite their proposals to limit weapons in space, China's 2007 ASAT test demonstrated for the first time China's ability to kinetically destroy satellites in orbit. While much of the world, including the Bush administration, condemned China's actions, Bush continued to assert that the US would not engage with more formal efforts at limiting arms in space. While the administration denied doing so, they too pursued an ASAT test the following year, successfully destroying an old NRO satellite with a ship-launched missile.

This series of tests from China and the US continued to signal not just the increasing importance of space but the increasing dangers in operating there. While the Obama administration eased Bush's hard line against arms negotiations, it soon became clear that in addition to China's increasing hostility in space, Russia was also in the process of revamping its own space capabilities. If the Obama administration did not express a significant concern about this progress, members of Congress were beginning to. In 2011, Congress included an amendment aimed at limiting American cooperation with China in space in response to China's continuing hostility and theft of intellectual property. What became known as the Wolf Amendment has meant that neither NASA nor the Office of Science and Technology Policy (OSTP) within the White House can engage with China on space-related matters without explicit authorization from Congress.

With the Wolf Amendment focused on limiting cooperation on civilian space matters, other members of Congress began to focus on the status of space within the military. When the space race began in the 1950s, there was little understanding or recognition of the military implications for operations there. As such, the Air Force was put in charge of most space-related matters with its own leaders creating the word "aerospace" to support such a move. Over the ensuing decades, other services also began to take an interest in space as the Navy helped to develop early versions of GPS and the Army began to investigate space systems for communications and missile warning and tracking purposes. The then-classified NRO, while independent, also worked closely with the Air Force on the most highly classified ISR missions. However, space, for the Air Force, remained a rather low priority especially compared to more high visibility and traditional missions such as air-to-air combat, air mobility, and transportation.

The Air Force's ignorance of the space mission was routinely criticized, and the Rumsfeld report drew particular attention to it. Noting that the government was not appropriately organized to deal with the increasing threat to American space assets, the report proposed the creation of first, a space corps that could develop both the processes and people needed to respond to threats in space with a space department to follow thereafter. At the time, like the rest of the report, the proposal gained little traction and the Air Force continued to treat space with secondary importance. This was reinforced by the disestablishment of US Space Command following September 11. As detailed in the previous chapter, USSPACECOM was first stood up in the early 1980s to reinforce the importance of space militarily as well as to reignite the Air Force's focus on space issues. Following 9/11, unified commands were reorganized with the establishment of a new US Northern Command, which led to the absorption of USSPACECOM under US Strategic Command in order to keep the number of unified commands to ten.

As potentially hostile acts against American assets in space began to increase under the Obama administration, policymakers once again began to revisit the notion of not just bringing Space Command back but also the notion of a space corps. Proponents in Congress, including Representatives Mike Rogers and Jim Cooper, began to float such an idea arguing that the Air Force had treated space like a "money pot" for its other priorities and had ignored the issue for too long.[7] While the Rogers and Cooper proposal to study a space corps gained little traction in Congress in 2017, a change in the political environment and the election of Donald Trump served to change its prospects.

At a rally in March 2018, Trump first mentioned the idea of a space force, seemingly out of nowhere. Even though the mention came and went during the rally, Trump continued to make statements in favor of it, directing the DOD in June 2018 to prepare recommendations on the matter. In August, the DOD recommended continued study of the matter along with the reestablishment of US Space Command. As debate over a new space force continued, some critics feared that by standing up a new independent military service, there would be more duplication of effort as well as an increase in bureaucracy, making it even harder to prioritize issues in space. Others also remarked that focusing on the military's role in space might also make it harder to deal with countries like Russia and China in space—if the US were to step up its own military efforts, it might lead them to do the same. Despite these and other concerns, legislation was signed in November 2019 that made the USSF a reality, albeit under the authority of the Department of the Air Force making the relationship between the USSF and Air Force akin to the one between the Navy and Marines.

Along with a stepped-up military posture in space has come a greater acceptance of rhetoric that acknowledges the prospect of warfighting in space. Where the idea of sanctuary had once reigned, US policymakers over the past several years have acknowledged that not only might warfighting and active conflict in space be possible, but it might be inevitable. This danger is only increasing at a time where the entire world is far more dependent on space-based assets and systems than ever before. Consider the role that American GPS satellites play in the global economy. Not only does GPS provide location data, it also provides ultra-precise timing via onboard atomic clocks. Banks and other companies use these timing signals to time stamp economic transactions around the world. It's not just the economy either as communications, emergency systems, and others also rely on the signals. Weather data from space is incredibly valuable not just in the case of incoming disasters but also to allow all of us to plan our day-to-day lives and activities. In short, at the same moment in time where the world has become more reliant on space than ever before, space has also transitioned away from sanctuary status to a place where conflict might occur at any time.

Civilian Space Policy in a Time of Competition

The quickening pace of China's development in space has also affected civilian space policy in the United States. With the adoption of a new approach to human spaceflight by the Obama administration in 2010, development of the Space Launch System (SLS) began, which would launch the saved Constellation crew vehicle on long duration space missions. The SLS would be based on space shuttle components but would also provide more power than the Saturn V rockets that propelled the US to the Moon. In the meantime, SpaceX and Boeing were chosen to develop crewed systems that could launch humans to LEO. Even still, during the Obama administration, NASA lacked a firm direction for where the SLS would go. The lack of destination was not NASA's biggest problem, however, as the SLS faced numerous delays in its development and construction.

Despite the problematic development of the SLS, the Trump administration also adopted a more aggressive posture in terms of human spaceflight. Specifically referencing China's actions in space and its long-term plans to go to the Moon, then-Vice President Mike Pence announced in 2019 a new NASA program, Artemis, that would use the SLS to return Americans to the Moon by 2024. The proposal was strongly reminiscent of Cold War proposals that aimed to demonstrate American leadership in the face of challenges from abroad. With NASA's full-throated support, Artemis received the go-ahead from Congress despite criticism of a too aggressive timeline. Even though international competition from China contributed to political support for the program, it was still not enough to convince Congress to provide significantly more resources. By the time Joe Biden entered office in January 2021, it had become clear that Artemis would not be able to make its 2024 landing goal particularly since the SLS had not been launched yet.

In the meantime, NASA has begun to reap the benefits of investment in commercial space companies. In 2020, SpaceX launched its Crew Dragon to the ISS for the first time with two NASA astronauts on board. The launch of

Demo-1 made them the first private company to ever launch humans into orbit. To date, SpaceX has launched several crews for NASA as well as two private space missions, Inspiration 4, which orbited Earth for three days and Axiom-1 a private mission to the ISS. Other companies including Boeing are continuing to develop their own crewed systems.

With transportation to and from LEO generally assured via commercial companies, space stations are set to be the next target of commercialization. While the ISS continues to operate well, its lifetime is quickly coming to a close with a planned decommissioning in 2031. In its place, NASA is starting to support the development of a private space station, which may utilize parts of the ISS in addition to building and launching new elements. Once again, however, the looming threat of China's advances in space have played a role in this. Some policymakers have specifically cited China's new space station and continuous presence in LEO as a reason for NASA to continue to have some presence there for the foreseeable future.

Conclusion

The end of this chapter on the post–Cold War history of US space policy, then, in many ways mirrors the state of space policy during the Cold War. The 1990s found the US in a position without challenge either on Earth or space. Over the ensuing decades, however, new and old challengers alike have developed new capabilities that potentially threaten the US position in space. The result is that drivers that initially influenced space policy in the 1950s and 1960s like international competition, military pressure, and economic benefits, are once again potent stimuli for what the US chooses to do in space. There are several differences, however, between the space environment now and the space environment 60 years ago. One difference is the increasing presence and power of private companies in space. Musk's SpaceX has grown from a small startup to possessing more space capabilities than most countries—this has consequences not just for what the US can do in space and how, but for the dynamics involved in space to begin with. Increasing numbers of satellites mean the orbits around Earth are increasingly congested and space debris is a growing threat. As more non-state actors can access space, there are a growing number of people and companies that must be considered, significantly complicating behavior in space.

A second difference is the world's reliance on space-based assets. Communications satellites in the 1960s were a novelty, something unique and interesting but not something most ordinary people might encounter. Today, much of our daily life utilizes something in space at some point or another. Whether it's going to the ATM, filling up a gas tank, accessing the internet, or calling a friend halfway around the globe, satellites play a fundamental role in facilitating what it is that humans do. Nothing the US does in space can be done without recognizing this pivotal dependence on space and much of what the US is looking to do is to safeguard that dependence. Thus, while competition with a surging China and a resurging Russia might seem familiar to Cold War audiences, the circumstances under which space policy is made today are significantly different from 60 years ago. Space today is not simply Cold War 2.0 but a far more complex, difficult, and important policy area that requires substantial attention.

Summary Points

- The First Gulf War demonstrated the new importance of space-based capabilities coming out of the Cold War, thus setting the stage for the increased use of space for military purposes. As other countries have also learned this lesson, there is now growing concern about how to protect American assets in space leading to the development of new and more advanced space systems as well as the USSF.
- George H. W. Bush's Space Exploration Initiative tried to move civilian space policy to a more expansive vision but NASA opposition and a high price tag meant that the space shuttle continued to dominate civilian space into the 1990s. The *Columbia* accident in 2003 opened a window of opportunity for policy change leading to the shuttle's retirement in 2011.
- Initially proposed by Reagan in 1984, the quest to build a space station was also transformed from a largely Western approach to an international one with the inclusion of Russia in 1993.
- Though it continues to be a secondary concern for NASA, space science efforts made headlines with the deployment of the Hubble Space Telescope and the launching of various Mars probes and rovers.
- The commercial space industry, after suffering from various stops and starts, came into its own in the early 21st century, advancing lower-cost launch options and demonstrating an increasing level of capabilities and power. With the failure of George W. Bush's Vision for Space Exploration, the Obama administration shifted toward a greater dependence on these companies for access to and exploitation of space.

Discussion Questions

1. What were the significant drivers of space policy in the post–Cold War era?
2. What difficulties has NASA faced politically in carrying out its major programs and why?
3. How did the legacy of the Cold War influence space policy in this era?
4. If space policy is path dependent, how might this recent history influence policy in the next 5–10 years?
5. What led to the shift away from space as a sanctuary? Is this a beneficial shift?

For Further Reading

Lambright, W. Henry, "Leading Change at NASA: The Case of Dan Goldin," *Space Policy*, vol. 23, no. 1 (2007): 33–43.

McCurdy, Howard E., *Faster, Better, Cheaper: Low Cost Innovation in the US Space Program* (Baltimore, MD: Johns Hopkins University Press, 2001).

Paxton, Larry J., "'Faster, Better, and Cheaper' at NASA: Lessons Learned in Managing and Accepting Risk," *Acta Astronautica*, vol. 61 (2007): 954–63.

Report of the Commission to Assess United States National Security Space Management and Organization, January 11, 2001, https://aerospace.csis.org/wp-content/uploads/2018/09/RumsfeldCommission.pdf.

Rogers, Mike, "Remarks of Congressman Mike Rogers, Chairman, House Armed Services Strategic Forces Subcommittee, Presented to the 2017 Space Symposium," *Strategic Studies Quarterly*, vol. 11, no. 2 (2017): 3–12.

4

Presidents and Space Policy

When it comes to thinking about American space policy and presidents, it's hard not to turn directly to John F. Kennedy and his ambitious proposal to send astronauts to the Moon by the end of the 1960s. As we will note in this chapter, this model is quite often held up not just as an example but as a lesson of what is required to make space dreams a reality. However, rather than begin with Kennedy's May 1961 speech to the Congress where missions to the Moon were first proposed or even his later Rice University speech where he explained that we don't undertake such things because they are easy but because they are hard, instead, consider two other presidents: George H. W. Bush and George W. Bush. In 1989, George H. W. Bush, on the 20th anniversary of the Apollo 11 Moon landing, made a speech in front of the Smithsonian National Air and Space Museum proposing a new human spaceflight policy. Called the Space Exploration Initiative (SEI), the plan called on NASA to develop a new generation of rockets and vehicles to return the US to the Moon and later to Mars. In 2004, George W. Bush announced the Vision for Space Exploration (VSE) which, like his father's SEI before him, proposed to retire the space shuttle and develop a new generation of space exploration technology to return American astronauts to the Moon by 2020 and then to Mars by 2030.

As you might have guessed by the similarity of their proposals, the SEI failed to catch on but the VSE in the 21st century achieved relatively more policy success. Why? Both the SEI and the VSE were largely similar in content, both were announced following major space shuttle disasters, and both were introduced by Republican presidents. The SEI, unlike the VSE, could have had a connection to Cold War global politics, which might have made its argument stronger. George H. W. Bush even had a much stronger interest in space policy and exploration than his son did (despite having been governor of Texas previously). For those space policy advocates who argue that, following the Kennedy example, presidents are vital to the space policy process, both plans were developed by the executive office. By all accounts, both policy proposals should have seen a successful adoption, if not implementation. What explains the divergence?

The answer is, like all things, complicated. In 1989, NASA's bureaucracy, driven by factors like identity, organizational culture, and preferences, were not particularly enthusiastic about the proposal just as they were recovering from the *Challenger* disaster. As such, they put an enormously high price tag on the project and did not mobilize to support the president's policy proposal.[1] Congress, faced with a massive cost in difficult economic times, was also hesitant. By the time Bush left office, the SEI was all but dead. In 2004, on the other hand, the *Columbia* disaster had made it clear to NASA, and others, that it was time to transition away from the shuttle. NASA not only quickly reorganized to provide support to fleshing out the VSE but worked with Congress to find a means of paying as you go for a project that would still be quite expensive, if not as high as the SEI was estimated to be.

While presidents are indeed important to setting American space policy, this comparison shows that they are not the only important actor. Nor are they always the one that has the most influence on policy decisions. As it turns out, the example set by Kennedy, of boldly declaring an ambitious goal and rallying the country to it, is an anomaly in the history of space policy. More often, the president is one in a set of actors competing to shape and influence US activities in space. In other words, while presidents may be necessary, they are not, on their own, sufficient to setting it. This chapter more fully fleshes out this idea. We begin first with a more general exploration of the importance of presidents to setting policy as well as the ways in which the institution of the presidency either helps or hurts presidents in doing so. Then, we turn to why presidents wish to influence space policy more specifically, as well as the tools that presidents have at their disposal to do so. Finally, we consider the larger debate over the role of presidents in space policy that continues to influence the field today.

Presidents and Policy

Presidents are the chief executive officer of the US government. Under the Constitution, they are charged with, among other things, faithfully carrying out and executing the country's laws. While this makes it seem as if they have no role in making the laws (that power being vested in the Congress, the legislative branch), that is not entirely the case. Rather than seeing the three branches of government as separate (but equal) branches as is typically taught, a better way of thinking about it is separate institutions that *share* power.[2]

Laws passed by Congress do not become a law until signed by the president (or when a presidential veto is overridden by two-thirds of both chambers of Congress, a very high bar). As such, Congress must consider what the president thinks about a potential law if they would like to see it signed.

Additionally, presidents are unique in that they are the only person who is voted on by the entire country, therefore, giving them the unique ability to claim that they are the country's representative. In other words, presidents can claim to better have the interests of the whole country at heart during policy discussions. When there is a time of crisis or national calamity, the president is who the public often looks to first not just for consolation or recovery but for a response that fixes or makes better the policies that led to such a situation in the first place. As the only nationally elected figure, the president is in a strong position to attract attention and make proposals before Congress can adequately respond. This first-mover advantage and the public attention that comes with it allows presidents to set the policy and legislative agenda, influencing what Congress considers and ultimately brings to a vote.[3]

Other features of the executive branch contribute to the idea that presidents are especially able to influence and make policy—presidents have an entire branch of government that reports to them and can provide expertise and information that Congress does not necessarily have at their disposal.[4] Where Congress might not have all the data at hand, presidents can argue that they have all the information necessary to make an informed decision. This includes sensitive or classified information that Congress might not have ready access to or discussions with allies and partners that might inform the global situation.

In addition to these general characteristics that influence the power of presidents to influence policy, there are some types of policies that are easier for presidents to influence than others. Americans typically understand that the legislative and executive branches (and sometimes the judicial branch) must work together to pass, implement, and enforce laws. However, when foreign countries and their leaders look to Washington, DC, they cannot possibly try to communicate with 535 members of Congress; instead, they turn to the president. The president's power in terms of foreign policy is further aided by their constitutional responsibility of being the commander in chief of the military and thus responsible for national security and defense. The result is that presidents tend to have more power and influence when it comes to foreign and defense policy compared to domestic policy where members of Congress have a much greater electoral incentive to be involved (discussed further in the next chapter). In terms of space policy, this means that presidents may be more influential when it comes to those aspects of US space activities compared to human spaceflight, which can be more domestically centered.

Even if presidents are inclined to think about space policy, there are institutional limitations to their ability to do so. Even though there is an entire branch of government devoted to the workings of government, the president is still only one person. The thing that makes them important in setting policy—being the only nationally elected figure—also significantly limits them. Where 535 members of Congress can divide the work among themselves, presidents must consider the wide (and growing) range of policy areas that concern the United States. The result is that presidential attention is necessarily limited to the few policy areas that are important at any given time whether to the good functioning of the country or to the public who has elected them. Consider the issues that tend to make the news—the economy, foreign policy, and other news cycle-specific items tend to take up the biggest chunk of presidential attention, leaving little time for less important issues like space. The result is that presidents usually only have enough time and attention to spend on space when there are problems with the current space policy or when space policy becomes relevant to the president's primary agenda.[5]

Finally, there are political factors that can make presidents more or less influential in public policy. One side effect of the separation of power is that politicians of different political parties can control the presidency and Congress. In periods of unified government where the majority of the House and Senate and the president are of one party, it is easier to come to agreement on major policy issues and pass legislation. This can make it easier for presidents to influence policy, especially if their own party sees it as important to help the president succeed and, therefore, strengthen their own political power.[6] On the other hand, periods of divided government are more difficult. During these eras, the president may be of a different political party than the majority of either the House or Senate or both.

Divided government may or may not be a problem depending on how polarized the parties are. Party polarization represents the ideological distance between Republican and Democratic parties—when polarization is low, their positions on issues are not that far apart from one another or even, sometimes, overlap. On the other hand, when polarization is high, as it has been for several years, policy positions are far more divergent. When there is overlap between the two parties, it can be easier to achieve compromise and influence policy. That can be much more difficult in periods of high polarization.[7] While presidents might not have much influence in Congress in divided governments with high polarization, they may have influence in other ways. As polarization has climbed in the United States, congressional productivity, as measured by the number of laws passed each term, has declined. This provides an opening for presidents to exert what independent policy influence they have in the form of things such as executive orders, discussed later in this book, because of congressional inaction.

All these institutional factors are important in terms of determining not just how and why presidents influence space policy but the extent to which they are able or encouraged to. Many of them, including the fact that presidents are the only nationally elected figures naturally looked to in challenging moments, gives them advantages in setting agendas and influencing policy. Political factors like divided government and polarization further contribute to their ability to influence policy more generally. All of this is aside from the unique skills that individual presidents bring to the office in terms of their ability to frame arguments, speak to the American public, and rally others to their cause.

Motivations

The institutional characteristics of the presidency have much to say about what role presidents play in policymaking more generally, but there is also much to be said about their motivations. *Why* do presidents seek to become involved in policy more generally and space policy specifically? In terms of the more general answer, there are a few reasonings. First, presidents must carry out and implement public policy. To the extent that this is their job and the public holds them accountable for it, they must be involved in the policy process to some degree; this is particularly true with national security as they are also the commander in chief.[8] Because of the role that space policy plays in national security, presidents may have additional motivations to be concerned about this policy area. Second, presidents have personal motivations to leave their stamp on policy, which is the output of the political process.[9] In a president's first term, this is often connected to an electoral motivation. As they campaign to be reelected (if they choose to run for reelection in the first place), they must be able to show the electorate what exactly they did in their first four years. Thus, the motivation to remain in power is a particularly strong one for a president at first. But what about in a second term when they cannot be reelected? In second terms, presidents tend to turn their attention to questions of legacy and reputation—what can they credibly claim to have achieved? What will they leave behind that will result in the country, and the world, remembering them? Presidents are often associated with space achievements (for example, Kennedy and the Apollo program) and often claim space achievements to be part of the larger legacy they leave behind (for example, Donald Trump and the creation of the USSF).

Presidents, then, obviously have reason to be involved in setting and influencing public policy, but what about motivations to be involved in space policy? As noted previously, presidents have little time, so they tend to worry about space when it rises to the top of their agenda in some way. This could be due to a failure or crisis like *Challenger* or *Columbia* or because space has become linked to other agenda priorities like foreign policy or the economy. Kennedy's Apollo proposal is an example of both factors aligning: not only did it seem like the US space policy was failing and that the Soviets were pushing the boundaries in space farther and faster but demonstrating scientific and technological prowess in space became a key contest in the larger Cold War. Soviet firsts like Sputnik and Yuri Gagarin's first orbital flight made space yet another front in the ideological struggle, thus Kennedy was compelled not only to respond to the Soviet achievements but do so in a way that advanced US foreign policy interests as well.

Even putting aside Kennedy's Moon proposal as an outlier in US space policy, space continues to remain connected to international politics as a means of demonstrating US prestige and as a form of global competition and cooperation. In the mid-1990s as the space station project was falling short of political support, President Clinton was convinced to support it by linking it to helping the Russians recover following the end of the Soviet Union. The Russian space industry would receive a much-needed financial lifeline while the US could take advantage of their expertise in space stations. As the ISS developed, it in turn became an important symbol of international cooperation in space, even among erstwhile adversaries. More recently, international competition in the space domain has returned as Chinese ambitions have led some to believe that a new space race or race to the Moon is occurring. The Trump administration often cited the specter of Chinese space accomplishments to support the Artemis program and its potential to return Americans to the Moon along with more expansive military efforts in space.

In addition to its linkages with international relations, presidents often involve themselves in space policy for reasons of economic development and investment. While not the primary motivator of the 1960s space race, enormous state investment in terms of space technology had significant economic benefits in terms of directing federal money to less economically advantaged states like Florida, Texas, and Alabama, but also in terms of spin-off technologies. Engineers around the country devoted themselves to developing the computers, electronics and microelectronics, and other systems that would be necessary to land people on the Moon. These developments did not stay in the space sector but have rippled throughout society contributing to almost every bit of digital products we use today. Space-based systems like satellites have proven vital to the American and global economy, underlying and enabling economic relations around the world. These economic motivations remain important in setting space policy. During the debate over whether to develop the space shuttle, then-president Richard Nixon came down in favor, partly because of the economic benefit that would accrue to the aerospace industry in California, a state that would be necessary to his reelection efforts in 1972.[10] Nixon's decision in this case also coincided with the electoral motivation. At the same time, however, Nixon was hesitant to spend Apollo-level sums on the shuttle, which demonstrates the waning importance of space policy by the end of the 1960s.

This brief discussion of presidential motivations certainly doesn't exclude other potential reasons for presidents wanting to influence space policy. Some presidents might be truly interested in the field. Others might see an opportunity to make positive change. However, even if that were the case, the structural and time limitations on the institution of the presidency can serve to impact the time available for presidents to do so.

Policy Tools

Presidents have a variety of tools at their disposal for influencing, and even setting, policy. This section will consider how presidents have influenced and set space policy following roughly along with the stages of the policymaking process outlined in chapter 1.

Presidential Proposals and Statements of Support

Recall from chapter 1 that the first stage in the policymaking process is agenda setting. As discussed, bringing space to the top of the president's policy agenda is difficult, but when it does happen, presidents are in an excellent position to use their "bully pulpit" to propose changes that range from very minor to very major. Presidents are offered a variety of means through which to communicate with the public, and it is in any one of these venues where presidents may push changes to space policy. However, not all presidential statements about space do not have a similar importance. Often, presidents will make mention of space issues or accomplishments in various speeches and statements. While this is helpful to the space policy community in demonstrating a president's knowledge of and interest in the area, it often does not signal anything beyond that. As such, here, it's useful to break down the various types of statements and proposals through which presidents might speak about space policy.

Proposals to make major changes to any aspect of American space policy are obviously the most difficult to achieve. As discussed in chapter 1, the more common policymaking pattern is incrementalism where policies are changed only a little bit from year to year. Wholesale changes or new approaches are very difficult to enact absent some sort of major policy crisis, particularly if things are going well with the status quo. When we consider major civilian space policy proposals specifically put forward by the president (and not first instigated by NASA), there are four: Kennedy's Apollo push, George H. W. Bush's SEI, George W. Bush's VSE, and Obama's shift toward commercialization. Thinking about these in terms of success, only two can be considered fully adopted (Apollo and the VSE) with only one carried out fully (Apollo). In the other two instances, SEI was largely doomed to failure and the shift toward commercialization did not come without significant compromise and changes to Obama's initial proposal. One additional presidential-level proposal might also be Donald Trump's Artemis program, the current NASA mission designed to return humans to the Moon. This program, adopted by Congress, builds substantially on the infrastructure and systems that NASA began to build under the Obama policy shift though it will still be several years until it has been fully carried out.

What about the record on military space proposals? While it's a bit more difficult to get a handle on presidential proposals in this area given the often-classified nature of these activities, we might consider a few examples. One is Ronald Reagan's SDI, nicknamed "Star Wars." At its heart, SDI proposed developing the technology and systems to create a space-based shield against missile attacks. While Reagan argued that such a system was inherently defensive, meant to protect the United States and its allies from attack, the Soviet Union saw it as offensive arguing that the same weapons that might protect the US could also be used to mount an attack against the Soviet Union. Further, Soviet leaders argued that, should the system be deployed, the US might be more likely to attack the USSR knowing that they were relatively safe from any counterattack. As such, the debate over SDI during the 1980s was high—while Soviet leaders sought to include prohibitions against it in various arms limitation talks, Reagan refused to consider the idea. Given US investment in SDI at this time, the Soviet Union also reciprocated by increasing their research and development into missile defense, an investment that some argue helped drive the Soviet Union into financial insolvency and its eventual collapse.

More recently, Donald Trump made the creation of an independent military service for space an administration priority. While this idea had circulated in policy circles for some decades and several members of Congress had been pursuing something similar for several years, Trump's support and the political push it gave the proposal did help make the creation of the USSF a reality. Proponents, including Trump, argued that making such a move not only recognized the importance and value of space both militarily and more generally, but was necessary to safeguard US national security. In both the SDI and USSF example, presidents appear to have had an easier time in pursuing policy change even withstanding domestic criticism of their approaches. This may reflect the president's tendency to have greater power and influence on foreign and national security issues compared to domestic ones.

While proposals are important indicators of presidential interest and support, there are other ways for presidents to signal intentions. Presidents often make speeches and statements about space, usually in the context of major achievements like space shuttle flights, anniversaries, or scientific advances. While these speeches might not contain any new proposals, they are important when we consider how serious choices must be made about how to allocate a president's time and speeches. Finally, presidents often make mention of space in other speeches or statements. In many of these instances, they remark on recent achievements or highlight their accomplishments. Space achievements also are used inspirationally to describe what it is that Americans can do when they set their minds to it.

To better understand how presidents utilize the bully pulpit that is their ability to attract media and political attention, figure 4.1 shows how presidents have invoked space in various forms since 1957. Taking advantage of *The Public Papers of the President*—a repository of all written and oral statements a president makes while in office—it shows the pattern of presidential remarks or speeches solely about space, written statements made solely about space, and brief mentions of space that are either written or spoken in other statements or speeches. The first thing this figure makes clear is that brief mentions of space issues are the most frequent type of activity in this area. Usually, these brief mentions take the form of aspirational statements or claims of credit. For example, in a 1983 speech in Indiana, Ronald Reagan remarked, "And advances in space travel will make the space shuttle Columbia look as old-fashioned as Lindbergh's plane, the *Spirit of St. Louis*."[11] More recently, Donald Trump often touted the creation of the USSF during his 2020

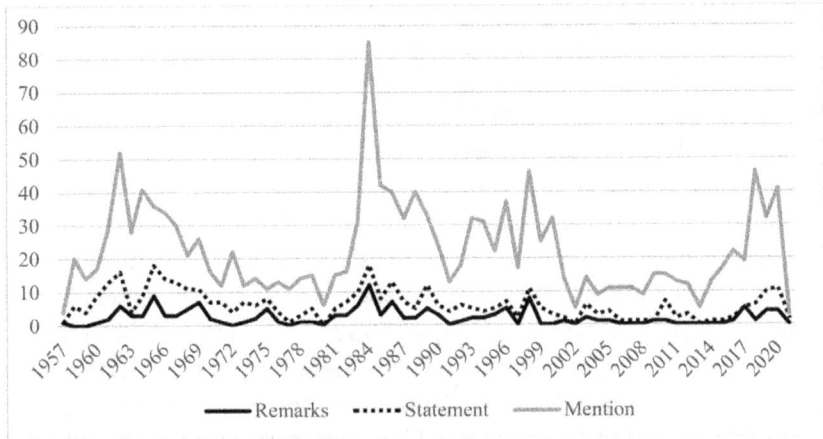

Figure 4.1. Presidential statements about space, 1957–2020. Figure compiled by the authors with data from The American Presidency Project, https://www.presidency.ucsb.edu/.

reelection campaign. Even though these types of remarks may sound trivial, repeated often enough by a president, they send a signal about the importance of space and how top of mind it is for a given president. While some of this might be cyclical, depending on major events going on in space, presidents also make active choices in their rhetoric. The pace of events in space might be considered relatively consistent between the Reagan and George W. Bush administrations given the two shuttle disasters as well as concerns over the military use of space, but Bush had significantly less mentions of space in his remarks compared to his 1980s predecessor.

Another trend that is obvious in figure 4.1 is that written statements and oral remarks solely about space policy are far less frequent and tend to rise and fall depending on outside factors and events. However, what also seems to be ironic is that as the importance of the space domain has increased in the 21st century, presidents have made ever fewer statements and speeches about it even compared to the 1970s. While those numbers rise toward the end of the series, most of the statements and speeches about space during the Trump administration concerned the creation of the USSF and activities around its establishment. While this data cannot tell us about why individual presidents may *choose* to invoke space more or less in their public statements or how successful presidents may or may not be in changing space policy, it does give us some indication about the level of importance to the individual.

Despite the attention that the president no doubt commands when speaking, research has often showed just how limited presidential influence is in this respect. George Edwards, in examining the limits of the president's ability to rally public and political support notes an important paradox: despite presidential belief that communicating with the public is an important source of power and prestige, only very rarely does the president move public opinion. Party polarization, intraparty division, divided government, public fatigue, and other institutional factors limit the extent to which presidents can use the bully pulpit.[12] In previous years, the bully pulpit also depended significantly on media coverage and whether media outlets chose to give the president's speeches coverage. While the advent of social media has made engagement with presidential speeches more frequent and significant, research still continues to show that a president's ability to shape public support and public opinion is rather limited.[13] These limitations combined with space policy's lack of saliency among the general public mean that, even though the president might be paying rhetorical attention to space issues, that may not be enough to move the needle to any significant extent.

One other aspect of presidential involvement and leadership that this data allows us to explore is the relative importance of different *types* of space policy, primarily civilian and military. Figure 4.2 combines the number of remarks, statements, and mentions but divides them on what kind of space policy they reference—civilian, military, or in some instances, both. This data shows that presidents have primarily spoken about civilian space policy issues, peaking in 1984 (this coincides with Reagan's reelection campaign in which he often mentioned the space shuttle and the recently approved space station project). These types of civilian mentions remained high throughout the Clinton administration. Clinton also, like Reagan, liked to mention the space shuttle and by then the ISS, which was under construction. While military space issues are clearly secondary in this data, their peaks are a reflection of global security concerns—the Cold War and the SDI in the 1980s, anti-satellite weapons in the 2007 time frame, and the USSF in the late 2010s.

Even still, the choice to focus on one aspect of space rather than another also tells us something about not just what presidents think matters but what they think the public wants to or needs to hear. Because understanding space can be difficult, it may be easier for the public to digest activities like human exploration and the space station compared to more complicated issues like space warfare. Just because discussions of military and national security space issues may not be prevalent in this type of data does not mean they are not going on behind the scenes. While this is a limitation of using this kind of data, it also tells us that presidents are often trying to reach a more general audience that may not know or be interested in specific and technical issues about space. This is also a reflection of presidential concerns about building

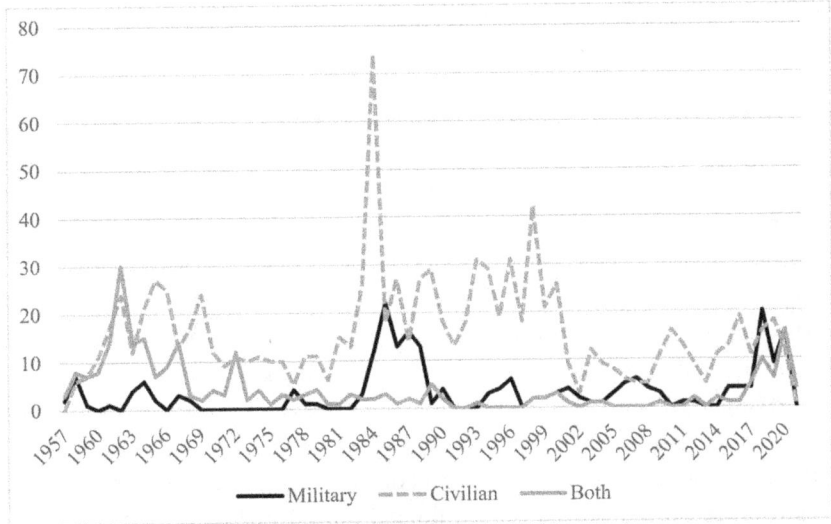

Figure 4.2. Presidential statements by type of space policy, 1957–2020. Figure compiled by the authors with data from The American Presidency Project, https://www.presidency.ucsb.edu/.

a legacy of their accomplishments in space. Thus, while public opinion may not be the most important thing in space policy, for presidents or Congress, presidents still need to ensure that there isn't a tidal wave of negative opinion against specific policy actions.

Executive Orders, Signing Statements, Policy Directives, and National Strategies

The types of presidential statements we have discussed so far can be helpful in setting policy agendas, shaping policy options, and convincing the public and other elected officials to adopt particular approaches. However, presidents do have some limited tools through which they can make policy unilaterally. To begin, executive orders (EOs) are tools that presidents sign on their own, without approval of Congress, that direct the various agencies of the executive branch to carry something out. While this definition makes it seem like the president could carry out much of their policy priorities in this way, EOs are still rather limited to what presidents are already allowed to do or direct legally. For example, one of the more well-known EOs during the Obama administration directed the Department of Homeland Security to *not* deport so-called Dreamers, young people who had been brought to the United States by their parents illegally. While Obama's authority to do so was challenged, the Supreme Court ultimately ruled that Obama did have the power to do so under previous legislation. As a result, EOs typically direct agencies to carry out specific actions, interpret law in a particular way, or institute limited new programs that might utilize funding that had been previously appropriated.

Even with these limitations, EOs are important tools for presidents particularly as the political environment has become more polarized and Congress has become less productive.[14] The result is that over the past several decades, presidents have issued increasingly more EOs. Even with this increase in frequency, space is not typically a policy area that is targeted. During the Kennedy administration, the president issued several EOs dealing with labor disputes in the space and missile industries while another EO during the Johnson administration renamed Cape Canaveral space facilities the John F. Kennedy Space Center. EOs also have been used to establish (or reestablish) the National Space Council (discussed further later).

Though related to an extent to executive orders, presidents have sometimes used signing statements to influence how legislation passed by Congress and signed into law by the president is implemented or interpreted by the executive branch. Signing statements, as the name implies, are statements presidents sign alongside a piece of legislation that detail how the president is interpreting portions of the bill or claiming that parts of the bill are unconstitutional. Though the practice originated with James Monroe, modern presidents have put them to greater use, with George W. Bush raising constitutional objections to laws in 132 signing statements over his eight years in office.[15] Since the opening of the space era, presidents have issued only seven signing statements in conjunction with space-related legislation: Eisenhower (1958), Reagan (1984), George H. W. Bush (two in 1992), Clinton (1998), George W. Bush (2004), and Trump (2020). The data suggest that, like executive orders, signing statements are not often used to influence space policy though they may be important in some minor cases.

While EOs are legal instruments presidents can use to make (mostly) minor policy changes, tools variously known as presidential or policy directives, which do not have status as an EO, can still be important for space. Policy directives or more general statements of administration position can charge agencies with making or changing things like regulations or how agencies operate and are organized. They do not necessarily need congressional authorization if agencies are already carrying out policy functions, but if these functions require additional funding or legislation, Congress will need to be involved at some point. During the Trump administration, the National Space Council developed seven "Space Policy Directives" (SPDs) that dealt with topics such as the Artemis program, establishing the USSF, space traffic management, and position, navigation, and timing systems like GPS. These

SPDs directed executive branch organizations to carry out certain activities such as developing a space traffic management system under the Department of Commerce or directing the DOD to develop plans for the USSF. While the use of specific SPDs during the Trump administration might suggest that this is a new tool in the president's policy arsenal, presidents have used directives previously, even if they were under a different name like "National Security Directive."

Finally, national strategies in areas like defense and national security can also influence space policy. While there are many different definitions of "strategy," documents such as these typically describe a president's priorities and the means through which those priorities can or should be achieved. These documents in turn provide governmental agencies with an overall goal to work toward, shaping policy and budgetary proposals going forward. While the National Security Strategy (NSS) serves as the keystone strategy document, there are various other national strategies including the National Defense Strategy (NDS), Nuclear Posture Reviews (NPRs), and strategies pertaining to space variously called National Space Policies, National Space Strategies, National Security Space Strategy, or Defense Space Strategies. These documents are important as they offer more detail and guidance than what is contained in the NSS. For example, the 2022 NSS only contains brief mentions of space with respect to improving upon technological development, integration, and a whole of government approach to dealing with competition. Additional guidance represented by the other aforementioned strategies details how these aspirations can be accomplished. Nested underneath the larger strategies like the NSS and NDS, each of these strategies is typically produced by individual administrations though timing can change and sometimes years can pass between the issuance of new strategies.

As an example, the 2011 National Security Space Strategy coined the description of space as increasingly congested, contested, and competitive, a phrase which has been used widely since in space policy debates. Following from that, it identified three national security space objectives for the United States: "strengthen safety, stability, and security in space; maintain and enhance the strategic national security advantages afforded to the United States by space; and energize the space industrial base that supports US national security."[16] Based on those goals, the document goes on to identify broad approaches to achieve these objectives and charges various agencies with implementing them. More recently, the 2020 Defense Space Strategy identified "desired conditions" in space over the following 10 years, stating: "The space domain is secure, stable, and accessible. The use of space by the United States and our allies and partners is underpinned by sustained, comprehensive US military strength. The United States is able to leverage our use of space to generate, project, and employ power across all domains throughout the spectrum of conflict."[17] Like the 2011 strategy, the document also identifies particular policies and approaches to help the US achieve the desired end state.

While these various strategy documents take different approaches and are issued under different names, the common theme is that they identify a president's ultimate goals in space and elsewhere and provide some direction to the US government on how to carry them out. Though they do not have the force of law, they influence bureaucracies and the presidential appointees who lead them to pursue certain types of activities over others. Particularly in the national security realm of space, these types of statements are quite important and have the potential of making somewhat significant changes to government policy. Unfortunately, most of these strategies are classified or have classified components making it difficult to assess exactly how successful a president is in influencing military space policy via these sorts of documents.

Budget

In addition to taking the initiative to make policy proposals, presidents can also set the space agenda and thereby influence policy via the annual budget process. Though Congress has the constitutional power of the purse (discussed further in the next chapter), presidents begin the annual budget-setting process by sending to Congress a proposed budget that encompasses all the president's priorities and what resources they believe are necessary to carrying it out. Presidents are assisted in this task by the Office of Management and Budget (OMB) whose analysts are often quite influential in determining how much money different programs and offices would receive under a proposed budget.

Given that money is limited, how presidents choose to divvy up the federal largesse is not just a reflection of policy but *priority*. In other words, presidents may think a policy area like space is important, but unless it is budgeted the appropriate amount of resources needed to fully implement it, it is not prioritized as such. The result is that the president's budget isn't just a reflection of the policy and policy changes that presidents are seeking but how important that policy is to them. George W. Bush's VSE is a good example of this problem. While Bush proposed and supported the VSE and congressional adoption followed, in later years of his administration, the budget proposal never provided enough funds to carry out the major components of it (e.g., building the Ares rocket, developing a crew vehicle) on the timeframe initially desired. When looked at this way, things like proposals and supportive statements may be just cheap talk—what signals an administration's priorities more often is where the money goes.

Budgets, however, do offer presidents the opportunity to pursue incremental changes to both substance and funding. Minor shifts in funding from year to year can lead to larger change over time. For instance, following the Obama administration's shift away from the VSE and toward a mission potentially aimed at an asteroid, funding for asteroid detection and study did slowly increase. Today, NEO research is being funded at substantially higher levels than prior to the Obama administration. Even still, however, research shows that while presidents tend to propose budgets that fund NASA at relatively high levels, the Congress almost always reduces what they eventually receive.[18]

Nominations

While the agenda-setting functions of proposals and budgets are important benchmarks, the president is assisted in several policymaking stages by the people they choose to appoint to agencies like NASA and positions within the DOD that have to do with space operations. People in these roles not only assist at the outset of the process, informing presidents of options, providing information, and highlighting potential troubles, but they are also important during policy formulation, adoption, and implementation. As Congress formulates policy, department leadership can similarly provide information and lobby members of Congress and their staff to adopt the administration's desired position. Once legislation is passed and policy adopted, nominees and their agencies become central in putting that policy into action.

Who presidents decide to nominate for these positions are important signals of presidential intention.[19] Nominees who seem close to the president may have more influence or access to the president whereas others may have to work through other people to get presidential attention. When George W. Bush came into office, he nominated a former White House budget official, Sean O'Keefe, to be NASA administrator. While O'Keefe did not have much experience with the space program, his role as someone who looked closely at federal spending helped to signal the president's intentions regarding civilian space policy. At the same time, Donald Rumsfeld, who had just finished a critical report on the state of national security in space, became secretary of the DOD, a quite powerful and influential position when it comes to military space affairs. While 9/11 ultimately scuttled many of his reform plans as well as a potential focus on defense affairs in space, this appointment also helped signal presidential intentions.

Presidential nominees, though appointed by the president and expected to represent the policy interests of the administration, serve many masters. To sit in an appointed position, they must not only be nominated by the president but must also be confirmed by the Senate. In this sense, members of the Senate can also be influential in determining members of the executive branch. Lori Garver, who served as deputy NASA administrator during the first years of the Obama administration, described how the person Obama initially wanted as NASA administrator was undercut by lack of support in the Senate, particularly from Florida Senator Bill Nelson who, not only had a constituency interest in space, but had, in the 1980s, flown on a space shuttle as a member of the House of Representatives. The result was that contrary to Obama's first wishes, he ultimately nominated Charles Bolden, a former astronaut who flew with Nelson because of Nelson's support.[20] Ironically, Nelson is currently serving as NASA administrator under the Biden administration.

Senators are not the only ones who may claim influence over executive branch nominees. Once in office, appointees must contend with their own agencies who have established a certain organizational culture along with missions that they deem to be most appropriate (described further in chapter 6). Sometimes, appointees can be "captured" by bureaucratic interests and pursue them more than any presidential ambitions. Garver also relates several examples of this regarding Bolden during the Obama administration—while Obama sought to include more commercialization in NASA's approach, Bolden was not always supportive of that initiative, instead appearing to prefer options that NASA bureaucracy was supportive of and that, they believed, protected NASA's mission and resources.[21]

Understanding these caveats to appointee power and interest, presidents still get the benefit of setting an agenda in choosing who to install in influential posts. If appointees move too far from the administration, they can be fired or pushed to resign. During the debate over George H. W. Bush's SEI, the then-NASA administrator, Richard Truly, was fired for his apparent lack of support of the president's initiative.

National Space Council

As we will discuss throughout this book, there are a variety of government agencies involved in carrying out space policy with actions that often need coordination. Even at the dawn of the space age, it was clear to policymakers that presidents might need a central body through which they can receive information, develop policy proposals, and coordinate policy implementation. The result was that, in addition to NASA, the National Aeronautics and

Space Act of 1958 also established the National Aeronautics and Space Council (NASC). The NASC was intended as an advisory body to the president to assist them in making space policy and coordinating it among different government organizations like the DOD and NASA. Originally, the NASC was to be headed by the president and comprised of additional members including the secretaries of state and defense, the NASA administrator, the chair of the Atomic Energy Commission, and other presidential selections.

Even though Congress thought such a body would be beneficial and Eisenhower signed the bill into law, he was not happy with the NASC, seeing it as unnecessary. As a result, under the Eisenhower administration, the NASC saw no real policy action. Like many other elements of space policy, though, the Kennedy administration took a different tack. Given then-Vice President Lyndon B. Johnson's interest in space stemming from his time as Senate majority leader, the administration pressed for legislative changes to the act, which made the vice president chair of the NASC rather than the president. However, despite some early action in the NASC during the Kennedy administration, the council also succumbed to lack of presidential and vice-presidential interest extending throughout the Johnson administration. By 1973, the NASC was disbanded in reorganization efforts led by the Nixon administration.[22]

Without the NASC, presidents chose to organize and coordinate space policy through various alternative means. Some administrations utilized ad hoc task groups as the Nixon administration did with the Space Task Group. Later, the OSTP was created in 1976 and used in the Carter administration. The Reagan administration chose, instead, to utilize a subcommittee of the larger National Security Council (NSC) called the Senior Interagency Group on Space (SIG-Space). SIG-Space was used heavily during the Reagan years, providing input on major space policy decisions including the SDI, support for development of a space station, and how to proceed following the *Challenger* accident.[23] Variations of this use of the NSC were adopted by both the George W. Bush and Obama administrations.

The modern National Space Council (often abbreviated NSpC to distinguish it from the National Security Council) was first created via legislation in November 1988 and brought into being shortly after George H. W. Bush was inaugurated.[24] The new NSpC was to be chaired by the vice president and include other officials such as the secretaries of state, defense, treasury, commerce, and transportation, the national security advisor, the NASA administrator, the head of the OMB, the director of the CIA, and several other science and technology advisers. Additional legislation passed while Bush was in office provided the NSpC with additional funding and authorities. The NSpC and Bush's vice president, Dan Quayle, played a vigorous role in the development and advocacy for the SEI. NSpC officials from the time, however, admit that while they were active in terms of civilian space policy, their involvement on space national security matters was far less substantial.[25] Clinton's election in 1992 ended this first reincarnation of the NSpC.

The Trump administration resurrected the NSpC in 2017 still with the vice president as chair and an expanded membership. This move highlights the renewed importance that space has played in international and domestic affairs over the past two decades. As such, the Trump NSpC played an active role in coordinating the space policy directives noted previously, helping develop and promote the Artemis program at NASA, and advocating for an independent military branch for space. When Joe Biden was elected and took office, the NSpC remained—the first time in its history that it survived into a new presidential administration.

The saga of the NSpC is illustrative of several trends. One is the presidential interest in space generally does wax and wane. Presidents who have tended to place a greater emphasis on space for whatever reason have tended to utilize this tool more than others. It doesn't only demonstrate interest but elevates space policy from one of many dozens of responsibilities of groups like OSTP or the NSC to the single responsibility of an organized and recognized group. The fact that the president has established it as a means of working on space policy issues lends it some of the president's authority and responsibility, allowing it to better exert influence and power over agencies like the DOD and NASA who might otherwise fight among themselves for influence and resources. Even with the aura of presidential influence and involvement, the NSpC can still have difficulty in adjudicating agency disputes particularly since it lacks legal, legislative power to overrule executive branch organizations.[26]

Second, since the Kennedy administration, the vice president's position as chair has given them a somewhat powerful perch to exert their own influence and authority. Even without a formal body like the NSpC, vice presidents have often been influential in space policy issues. During the Nixon administration, Spiro Agnew (Nixon's first vice president before he was forced to resign) headed the Space Task Group charged with providing proposals for future American space plans following the conclusion of the Apollo program. George H. W. Bush, as Ronald Reagan's vice president, often spoke with astronauts during their missions and visited various space installations. He was the first person sent to console the *Challenger* families in 1986 as Reagan had been preparing to give the State of the Union address. Al Gore continued his interest in space, particularly for better understanding Earth, from his time in the Senate to his years as Bill Clinton's vice president.

While some vice presidents seemed to have a genuine interest in space, over the years, space has become a more strategic portfolio for vice presidents to undertake. Constitutionally, vice presidents have little formal power beyond breaking ties in the Senate and stepping into the presidency when needed. Historically, they've also had little influence—as FDR's vice president, Harry Truman did not even know that an atomic bomb was being developed. When Truman assumed the presidency following FDR's death, Truman took his experience as a lesson, understanding that vice presidents did need to be given expanded responsibilities and greater insight into the workings of the government. Since then, vice presidents have slowly accumulated greater power and influence although this is often dependent on how much power presidents want to hand over in the first place. The result is that space is viewed as a policy portfolio that vice presidents can be given without taking away too much attention from the president since space issues typically have a lower priority. And yet, space is interesting enough with adequate public attention to satisfy the need of vice presidents for something a bit more substantial to work on in addition to giving them something to claim credit for should they run for the presidency.

Finally, the expansion of the NSpC from certain core members of the president's cabinet to a larger group that now includes the secretaries of education, interior, agriculture, labor, and several others demonstrates just how important space has become to our daily way of life and the vast amount of people who depend on or use it in some way. Space has moved beyond issues of prestige and defense to include issues of education, access to communication, farming, and monitoring global climate change. While this is undoubtedly a good thing, the expansion of interest in space issues further complicates its development and implementation. This is only compounded with an NSpC that continues to rely on informal rather than formal authority.

Conclusion

As discussed in chapter 1, public policy tends to be path dependent—in other words, the initial starting point and course of a particular policy rules out some future changes and makes it harder to choose alternative options. Given this, the early years of space policy and the patterns through which it was made, provided examples of what appeared to be "good" illustrations of how major space policies are created and adopted. In this case, it was the Kennedy approach of making a broad national goal, linking it to national security and international concerns, and rallying the support of the public and the Congress behind him. Not only was the Apollo program congressionally endorsed following Kennedy's announcement, but its eventual success in landing humans on the Moon reinforced the notion that this type of policymaking is desirable and successful.

Since the 1960s, then, proponents of more expansive space policies have often argued that a similar presidentially led approach is necessary in order for it to succeed. If only they would adopt the Kennedy model, the argument goes, then the US would be able to do more in space. Unfortunately, Kennedy has proven to be the anomaly rather than the gold standard specifically because of the unique circumstances in which he was engaged in making space policy. In addition to the pressures of the Cold War (including the recent failure of the Bay of Pigs and the spate of Soviet space successes including Yuri Gagarin's flight), there was no competing policy agenda or alternative to what Kennedy was offering. Public support for space exploration was still rather high and the war in Vietnam and efforts to reduce poverty had not yet ramped up and were not yet competing with space for scarce budgetary resources. Few members of Congress had entrenched space interests in their states and districts, and while the general military-industrial complex was growing in importance, there was not much of a space-specific industrial base.

These circumstances have changed and thus has the role that presidents play in this process. In addition to ideas about space policy that have become accepted and entrenched, there are a greater number of actors with greater amounts of power that must be involved in the policymaking process diluting the power of a unilateral president. The growth of the space domain and its importance to multiple areas of everyday life have increased the number of actors that want and need to be involved in policy initiatives. The growth of the commercial space sector has introduced new players with significant economic priorities and power. More countries than ever before have access to space and space technologies and thus must be considered. The conclusion to take from this is that while presidents are certainly necessary to modern space policymaking, they are by no means sufficient on their own. For presidents to be successful in making and changing space policy, they must work within the system of separated institutions sharing power *and* be willing to invest the political capital necessary to do so. While Obama was willing to propose ending the Constellation program altogether, he was not willing to put so much effort into it to get his own way, thus necessitating compromise with other political actors. Trump, on the other hand, in seeking the establishment of the USSF, was willing to trade it for paid family leave for federal employees. The lesson, perhaps, is that proposing is easy; the politics of moving the ball forward is the harder choice to make.

Summary Points

- Institutional characteristics of the presidency such as being nationally elected, having a vast executive branch to support them, and the ability of presidents to "go public" and set the policy agenda, influence the ways in which presidents can influence space policy.
- Presidents have various motivations in seeking to influence space policy from being expected to intercede during periods of policy crisis, to wanting to improve it, to using it as both an economic and geopolitical tool.
- Presidents can influence and set space policy in a variety of ways including in policy proposals, public statements, executive orders and agreements, policy directions, national strategies, budgeting, and nominations to agency leadership.
- The National Space Council has been variously used to support and coordinate space policy across the federal government and is headed by the vice president who is often the administration official tasked with general space issues.
- Though the "Kennedy myth" of presidential dominance of space policy continues to exist, the president does not have absolute authority to set space policy.

Discussion Questions

1. Given the increasing importance of the space domain generally, why do you think it continues to be considered a lower priority issue for most presidents?
2. Which of the president's tools do you think is most effective for influencing space policy? Why?
3. What is the role of vice presidents in space policy? Is it substantive or more symbolic?
4. Is a mechanism for coordinating space policy across the executive branch necessary? Or does it add additional complications to the policymaking process?
5. Under what kinds of circumstances might presidents be successful in changing and/or influencing space policy?

For Further Reading

Handberg, Roger, "Human Spaceflight and Presidential Agendas: Niche Policies and NASA, Opportunity and Failure," *Technology in Society,* vol. 29 (2014): 31–43.

Launius, Roger D., and Howard E. McCurdy, eds., *Spaceflight and the Myth of Presidential Leadership* (Urbana, IL: University of Illinois Press, 1997).

Logsdon, John M., *After Apollo? Richard Nixon and the American Space Program* (New York: Palgrave Macmillan, 2015).

5

Congress and Space Policy

In learning about the US government, students are often taught that the Constitution sets up a system of checks and balances among three separate, but equal, branches. While this is true, a better way of thinking about the three branches of American government is that they are a series of separated institutions that *share* power. The president can do little without Congress, Congress needs the assent of the president (for the most part) to carry out its function, and the judiciary relies on the other two both for its membership and to carry out its orders. While this is absolutely a system of checks and balances, the cooperation of all three is necessary for the success of public policy.

The same is true of space policy, as Congress considers space more often and more deliberately by a greater number of individuals compared to the singular president. There are any number of instances where the Congress has exerted this power: in the creation of NASA in the late 1950s, in the funding of a new space shuttle after the *Challenger* disaster in 1986, the continuation of the SLS in the 2010s, and most recently, in the creation of the USSF. This list, however, only hints at the immense influence that Congress can and does have; while these are big policy moments, Congress is a continuous monitor of all sorts of US space policy that has the tools and resources at its disposal to continuously consider some aspect of space policy. In studying Congress in the 1980s, Matthew McCubbins and Thomas Schwartz proposed that they engage in two types of oversight: police patrols (continuous monitoring) and fire alarms (waiting until they are alerted that something is wrong).[1] The metaphor can also be applied to the relationship between the president and Congress regarding space policy: where the president may only come to consider space issues when a fire alarm goes off, the Congress serves as a continuous police patrol, alert to any potential issues and at work trying to stave them off.

Unlike the presidency, however, Congress is made up of a series of different actors and organizations. As such, this chapter begins with a brief discussion of the different elements of Congress that influence who considers space policy and how they do so. Following this discussion, the chapter takes up the role that Congress has along with the tools it uses to affect space policy including legislation, budgeting, and oversight. More importantly than knowing how members of Congress can affect space policy, however, is understanding what motivates them to do so. Although political scientists consider members to be "single-minded seekers of reelection," space policy rarely is a salient issue for most voters. In such a case, why else might members want to influence its direction, and does reelection play any role at all? Finally, this chapter considers Congress's overall significance to space policy examining the patterns that the two chambers engage in between each other and with the president.

Institutional Pressures

The term "Congress," signifying a unified actor, is something of a misnomer. In reality, what seems like one actor is actually many: 535 to be exact, divided into two chambers that are then subdivided further into committees. The ability of the chamber to influence policy, any policy, depends to a great extent on all these parts working together as a whole. However, the fact that they are a significant number of individuals, with each chamber working under different institutional constraints, more often than not frustrates the policy process. This is truer when it comes to space policy, an area that is not high on most people's agendas most of the time.

To better understand the dynamics that play out in congressional consideration of space policy, we must first take a brief detour into the structure of Congress as the differences between chambers, members, and committees are important. There are three differences between the House and Senate that influence how each chamber considers space: size of the membership, style of representation, and constitutional responsibilities.[2] First, the biggest difference between the two chambers is sheer size—435 members represent slices of states in the House and 100 members represent whole states in the Senate.[3] One of the major patterns of behavior that results from this difference is that the House has more members to work on more policy issues at any given point in time. This size allows for less salient issues, like space policy, to be considered more frequently and at a greater depth, typically, than time and personnel allow for in the Senate.[4] This is furthered by the notion of representation in the two chambers—members of the House represent districts, small parts of states that are roughly equal in population whereas senators represent whole states. Because space policy tends to affect very localized areas where major NASA centers, military bases, or industrial contractors are located, individual House members representing those areas tend to have a disproportionate interest in space compared to members who do not. Finally, the Constitution requires that government budgets originate in the House. Because this is done (usually) on an annual basis, the House has a policy tool at its disposal—the budget—that gets looked at yearly by members who have a distinct interest in space.

These differences lead to starkly different patterns of behavior between the House and Senate. Representatives tend to be hyperlocalized, focused on issues that affect their districts since they must be reelected every two years.[5] While this might seem like it gives the House an advantage in setting space policy, it can also be a hindrance—acting in the space domain is inherently a long-term, expensive project that typically will not show results in the short term. Thus, even when members of the House might have a direct local tie to space issues, they might not be able to generate too many benefits from it over the course of their two-year terms. On the other hand, the Senate's focus on broader, more national issues, assisted by their six-year terms, give them a bit more influence when it comes to space, especially when space is linked to matters of national security or international competition as it was during the space race.[6]

Another important distinction when it comes to considering the role and power of Congress is its committee-based structure. Trying to get 100 people to work together all at once on a single piece of legislation or issue is problematic; this is only magnified in the House with its 435 members. As such, very early on in American history, both the House and the Senate determined that they would function more smoothly by instituting a series of committees with designated jurisdictions over specific policy areas. In addition to helping manage the congressional workload, committees also allowed individual members to gain expertise in a limited number of policy areas. When these committees reported out legislation, the rest of the chamber tended to trust the committee and its members on what the bill contained. Committee chairs, who had significant control over their committees, historically achieved that position by dint of seniority in both the chamber and on the committee.[7]

In the early to mid-20th century, committees and committee chairs were fiefdoms unto their own. This was still the case as issues of space policy first came to the fore in the 1950s. As mentioned in chapter 2, after NASA's establishment in 1958, the organization quickly recognized the enormous power that committee chairs held, leading it to make the strategic choice to locate certain new installations (aside from the National Advisory Committee for Aeronautics centers NASA inherited) in the home states of chairs that oversaw them. The result is that most of NASA's field centers are today located in the southern United States. Even as this was occurring, however, political winds were shifting as a realignment among voters, particularly in the South, began and institutional changes in the House and Senate followed. The result is that since the 1960s, committees and their chairs have seen their power decrease in favor of centralized power in the hands of party leadership. This is not to say that party leaders in the committee era had no power whatsoever—Lyndon Johnson when he was majority leader of the Senate led the way in early space policy. But overall, their power has been weakened and replaced not only by political parties but by empowered policy entrepreneurs who take a keen interest in promoting one policy vision or another.

The result has been that space policy today can now be influenced in the Congress from various directions. While committees are still important, the support of party leadership is now all but required to advance any significantly new pieces of legislation or policy. Additionally, members of Congress who are not a committee chair can still have significant influence. For instance, Representatives Jim Cooper (Tennessee) and Mike Rogers (Alabama), despite not being chair of either the House Science, Space, and Technology or Armed Services Committees, were the main congressional proponents of establishing an independent space force in the US military. Their efforts eventually resulted in the establishment of the USSF in 2019. Of note, however, is that the USSF's establishment resulted partially from a political bargain entered into by party leadership in return for allowing paid family leave for government employees.

While this discussion of Congress as an institution may seem like a digression of sorts, understanding these differences contributes to understanding the rules of the game in which space policy must compete. Since the rules of the game influence the ultimate outcome, we must understand the conditions under which policy is created.

Congressional Motivations

Beyond the vague notion that members of Congress are in Congress to "represent" their constituents (whatever that means to them), why do members of Congress behave the way they do generally and why do they behave the way they do specifically as it pertains to space policy? For some, their behavior may be determined by what they believe to be the "right" thing to do or, in other words, to make good policy, to make changes in a positive direction. Others might have more nefarious motivations in the direction of gaining power, prestige, money, or future job opportunities that might come their way because of a congressional career. While all of this may be true, congressional scholar David Mayhew argues that members of Congress are "single-minded seekers of reelection" because it must be the "*proximate* goal of everyone, the goal that must be achieved over and over if other ends are to be entertained."[8] In order to be in a position where a member can make good policy, gain power, get a better job, or earn more money, they must be concerned with being reelected.

Since Mayhew first outlined his argument, political scientists have continued to find evidence of this electoral motivation in how members vote on legislation, what types of committees they seek to join, the types of policies they pursue, and the types of statements that they make. (Mayhew analyzed many of these actions in a later study.) Reelection as a motivation has proven time and time again to provide powerful analytic leverage in explaining the behavior of Congress members, but there is a catch: does reelection matter in an area of policy like space that not many people pay attention to?

Aside from great events like various space launches, landings, or even disasters, the public pays stunningly little attention to space (we will discuss this further in chapter 9). Particularly when compared to perennially significant issues like the economy, taxes, and health care, space ranks relatively far down the list. If voters either do not care or do not pay attention to space policy, does it matter what their representative or senator does on that issue? In other words, because of the low saliency of space policy, the reelection motivation may not be the biggest factor in understanding how members behave. Without voters to pay attention to it, there is little incentive to do things that would benefit them. Research carried out by Alan Steinberg has shown this to be the case regarding space. He found very little correlation between the general public's attitude toward space spending and NASA's budget.[9] Using the same data, other analyses have shown that the population that does support more spending for space exploration makes up a small proportion of the public, further diminishing the possibility that reelection drives congressional behavior in space policy.

If the public in general cares very little about space issues, it may instead be the case that more localized populations do. For example, communities surrounding major NASA or USSF installations, or space companies may actually care a great deal about space policy and pay attention to it. Government funds that go into supporting such locations provide not just jobs, but often high-paying jobs that attract a well-educated workforce. This in turn can further stimulate the local economy surrounding these installations as support services, food and beverage, housing, shopping, and other needs must be met. For members representing such places, they may have a greater motivation to do things to please their voters because their voters and the community they represent have a direct stake in it. Steinberg, working with Martin Machay, found that, regardless of public opinion, members with a NASA center or significant air and space manufacturing interests were more likely to vote for NASA authorization.[10] By being able to claim that the member brought money, jobs, and opportunity to their districts or states, members can further bolster their reelection prospects. Thus, in space's case, the reelection motivation does play a role but only indirectly through the presence of government or industry.

This indirect influence of reelection also suggests that space policy tends to take on more of a localized concern; in other words, although space has national, international, and perhaps galactic implications, what drives policy behavior in Congress is the local. One of the leading justifications for supporting space activities has often come from its economic impacts both locally and nationally in terms of how technological developments can spur further economic growth and development. It is not unusual to see NASA touting the local economic impact in regions where it has major installations.[11] These types of arguments are often easier to make to voters who may not be able to see the value of investments in space because of scientific, exploration, or educational ideals.

While one face of space policy might be locally focused and economically driven, the other image often associated with space is that of international prestige and competition and national security. Recall from chapter 2 that the original space race was driven largely by concerns that the Soviet Union was progressing faster and farther in terms of technology and weapons development than the United States. As a result, the United States engaged wholeheartedly in a race to speed up space development, ultimately resulting in the Apollo program and the 1969 Moon landing. Since then, national security and international concerns have continued to play a role in the development of civilian programs like the space shuttle, the ISS, and more recently the Artemis program. Similarly, military space developments have been driven significantly by these security concerns, particularly as other countries step up their own space development.

Whether space is viewed primarily as a domestic policy issue or a foreign policy issue has the potential to change congressional behavior and thus a member's motivation for action. When space is seen as a local concern, unless a member has a direct connection to it in their district, there is little reason for them to support significant changes in space policy or even significant funding increases. On the other hand, when space is seen as an extension of international competition and intrinsic in securing the safety of the United States, members may tend to be more supportive of extensive activities. After all, if US national security is at risk, that will be a concern to their voters come election time.

Policy Tools

The previous discussion alluded to several different roles that Congress plays in setting policy and the tools it has at its disposal. As the legislative branch, the primary role Congress serves is the passage of legislation. The structure of Congress discussed previously, though making legislation doubly difficult, reflects the beliefs of the Founders, particularly their skepticism of government and their desire to have different populations represented differently (states versus individual people). Legislation was not supposed to be easy to pass, instead, it was to go through different "filters," ensuring that what was eventually passed was well considered and not hastily done. Additionally, the framers of the Constitution included other powers for Congress to use to better carry out its function to check and balance the other branches of government.

Legislation

Legislation can be written with all sorts of ends in mind—bills can establish organizations and agencies (enabling legislation), they can direct the executive branch to do certain things or direct them *not* to do something, they can establish certain acts as crimes and thus prohibit them, they can rename buildings and streets or declare federal holidays.[12] Of course, there are also some things that bills cannot do, namely declare someone guilty of a crime (bills of attainder, which are specifically prohibited by the Constitution). For space policy, two types of legislation are particularly important: enabling legislation and authorizations. As stated previously, enabling legislation establishes organizations. For example, the 1958 National Aeronautics and Space Act created NASA from an amalgamation of different agencies including NACA. The legislation sets forth NASA's broad operating principles and purposes, many of which continue to be referenced today to justify actions taken on the part of the agency. For instance, the act notes in several places that the newly created space agency should contribute to international cooperation and the facilitation of the peaceful uses of outer space. One of NASA's major justifications for cooperation with other countries on projects like the ISS date to this law. Beyond setting out general organizational parameters, however, the act broadly allows NASA to determine its own best internal structures.

Once agencies are established, their activities must also be sanctioned through authorizations. This type of legislation gives permission to the executive branch to carry out particular policies and sets a recommended budget level for them. Historically, authorization legislation for domestic agencies like NASA was passed on an annual basis. However, as polarization has increased in the United States and the productivity of the Congress has decreased, the Congress simply does not have enough time every year to pass a new authorization as frequently. The result is that NASA authorizations are typically passed every three to five years. On the other hand, military space activities are authorized in an annual National Defense Authorization Act, which continues to be passed on a yearly basis (though not without significant difficulty). Authorizing legislation is typically drafted by committees that oversee the topic area. It is important to note that while authorizations tell an agency what it can and cannot do, they do *not* give it the funds to carry those activities out. Those funds come separately from budgeting legislation.

Following larger trends in legislation, authorization bills have gotten longer over time. This reflects the more difficult nature of policymaking and legislating as issues become more complex (as space certainly is), but there are other reasons for this as well.[13] Congress, even with members who specialize on certain committees and knowledgeable staffers who provide further support, is at a considerable information disadvantage compared to executive agencies. For example, NASA has a body of technical and subject-specific knowledge about what is both possible to accomplish, how much it will cost, and how long it will take. Understanding this, Congress has come to include more specific language about what agencies should and should not do to try to overcome this disadvantage. One example of this was in the 1980s when Congress included a provision in a bill forbidding the military from carrying out anti-satellite tests in space, which they had been planning.[14]

Even beyond authorization bills, other pieces of legislation can include provisions that relate to space policy. For instance, several pieces of legislation have been passed dealing with launch regulations since the 1980s. These bills have required the Federal Aviation Administration (FAA) to create and enforce regulations for commercial space companies to ensure that their launches are not only safe but do not violate any international treaties to which the United States is party. At the same time, such legislation also has restricted the FAA's ability to regulate things like human spaceflight undertaken by private companies so as not to stifle a newly developing industry. Additionally, amendments pertaining to space may be included in other pieces of legislation that deal with a variety of policy areas like infrastructure, science, military research and development, and economic development.

Budgets

While authorizations and other types of legislation can originate in either chamber of Congress, according to the Constitution, budgets must originate with the House of Representatives as the "people's chamber." Though authorizations are important, budgets may be even more so. Endowed with the power of the purse, the Congress must agree on a government budget each year or extend the previous year's budget; if they do neither, the government, aside from critical functions, shuts down.[15] While authorizations are written by subject-specific committees, each chamber has an Appropriations Committee that oversees this process with subject-specific subcommittees to split up the work. This division of labor, while it makes the process more difficult, is intended to account for the whole of government. While subject committees authorize policy and activities and recommend appropriate budget levels, they might not consider all the needs of government in doing so. The Appropriations Committees on the other hand, seeing what policy has been authorized and weighing the needs of the whole government, decides how to spend the limited resources. The result is that agencies and policies may be funded at lower levels than authorizations allow. For a policy area as expensive as space, this appropriations process is critical and the level of funding that is granted to NASA, the military, and other space activities is often indicative of the amount of support and/or interest Congress has in the area.

Appropriations bills sometimes also become a place where policy is effectively written into law in the form of riders. By congressional rules, appropriations bills cannot enact new policy so to get around this, members often use riders that stipulate what appropriated monies can and cannot be used for. The result is that the appropriations bill becomes a de facto policy. Jason MacDonald argues that these policy riders effectively give the Congress a greater hand in bureaucratic action than is commonly appreciated particularly because appropriations bills are often considered "must pass" legislation.[16] One major aspect of American space policy has been limited by a policy rider over the past several years, and that is the potential for cooperation with China on space matters. As a result of concerns over Chinese spying and theft of intellectual property that could occur through scientific cooperation with China, Representative Frank Wolf inserted a provision in a 2011 appropriations act stating:

> None of the funds made available by this division may be used for the National Aeronautics and Space Administration or the Office of Science and Technology Policy to develop, design, plan, promulgate, implement, or execute a bilateral policy program, order, or contract of

any kind to participate, collaborate, or coordinate bilaterally in any way with China or any Chinese-owned company unless such activities are specifically authorized by a law enacted after the date of enactment of this division.[17]

As a policy rider, the so-called Wolf Amendment has continued to be included in annual appropriations legislation, halting any attempts at cooperating with China in a civilian or scientific manner in space.

Oversight

While the legislative branch's main policy tool is legislation in its various forms, it does have other items at its disposal. As a means of checking the activities of the other two branches, Congress can hold hearings and conduct investigations into various policy areas, government organizations, and societal issues. The most public and visible form is a congressional hearing that is most often held at the committee (sometimes subcommittee) level. These forums feature opening statements from committee chairs and witnesses and sometimes intense questioning. Often public, hearings can be used to gather information and highlight major problems for media and public attention. While hearings are an opportunity to make public different perspectives or bring to light major issues, they have been criticized in recent decades for becoming mere media stunts with little substance attached beyond trying to attract attention. Because committee chairs control the agenda of their committees, they have full power to schedule hearings, determine the topic, and invite witnesses in turn giving the majority party a great deal of power. While many committees and committee chairs operate with cooperation from the minority party, this contributes to the critique that congressional hearings have simply become an opportunity for political grandstanding.

Congress has often utilized this tool in examining space policy, particularly in the wake of policy failures. Figure 5.1 displays the frequency of congressional hearings on space broken down by chamber. Overall, you can see that the number of hearings is somewhat cyclical depending on the salience of space in the country. After holding relatively fewer hearings in the 1970s, the number of hearings picked up in the 1980s. While this might be expected in the wake of something like the *Challenger* in 1986, the data show the increase dating to the early 1980s. Several of these hearings related to the Reagan administration's SDI, thus showing Congress's interest in overseeing several aspects that came to be somewhat controversial. Hearings remained frequent into the early 1990s as Congress debated, and came close to canceling, the ISS. Another dip in the 1990s reflects several trends: space in general became less

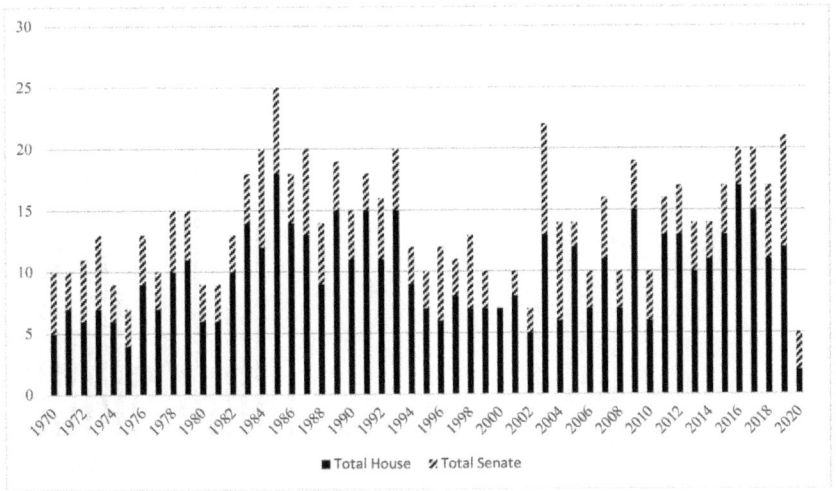

Figure 5.1. Congressional hearings by chamber on space, 1970–2020. Figure compiled by the authors.

newsworthy as NASA gained steam and momentum with several successful missions and the start of construction on the ISS. Additionally, with the Cold War over, military space issues like the SDI also became less of a concern. Hearing frequency once again picks up in the 2000s following the *Columbia* accident and increasing military and national security concerns having to do with space.

Hearings can also tell us about concerns in Congress on military and national security issues in relation to space. Figure 5.2 shows the frequency with which Congress (as divided by chamber) has held hearings on military and national security space-related issues. While the series shows no hearings held between 1970 and 1981, this is somewhat misleading. Because of the sensitive nature of space capabilities, much of what Congress might be doing in terms of oversight is likely classified and unpublished. Similarly, hearings on military and national security space spending would not be available either. Given this hesitance to discuss these issues in the open, that Congress would hold these open hearings at all is indicative of the seriousness that they took their oversight responsibility. Even though the more sensitive issues would still likely be taken up behind closed doors, these data do suggest some larger findings.

First, like figure 5.1's overall pattern, the number of military/national security space hearings begins to tick up in the 1980s as Reagan's SDI program got underway. Aside from the peak of the series in 1998, the Congress does not

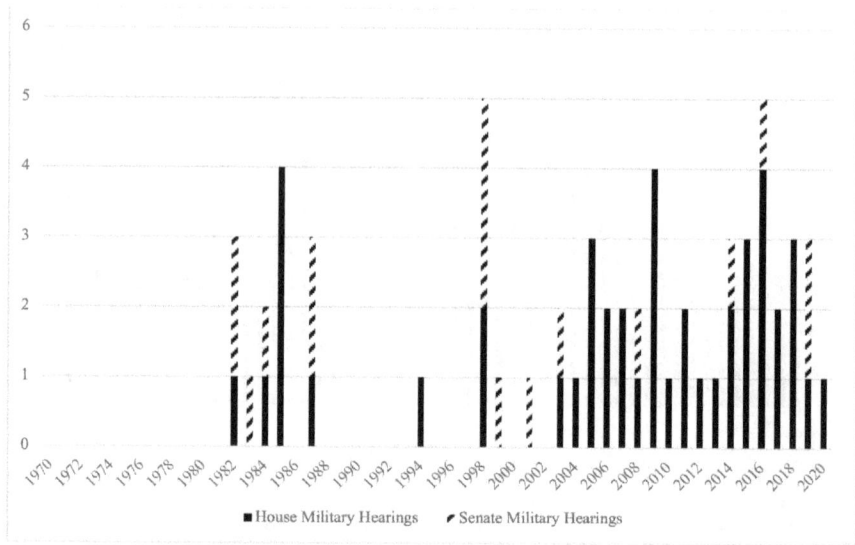

Figure 5.2. Frequency of military/national security space hearings, 1970–2020. Figure compiled by the authors.

largely take up military/national security space issues again until the 2000s. Given the establishment of the USSF in 2019 and continuing concerns about both Chinese and Russian actions in space, it is likely that these types of hearings will continue reflecting Congress members' concerns.

Sometimes, in concert with hearings, members of Congress will request or undertake investigations of certain policy matters. While committees certainly have the power to do this (with hearings being the opportunity to hear from witnesses and gather information), Congress has another means of investigating issues as well: the Government Accountability Office (GAO). The GAO, first established in 1921, provides nonpartisan advice, investigations, and advice to members of Congress. While we revisit the GAO's role in space policy again in chapter 7, a few points are relevant here with respect to Congress. For example, members of Congress can request that the GAO undertake a particular examination of a topic, or GAO may do one as required by law. The GAO also serves as an appeals court of sorts for government contracting decisions and can provide expert testimony in hearings before Congress. Because individual members request many investigations and reports, the frequency with which the GAO reports on an issue or agency can be indicative of increased attention being paid by either an individual member, committee, or chamber. The results of such investigations are often then used in the writing of legislation as members of Congress seek to remedy or mitigate potential policy problems.

In the past, other similar organizations have existed to provide expert advice to Congress including one important to space: the Office of Technology Assessment (OTA). Between 1974 and 1995, this office provided nonpartisan advice and information to Congress on critical science and technology issues. The OTA was supposed to respond not only to the problem of the executive branch often having more information and insight on issues than Congress but also to the growing complexity of science and technology. The OTA was closed in the mid-1990s when Republicans retook control of Congress.

Informal Tools

In terms of the role of Congress in space policy and the tools that they have at their disposal, legislation, appropriations, and hearings and investigations are formal, institutional options. They are outlined by the Constitution and previous law and are all conducted within the confines of Congress. This does not mean that members do not have other ways of influencing space policy. One of the major means at their disposal, particularly over the past several decades, is the media. Members of Congress, depending on their status, seniority, or importance to the policy issue at hand, can command a great deal of media attention that can be used to highlight policy problems and advocate for specific solutions. While this has mostly been in the form of television, newspaper, and radio, the growth of the internet and social media has amped this up considerably. With traditional forms of media, members were dependent on the willingness of the media to publicize them and their issues; with the internet and various forms of social media today, no such gatekeepers exist. While it is certainly helpful to have traditional media also covering a story, having something go viral on the internet can help jump-start that attention. In addition to media attention, members can also give speeches, participate in policy events, or write essays or editorials as a means of driving attention.

A final informal tool members can exercise is simply lobbying the executive branch. Much like traditional lobbyists who try to convince policy leaders to adopt a certain policy or enact it in a certain way, members of Congress have a similar power in that they can speak to executive branch officials from the president on down and try to press them to behave in a manner consistent with their wishes. Because the executive branch and different government organizations have an inherent interest in pleasing Congress (they do provide the budget after all), members will often find an audience willing to do what they can to help, within reason and within legal bounds. This relationship is reflected in the idea of an "iron triangle"—the notion that members of Congress, relevant businesses and contractor groups, and executive branch agencies dominate policymaking. The iron triangle has often been used to analyze defense and national security policy so the overlap with space policy means that similar dynamics can be at play. While significant attention has been paid in recent years to conflicts between the executive and legislative branches, most of that has been centered on high-level, major policy issues. The reality is that the US government has a great many things to oversee and do, many of which often escape media or public attention—space is often one of them. This gives those government agencies involved with space issues as well as members of Congress a bit more leeway knowing that the eyes of the public are not always on them.

Constitutionally, beyond representing the voice of the people, the role of Congress is to write and pass legislation that essentially runs the country. In the end, the executive branch can do little if there is no money coming into the government and none to be spent. Even if it did have those resources, the executive branch must be told what to implement and execute. Thus, when it comes to space policy, the role of Congress is to provide those organizational means, the resources, and a direction whether that is in the realm of the military, civil space, or commercial space. Beyond legislating, Congress is endowed with additional tools that it can use to ensure that the executive branch does what the people's branch has charged it with doing as well as setting the agenda and bringing attention to little known problems or questions. That stated, understanding what means the Congress has at its disposal to influence policy means little unless its motivations for using those tools are also properly understood.

Congress's Significance in Space Policy

Over the past few decades, Congress's productivity has severely decreased—when looked at in terms of how many bills are passed in each two-year congressional session, the number has indeed plummeted. However, examining only the number of bills passed can be misleading particularly because of the other tools Congress has at its disposal to influence policy. There is ample evidence to support the argument that Congress is a significant influence in American space policy.

Because of the high costs of almost any activity in space, the budget is perhaps the most important policy tool. If NASA or the DOD do not have the funds to undertake something, they simply cannot do it. Because of Congress's power and influence in setting the budget, this makes the institution quite influential for space policy. One objection to this might be that the president does have an agenda-setting power in proposing a budget each year prior to Congress's consideration. However, at least with respect to NASA, Congress rarely accedes to the president's request. In an analysis of NASA's budget between 1958 and 2008, Congress only appropriated more money than the president requested in one year. In all the others, it appropriated less. The single exception occurred following the *Challenger* disaster when Congress, over the president's objection, appropriated funds for a replacement orbiter.[18]

In terms of military space, the budgetary influence is a bit less clear. The major reason for this is that many military space activities take place in the "black," or rather, are classified and therefore not publicized. While some members of Congress (like those who are chairs or ranking members of space- and defense-related committees) might be fully informed of these classified programs, most members are not, and the specific budget figures are not publicized. Given the increasing amount of money spent on defense in general, we can assume that money for military space activities has also increased. However, that does not mean that Congress has completely acquiesced to all military and classified activities in space.

While interest in space weapons has waxed and waned over the years, increased Soviet activity in ASAT weapons in the 1970s caused the United States to also increase ASAT development. As President Ronald Reagan came into office and started the SDI in the early 1980s, policymakers began to be increasingly concerned with the overall effects that weaponization of space and ASATs could have. As a result, Congress included in legislation a stipulation that the DOD was not to conduct ASAT tests in space unless the Soviet Union did so first. Thus, even though ASAT capabilities were being contemplated and developed with the funds provided by Congress, Congress still acted to limit how such weapons might be tested and/or used. The Wolf Amendment, described previously, has similarly limited the potential for American and Chinese cooperation in space even if the executive branch sought to do so.

Aside from legislating what can and cannot be done in space and providing the funds (or not) for it, Congress also has informal mechanisms of policy influence via hearings and behind-the-scenes discussions. Following the *Columbia* disaster in 2003, the Bush administration decided, among other things, to retire the space shuttle by 2010 and initiate a new program called Constellation. Because of the danger that was now apparent in the space shuttle vehicles, NASA leadership decided that all remaining shuttle flights would go to the ISS so that if detailed inspections of the heat tiles on the shuttles showed any sign of damage, the crew could stay safely aboard the station. As a result, a final planned servicing mission to the Hubble Space Telescope was canceled because the shuttle would not be able to reach both the Hubble and the ISS. There was immediate public and congressional backlash to the decision given both the time and effort that had previously gone into building and maintaining the telescope as well as the valuable science that had emerged from it. In congressional hearings and other venues, members of Congress including Senator Barbara Mikulski strenuously lobbied NASA to reverse the decision, which they ultimately did. To mitigate any potential danger to the shuttle being used for the mission, a second shuttle was prepared for launch and moved to the launch pad to be launched in case the primary shuttle did suffer damage.

More recently, congressional influence has helped to keep the SLS alive at NASA. By the time Barack Obama became president in 2009, Bush's Constellation program was badly behind schedule and significantly underfunded. As a result, Obama proposed canceling Constellation altogether, moving to a reliance on commercial companies for launches to LEO, and directing NASA to develop a new heavy-lift vehicle. Congressional opposition became immediately apparent because, if the program was canceled, companies across the United States stood to lose valuable dollars in terms of supporting contracts. As a result, a compromise was agreed on that saved the Orion crew vehicle and started development of the SLS as a heavy-lift vehicle with a mandate to include as much shuttle-era technology and systems as possible. Much like Constellation, however, the SLS, developed by NASA's Marshall Space Flight Center in Alabama from the main components of the space shuttle, quickly fell behind schedule with dramatically increasing costs. While the requirement to base the SLS on space shuttle technology helped to preserve jobs and money that members of Congress were worried about losing, it also meant that NASA was not able to innovate new systems and technology. With commercial options from companies like SpaceX coming online, some called for the outright cancellation of the SLS in favor of commercial options. However, members of Congress including Alabama's Senator Richard Shelby worked to protect the SLS because of the money and economic benefits it brought to their constituents.

The President and Congress

If we consider Congress alone, the argument can be made that they are quite influential in the setting of space policy. However, what about their significance in comparison to other actors such as the president? Or even the power of the House versus the Senate? To the first part, careful readers will note that several of the instances discussed previously that are used to demonstrate Congress's influence involve the president. Reagan's "Star Wars," Obama's cancellation of Constellation, even George W. Bush's decision to retire the space shuttle are all examples of presidents exerting influence. In each of these cases, however, presidents mostly served as agenda setters and policy initiators. As discussed in the previous chapter, this is a powerful role for the president. Setting the agenda and making the first move confers a powerful first-mover advantage—in doing so, presidents can exclude some otherwise feasible policy options and therefore limit the choices that Congress might eventually make. Further, even though many of a president's policy pronouncements might be rolled back eventually by congressional action, some may argue that the president anticipated the rollbacks and therefore staked their own position further out than their own ideal point, understanding that the outcome would eventually be closer to what they wanted.

While this analysis might speak to the power of the president, what also stands out in these examples is the momentary nature of presidential involvement. Looking at a history of presidential influence on American space policy, one is struck by the infrequency of direct presidential influence, in other words, presidents are not *consistently* involved in setting and enacting space policy. This is only natural—the president being only one person and unable to consider all policy problems all the time, must make choices. Since space is a second- or third-tier policy item that is not all that salient to the public, it is not something that often appears on the president's radar. When it does, it is often because there is some sort of policy crisis or breakdown in the expected course that brings space to the fore of the national agenda and forces attention from the president. Kennedy's decision to go to the Moon, George W. Bush's VSE, Barack Obama's attempt to kill Constellation all occurred after significant policy failures and proved to be quite episodic in nature.

Compare this pattern of behavior to an institution like Congress. The size of Congress gives it more people who can look closely into less salient issues. Further, individual members are intrinsically motivated to be interested given connections in their states or districts to space activities. This creates a deeper and more permanent motivation for members to be interested in space policy. Further, while presidents may be able to set the agenda and set a rather amorphous goal (e.g., go to the Moon, go to Mars, etc.), Congress is the institution that writes the legislation that makes it happen, granting it the power to decide on the specific avenues that are taken to make those goals reality. It can decide how much money goes where, in essence determining the scope and pace of the policy in action. This pattern of consistent interest and involvement motivated by direct electoral connections often gives Congress a leg up on the president in exerting influence on the course of space policy.

The House and the Senate

Finally, as noted at the beginning of this chapter, what we often consider as a single institution, actually is two different ones with quite different rationales and motivations.

One way of seeing the difference is in terms of which chamber is more *influential* in driving policy. To answer this, we must first consider that space policy tends to fit the pattern of PE described in the first chapter. Recall that PE proposes that policies do change but tend to do so incrementally over time, establishing something of stable policy period. There are periods, however, where policy, rather than changing incrementally, changes rather dramatically in what are called "punctuations." Human spaceflight is a good example of a PE. Throughout the 1960s, the US had a general policy direction—get to the Moon before the end of the decade and hopefully before the Soviets. While there were minor changes along the way in terms of approach or plan, that was the general aim. Following that, the US built and flew the space shuttle. Again, it continued on a rather stable path until the *Challenger* in 1986. According to PE theorists, policy crises, or periods in time where policies may fail or show significant weaknesses, often serve as an opportunity for change. In the case of the *Challenger*, there was little change for several reasons, which are discussed elsewhere. The space shuttle continued as the dominant US human spaceflight program until 2003 and *Columbia's* breakup when the Bush administration decided to initiate Constellation. Finally, once that proved unworkable, the Obama administration shifted into the current human spaceflight strategy.

What does this matter for policy influence? While both chambers must ultimately agree to all pieces of legislation, including appropriations, there are quite different patterns of behavior with respect to space policy in looking at the House and the Senate separately. In considering NASA authorization bills, the Senate's version of the bill tends to be adopted more often in the wake of policy punctuations.[19] In other words, when space policy changes significantly, the Senate tends to have a greater influence. When space policy changes only slightly, periods of equilibrium, House-authored versions of legislation are more often built upon. This is to be expected simply based on the makeup of the chambers. As stated previously, the Senate tends to have broader, more national concerns than the House.

You can see another significant difference between chambers in figure 5.1—committee hearings. It is obvious from the figure that the House tends to hold more hearings on space than the Senate—a finding that is to be expected given the size differences between the chambers. The topics of these hearings are also somewhat different. While both chambers regularly hold hearings on things like authorizations, appropriations, and major space challenges, the Senate, reflecting a membership that is broader, tends to focus on more national concerns. As such, it's no surprise that Senate hearings often cover things like "NASA's Relevance to the US Economy" (1993) or "Contributions of Space to National Imperatives" (2011). On the other hand, and reflecting a membership that is more locally focused, the House often has hearings that focus on specific local topics. For example, in 2002, the House Science Committee held a hearing in Kansas City, Kansas, entitled "How Space Technology and Data Can Help Meet State and Local Needs." Rather than considering which chamber is more influential overall, then, a better question is, *When* is each better able to exert its influence?

Polarization, Partisan Politics, and Space

Congress is an inherently political institution, one which has seen significant changes in its makeup and partisan behavior since the 1960s. Political scientists have regularly shown the degree to which political polarization has increased and seriously split Republicans and Democrats in both chambers. The results of this polarization are as apparent as ever: two political parties that do not get along well with one another, increasingly frequent periods of divided government, and increased difficulty in passing legislation. To what extent has this larger societal and political trend impacted the consideration of space policy within Congress?

There are two ways in which this question can be considered: one, the ways in which polarization, insofar as it affects the operations of Congress, impacts the consideration of space policy, and two, whether space policy is partisan. In terms of operational effects, trends like a decrease in passage of legislation have had a significant effect on space. Authorization bills are no longer passed on a yearly basis and bills concerning other space-related topics such as regulation and commercialization are far more difficult to pass. Because fewer and fewer bills are voted on overall, provisions concerning space are often tucked into larger pieces of legislation. In one way, this might be a good means of achieving legislative objectives; if a bill under consideration has a lot of policy provisions included that appeal to a large majority of Congress, it can be easier to pass the bill (often called logrolling). On the other hand, each policy area, space included, may not be getting the attention it deserves or needs if considered on its own.

While passing authorization bills less regularly may be acceptable for space, an area in which projects tend to take a long time to plan, design, build, and carry out, what can be more impactful are the difficulties that Congress has in budgeting. Over the last few decades, Congress has routinely either been late in passing a federal budget or has not passed one at all, instead choosing to fund the government through continuing resolutions (CRs), which stipulate that government agencies continue to receive the same funding levels as they had previously for a set time period. While this mechanism is good in that the federal government can stay open and continue operating, it does not easily allow for changes in policy or direction. In other words, if a policy or program needs to change, it cannot while it is being funded under the CR. Similarly, even if Congress would like to increase funding for a particular area, the CR makes this generally impossible. The lack of annual flexibility in budgeting can impact space projects that need further funding or might need to be canceled.

The second question, What is the extent to which space is a partisan issue, is a bit more nuanced. While John F. Kennedy, a Democrat, was one of the most significant supporters of the space program in the 1960s, since then, it has generally been Republican presidents that have supported great expansions of space activities (Ronald Reagan, George H. W. Bush, George W. Bush, and Donald Trump). However, Democratic presidents have also been important in the setting of space policy as discussed in the previous chapters. In attempting to identify the partisan dynamics of space, scholars have found mixed results. While Wendy Whitman Cobb found that people who supported more funding for space exploration tended to be Republicans, other analyses have found less of a partisan relationship.[20] David Burbach, for example, finds that, among the general public, both Democrats and Republicans support space exploration but for different reasons. For Republicans, it is the connection with national security whereas for Democrats, it is environmental protection.[21] While this might be seen as good news in that people from both parties support space activities, the exact capabilities and activities needed to support environmentally oriented and national security-oriented space missions are quite different. As such, this split could cause just as many partisan fights as the question becomes not so much whether certain space activities are needed but exactly what those activities are meant to achieve.

At the congressional level, partisanship and space are similarly nuanced. In an analysis of roll call votes on legislation pertaining to NASA, Martin Machay and Alan Steinberg found partisan influence on a NASA authorization bill but not on appropriations. When money and economic impact was involved, the potential economic effect on a member's state or district was more important. Based on this, they argue that when economic interests are not at stake, space is susceptible to partisan influence.[22] This suggests that, given the low saliency of space as an issue, members of Congress may take party as a cue when voting on non-appropriations bills, but that, when important to a member's district, the electoral motivation wins out.

Conclusion

In thinking about the overall relationship between space and Congress, we can return to the fire alarm and police patrol analogy mentioned at the beginning of this chapter. Congress, playing the role of a beat cop, represents a continuous presence and interest in space policy. Not every member will have a strong stake in various aspects of space policy, but those who do have tools at their disposal that allow them to keep a continuous watch on space policy. Committees can hold hearings highlighting what issues they see as important, and members can request investigations and speak out to advance their own positions. However, because space policy often holds a lower place on the political agenda, many of these actions occur without a large audience and do not result in major change. Incremental change occurs through appropriations and authorizations and are largely driven by economic drivers influencing the chances of a member's reelection to Congress.

That stated, when a crisis or failure of policy does occur and the fire alarm goes off, it attracts the attention of both the Congress (the police) and the president (fire alarm). In the case of fire, those who have the means to beat it back become a primary focus; similarly, in times of crisis over space policy, presidents tend to take center stage. That does not mean there is no role for Congress to participate in the policy debate. While presidents can and do make major policy proposals (for instance responding to events like *Challenger* or *Columbia* or national security threats like increased weapons testing in space), Congress can either assent to them, modify them, or axe them altogether.

When looked at through the lens of this analogy, Congress is clearly poised to offer consistent influence on space policy over time, even if it is at a lower intensity than some might like. Because of this, it is important to identify and acknowledge the variables that, in turn, influence how Congress views space and what it does about it. This includes not only the views of their voters but also the economic impact that space activities offer directly to their constituents but also its relationship to national security issues. Major institutional differences between the House and the Senate further contribute to the ways in which space questions are considered. To be sure, the unique problems of the space domain present novel challenges to politicians—space is a long-term, high-cost, high-technology endeavor that does not warrant a lot of public attention despite the importance of space to both our daily lives and economic well-being—but this view of congressional influences reaffirms just how important institutions and the ways in which they function are for understanding space policy.

Summary Points

- Institutional features of Congress such as its bicameralism, different sizes, different constituencies, different term lengths, and different constitutional responsibilities affect how, and how significantly, Congress influences space policy.
- Congress has several tools through which to make space policy including legislation, appropriations, oversight, media access, and lobbying.
- Members of Congress are "single-minded seekers of reelection" meaning that the interests of their voters often drive their behavior in Congress, including on matters of space policy.
- The secondary nature of space policy also means that only certain members of Congress are heavily interested and involved in the matter.
- The House and the Senate can have varying influence on space policy depending on several factors such as the success or failure of various policy actions, the willingness of the president to be involved, and the larger political environment.

Discussion Questions

1. How do the institutional differences between the House and Senate impact space policy?
2. In your opinion, which is the most important policy tool Congress has at its disposal to make and influence space policy? Why?
3. Is reelection an important motivator for members of Congress when it comes to space policy? What else may influence their behavior?
4. Considering this chapter and the previous one, which branch do you believe has the most influence on space policy, the executive or legislative?
5. How does the overall political environment influence the consideration of space policy in Congress?

For Further Reading

Conley, Richard, and Wendy Whitman Cobb, "Presidential Vision or Congressional Derision? Explaining Budgeting Outcomes for NASA, 1958–2008," *Congress and the Presidency,* vol. 39, no. 1 (2012): 51–73.

Machay, Martin, and Alan Steinberg, "NASA Funding in Congress: Money Matters," *European Journal of Business Science and Technology,* 6, no. 1 (2020): 5–20.

Whitman Cobb, Wendy N., "Snapshot of a Shifting Senate: Senator Robert Kerr and Space, 1961–62," *Quest: The History of Spaceflight Quarterly,* vol. 25, no. 1 (2018): 29–40.

Whitman Cobb, Wendy N., *Unbroken Government: Success and the Illusion of Failure in Policymaking* (New York: Palgrave Macmillan, 2013).

6

NASA

The cultural impact of NASA is quite unique—rarely do you see the logos of federal bureaucracies worn on clothing like you do with NASA's "meatball." This cultural relevance speaks to the deep impact that NASA and its undertakings have had, even if some people don't know it. NASA has brought the world images of Mars, discoveries of water on the Moon, stunning pictures of our universe, and probes that are on track to leave the solar system. Even if it is not as many as before, people still crowd the Florida shorelines to watch people being launched atop massive rockets into outer space. Movies have been made that tell the story of astronaut heroes, both real and fictitious with NASA in the starring role. It is hard to deny that NASA just does "cool" things.

NASA's cultural relevance, however, belies its role and status as a government bureaucracy. For all the amazing things that it does, it does not do so in a vacuum. Being a bureaucratic organization and having to work with and abide by political leadership in Congress and the White House carries with it limits and specific responsibilities. To understand the role of NASA in US space policy, it is important to understand what being a bureaucracy entails. This chapter begins with a discussion of bureaucracies in general including their defining characteristics and general motivations. Then, we briefly detail NASA's founding with a particular focus on the entities that were brought together in the late 1950s that contributed to NASA's early organization and mission. We then examine NASA's ability to influence policy across the various policymaking stages, followed by an analysis of their significance in space policymaking. The technical and scientific nature of space activities confers on NASA a special ability to influence policy, but even it must behave within the context of the American political system.

Role and Importance of Bureaucracies

For all the complaints that are often directed toward bureaucracy in general, these types of organizations play an important role in policy generally. Before turning to NASA specifically, this section introduces some key concepts that help us think about how bureaucracies behave and what kind of influence they have on policy. In terms of the policymaking process discussed in chapter 1, the formal role of bureaucracies is to implement policy, but they play a large role throughout the process. They can provide information and policy ideas based on previous experience that in turn shapes how policy is crafted into legislation. Bureaucracies may also lobby political leaders for particular policy paths and help rally stakeholders in support (or against) potential policy changes.

To begin, what is a bureaucracy? A bureaucracy is any sort of organization with the following characteristics: it is hierarchical, it is organized based on specialization, populated by experts rather than political partisans, and operates based on certain formal rules. Many of these characteristics are intuitive—organizations require a set chain of command and people empowered to make decisions. Employees should be people well versed in their field who can make good decisions on policy and how it should be implemented, not people appointed for political reasons who may not have the same knowledge. A bureaucracy should not only be specialized in the sense that it provides a particular service or defined set of services, but its workforce should also be organized in such a way that different tasks are performed by different people. Finally, standard operating procedures (SOPs) not only clearly lay out how a task should be accomplished but they make this transparent and predictable for the population a bureaucracy serves.

While organizations that fit this definition have been around for thousands of years, the growth of government bureaucracy in America really took off after the Civil War as the responsibilities of the federal government grew. The role of bureaucracies and their importance was further advanced in the early 20th century as Max Weber, a German sociologist, argued that bureaucracies were in fact the most efficient way to run a government. Weber argued that these sorts of depersonalized and depoliticized organizations would be able to operate rationally, without politics or personality getting in the way. As expectations have grown for what the public expects the federal government to do, bureaucratic organizations have proliferated as a means of implementing increasingly complex legislation.

If we consider just how complicated the world has become over the past century as technology has advanced and the world has seemingly become smaller, the turn toward bureaucracy makes sense. Bureaucracies help to ensure that people who are experts in their fields carry out policies in a fair manner. However, the same fast-paced change that bureaucracies are created to respond to also highlights some of the problems inherent with them.

By the mid-20th century, several problems with bureaucracies began to be identified—the amount of paperwork was increasing as was the time it took to complete tasks. It began to seem that bureaucracies were unable to confront rapid change. As a result, political scientists began to try to understand these bureaucratic pathologies and what caused them. Anthony Downs suggested a particular life cycle to bureaucratic organizations: the initial rush of establishment brings about resources and eager new employees with particular motivations to make change happen. Over time, however, as resources dry up and enthusiasm over the new organization falls, those more eager employees exit the organization, leaving a core group of more conservative administrators. The bureaucracy's SOPs become more embedded. The organization overall becomes more resistant to change and more anxious to protect itself and what it views as its mission. The result for Downs is the "Law of Increasing Conservatism": "All organizations tend to become more conservative as they grow older, unless they experience periods of very rapid growth or internal turnover."[1]

In many ways, this search for stability is also a search for political survival. While it is rather difficult for us to identify a single government organization that has been eliminated in our lifetimes, organizational leaders appear to be continually concerned not just with ensuring that their organization has a mission and continues to have a mission but that they continue to receive more resources year in and year out. In this way, behaviors that reduce uncertainty and increase stability for the organization and its employees are encouraged. This continuing search for resources and survival has led some scholars to analogize bureaucracies to living organisms that continually seek out life-sustaining resources.[2]

These two views of bureaucracy, one as a rational and efficient means of implementing policy in a rapidly changing world and another as a self-seeking, bloated, inefficient barrier to change highlight both the benefits and consequences of bureaucracy. While it might be necessary to government functioning in a large democratic country, it is not always efficient and flexible. However, in a democratic society that seeks to treat each of its citizens equitably and fairly, efficiency might not be the primary value that needs to be upheld. Similarly, the predictability that bureaucracies offer helps to eliminate political or personal influence and ensure that citizens are not surprised by how different policies are implemented and enforced. Finally, while we all likely have our own complaints about the problems with bureaucracy in the United States, we must still acknowledge that it works generally well. Mail is delivered, social security checks are disbursed, and laws are enforced. American bureaucracy does not suffer from problems of endemic corruption as many other countries do. And bureaucracies, like NASA, can still do amazing things. With an understanding of the larger underlying characteristics of bureaucracies, we now turn to examine how NASA was formed and the role it has in US space policy.

Forming NASA and Developing a Culture

While the National Aeronautics and Space Administration was not formed until 1958, its roots go back earlier in the 20th century and the dawn of aviation. Established in 1915, the National Advisory Committee for Aeronautics (NACA) was charged with the promotion of the quickly developing fields of aviation and aeronautics. It was also charged with conducting and carrying out aeronautical research—a valuable resource just as planes were "taking off." Utilizing its series of wind tunnels, NACA's research proved vital during and after World War II and produced the first planes capable of supersonic flight. It had facilities and personnel located across the United States that would go on to form the core of the new space organization.

As discussed in chapter 2, the Soviet Union's launch of Sputnik greatly changed American attitudes toward space. Following the launch, the DOD responded by creating the Advanced Research Projects Agency (it would eventually go on to become the Defense Advanced Research Projects Agency [DARPA]) and NACA's then director Hugh Dryden formed a Special Committee on Space Technology to help coordinate the federal response in terms of space technology. It was the hope of the Eisenhower administration that these sorts of ad hoc organizations might negate the need for a larger governmental reorganization given their reluctance to mount a significant response to the Soviet achievement. However, public and political reaction to both Sputnik and Eisenhower's seeming lack of urgency overwhelmed these efforts. By early 1958, it became clear that the administration would need to take larger steps. Military leaders, predicting the military impact and potential use of space, believed that the military should have leadership of space. Eisenhower, on the other hand, preferred a split of military and civilian space activities to emphasize the largely peaceful side of space that might be seen more favorably around the world.[3] Congress agreed and by mid-1958 passed the legislation authorizing the creation of NASA.

While ARPA remained within the purview of the DOD, the next major question was, What current government organizations might be absorbed within NASA? Though NACA would constitute the core of the new NASA, Eisenhower's decision to split military and civilian space activities meant that some of the military organizations that were involved with developing missiles and rockets would also come under NASA's control. The first and foremost of these were the Army Ballistic Missile Agency (ABMA) located in Huntsville, Alabama. The ABMA's history dates to the end of World War II when Soviet and American officials sought out the builders of Germany's V-2 rockets, which were launched from Germany to attack Britain. Realizing the potential value of this emerging technology (among others), the United States recruited more than 1,600 scientists, technicians, and engineers, bringing them to the United States. Among this number were Wernher von Braun, the V-2's chief designer. Despite his membership in the Nazi Party, the use of concentration camp labor to build the rockets, and the destruction they ultimately wrought on Britain, von Braun and his team soon found themselves in the United States and made citizens. By 1950, he was at the ABMA working on rockets for the US Army.

Von Braun's work on rockets stemmed from an early infatuation with the possibility of spaceflight and space exploration. As such, he often justified his work for both the German and American militaries as advancing technology for this more expansive purpose since the militaries were the ones willing to fund and support the work. As part of his work for the ABMA, von Braun contributed to the development of the Redstone missile that drew on the V-2 design. As the United States contemplated its contribution to the 1957 Geophysical Year, both von Braun and the ABMA and the Navy's Vanguard program competed to provide a launch vehicle and satellite payload. Although the Eisenhower administration ultimately chose Vanguard, von Braun's team continued work on a more advanced Jupiter-C missile. Following the failure of Vanguard and amid post-Sputnik pressure for an American response, von Braun's team was given permission to attempt a satellite launch on their Jupiter-C, which successfully launched the first American satellite, Explorer 1, in January 1958.

Because of Eisenhower's decision to split military and civilian space activities, ABMA was transferred to the nascent NASA and, along with NACA, helped form the core of the new organization. Shortly after its establishment, the Jet Propulsion Laboratory (JPL) at the California Institute of Technology also transferred its working relationship from the military to NASA. In addition to these previously existing centers, NASA soon added additional facilities in Maryland (Goddard Space Flight Center), Florida (Cape Canaveral, now Kennedy Space Center), Texas (Manned Spaceflight Center, now the Johnson Space Center), and the Mississippi Test Facility (now the Stennis Space Center) in Mississippi. NASA headquarters was also established in Washington, DC. The space program would also require support from private industry, which in turn had facilities across the United States.

The way that NASA was formed and the placement of NASA field centers in different parts of the country had important consequences for NASA's culture and outlook on operations. First, while being a new organization, NASA was made up of several other agencies with long histories and particular ways of viewing problems and their solutions. NACA's deep expertise in aeronautical research and building experimental aircraft would be immensely valuable in designing and operating early spacecraft. The German rocket team at the ABMA also had experience in designing and building rockets, which predisposed them to emphasize engineering expertise and hands-on experience in building rockets as opposed to allowing contractors to do so. While these organizational cultures were not necessarily in opposition to one another, they helped lead the new NASA to favor particular skills and ideas, primarily engineering, research, and direct experience in building space systems over others. Second, the way in which NASA became spread out across the United States also had consequences for organizational culture. Because the different NASA centers and contractors came to embody different elements of design, building, and operations, strong coordination and planning was required to ensure that rockets and vehicles that were being designed in Alabama, built in various other places across the country, and launched from Florida with astronauts that were largely trained in Texas all worked seamlessly. Additionally, even though each center had a particular type of expertise, the centers also competed in terms of resources and power—the center that was given control over a program would ultimately receive significant increases in funding and attention, making these types of decisions important to center leadership. Thus, while centers largely had different roles and responsibilities, the distributed nature of their location complicated the relationships among them.

A final element of NASA's founding and early years also strongly contributed to its emerging organizational culture—the space race and its focus on human spaceflight. Although competition between the Soviet Union and the United States started with satellites and uncrewed craft, it soon became clear that the emerging space race would require ever greater technical achievements. If animals could be launched into space, then the next great achievement would be humans. Although Eisenhower tried to tamp down on the development of a human spaceflight program, NASA soon embarked on Project Mercury, the first attempt to launch an American into space.

By the early 1960s, NASA's main organizational elements had been solidified. Its system of field centers and division of labor among them has largely persisted though some responsibilities have shifted over time. Competition remains fierce for funding and additional opportunities that might see each center's resources increased. For instance, Marshall Space Flight Center has actively sought out additional responsibilities in terms of the ISS; several of the modules that now currently make up the ISS were tested and integrated in Alabama and a second mission control that assists with scientific experiments on the station continues to operate there. Members of Congress representing these states and districts contribute to this ongoing competition as they seek out greater appropriations and duties for the centers they represent. While efforts have been made over time to reinforce and strengthen the centralized control of NASA headquarters in Washington, DC, the fact that NASA is so spread out geographically means that these efforts have typically failed.

The result of these early years was to impart on NASA a particular organizational culture. Organizational culture refers to the assumptions and beliefs that members of an organization share about the way they believe their organization should operate and behave.[4] The types of things that an organization decides are important and become embedded as part of its mission will have a role in shaping how an organization views potential policy change. Howard McCurdy, in his study of NASA's culture and changes to it over time, identifies four main elements: a focus on research and testing, valuing in-house technical capability, having hands-on experience in building and operating spacecraft, and a belief in the value of exceptional employees and what they could accomplish.[5] While NASA did accomplish extraordinary things under the banner of this ethos, McCurdy details just how this organizational culture has changed over time as the agency has become arguably less flexible, less technically able, and more risk adverse.

Organizational Tensions

As this brief overview of NASA's formation suggests, while the organization shares similar ideas and values, there are significant differences across NASA in which of those goals should take precedence over the others. First and foremost is the preference for human space exploration over robotic and uncrewed missions. The preference for human spaceflight, as noted previously, goes to NASA's early formation and goals. However, even in the 1950s, several scientists argued strongly that humans were not necessary to advance scientific knowledge. Several noted that the presence of humans on space missions made it more complicated, more expensive, and perhaps detracted from the overall mission. Satellites and robots could explore the solar system without the added expense of life support systems and other necessities to sustain human life. Experiments could be automated and undisturbed by human activity. On the other hand, proponents of human exploration argued, things could go wrong with experiments or spacecraft that only humans might be able to respond to and fix. Robots would not be able to adapt to changing circumstances and refocus missions if something else looked promising or intriguing. Finally, there was the idea that human nature drove people to explore *in person* rather than through the camera lens; the logic of exploration and discovery drove human beings from their core and space would simply be the next frontier in which humans could do so.

While these arguments lent themselves to worthy academic debates, the practical reality of geopolitics at the time dictated that human spaceflight would be the currency by which the United States and Soviet Union not only competed but generated international prestige. Sending a robotic probe to the Moon, no matter its scientific worth, was simply not as impressive as sending a person. Partially because of this focus, human spaceflight quickly became the primary and largest part of NASA's mission and in turn became embedded in NASA's developing organizational culture. The result has been that human spaceflight endures as a major element of NASA's identity, gaining it attention, resources, and added interest. The dominance of human spaceflight at NASA continues today. Typically, 50 percent (or more) of NASA's budget goes toward human spaceflight with approximately 30 percent reserved for space science.[6] Further, science budgets are often targeted for cuts when human spaceflight programs almost inevitably run over their allotted funds. This means that there is often fierce competition within the scientific community over what types of projects NASA should take on, particularly as some science missions like the James Webb Space Telescope cost many billions of dollars themselves.

With roots in another element of NASA's early organizational culture, a second competing tension has been between a preference for hands-on experience in design and building versus those who believe that contracting those elements out to private industry is a better idea. While numerous private contractors ultimately built Mercury, Gemini, and Apollo spacecraft, including the massive Saturn V rocket, NASA engineers and even astronauts were deeply involved in their design and often oversaw construction directly. In addition to wanting to personally have a hand in building spacecraft, they believed that overseeing private contractors required NASA employees to have a deep well of experience and knowledge to be able to know whether the contractors' performance was up to par.[7]

But even as early as the 1960s, this technical desire was under threat. A competing Air Force model of relying heavily on private industry and contracting quickly gained influence as more members of the Air Force began to join NASA. As NASA's funding began to dry up and the Apollo program ended, larger economic and political forces combined to make contracting a more promising approach.[8] As the opportunity for hands-on work decreased as well, NASA lost many of its young, eager engineers and instead grew in terms of administrative professionals. This further reinforced the tendency toward contracting. While there is still a large contingent of engineers and rocket scientists working for NASA today, over time, the means by which spacecraft have been constructed have shifted further toward the private sector, which has led to a new tension over the role of commercial spaceflight companies.

Recall from chapter 2 that the development of the space shuttle depended on support from the DOD, a commitment from it to launch its payloads only on the shuttle. Given this, commercially developed rockets had little economic case and were not pursued. The 1986 *Challenger* accident challenged the dominance of the shuttle and encouraged the DOD to look elsewhere for launch services. As a result, the DOD once again supported the development of expendable launch vehicles, and some companies began to develop rockets on their own. The first of these, Orbital Sciences Corporation's air-launched Pegasus, was launched in 1990. While uncrewed rockets made a comeback, commercially developed crewed vehicles were slower. Not only were such systems expensive and out of reach for all but the richest countries, NASA did not actively encourage their development to protect the shuttle's monopoly. Even early efforts at shuttle replacements throughout the 1990s suffered from NASA's lack of priority on them.

The situation changed once again with a shuttle accident in 2003. *Columbia*'s breakup on reentry forced NASA to consider ending the space shuttle program and replace it with another NASA-backed and operated crewed system. By this time, however, several new commercial companies were beginning to emerge that had private commercial spaceflight as their ultimate goal. Thus, the tension within NASA became whether the shuttle replacement should be developed and operated internally to the agency or whether private industry could provide the service *to* NASA. While private companies had clearly established their ability to provide uncrewed rocket launches, there was still deep skepticism whether they could provide crewed launch services—not only had they never done so previously, but all previous efforts at human spaceflight were so expensive as to only be feasible with state-backed development. NASA strongly resisted the possibility that it could lose its purview over human spaceflight.

With the announcement of the Vision for Space Exploration in 2004, NASA found itself with a new mandate but uncertain funding levels. With SpaceX and other new space companies beginning to mature, then-NASA administrator Mike Griffin made a small bet on them with the COTS program. COTS provided some developmental funds to private companies that might be able to provide uncrewed launch services to the ISS in coming years with the hope that, once successful, the companies might be able to advance their technology to crewed services. Then-president Barack Obama made this shift official policy when, in 2010, he proposed that NASA allow private industry to provide launch services to LEO to allow it to shift its focus to deep-space exploration. Even still, a deep tension within NASA remained because of its lack of control over independently developed and operated launch services. Lori Garver, in her account of her time as deputy administrator of NASA during the Obama administrator relates that NASA deeply distrusted the upstart commercial providers and resented handing over control of crewed flight (albeit only to LEO) to commercial providers.[9]

As we know now, SpaceX and others have demonstrated an ability not only to successfully launch cargo and humans into space but to do so reliably and cheaply with largely reusable rockets. Considerable tension remains as the organizational cultures and procedures of NASA and private companies do not always mesh well. What SpaceX or Blue Origin may see as an acceptable procedure, rationale, or plan may not to NASA and vice versa. While private companies are beginning to prove their mettle in terms of spaceflight, NASA is still trying to understand how to balance its need for control versus trust when it comes to private industry.

It is obvious that many of NASA's organizational tensions result from its deeply ingrained set of values. The focus on human spaceflight above all else has led NASA to value it higher than uncrewed, scientific missions that might provide more and deeper understanding at a lower cost. It has also contributed to NASA's desire to control and provide human spaceflight rather than allow private industry to provide it as a service. The organizations that came together to make up NASA and their deep experience in research, development, engineering, and building contributes to the tension over in-house experience versus contracting, which again rears its head in trusting private industry to provide launch services independently. Over time, both NASA's organizational culture and the resulting tensions help to explain its behavior and its preferences for particular policy paths. Because NASA will always have more knowledge and experience about space exploration than its political bosses, it can use both its knowledge and preferences combined to deeply influence the policymaking process even before certain policies are adopted.

NASA's Policy Tools

As we consider how NASA influences space policy in the United States, recall from chapter 1 the different stages of the policymaking process: agenda setting, policy formulation, adoption, implementation, and evaluation. Although NASA's main function as an executive branch bureaucracy is to implement and execute public policy decisions, it influences the policy process long before policy is adopted. Thus, in examining the tools that NASA has at its disposal, we can consider them based on the stages of policymaking where they might have influence.

Agenda Setting, Formulation, and Adoption

How do policymakers know that there is a policy problem and what solutions might be deployed to solve it? While they might be independently knowledgeable and informed about any given policy area, it will be difficult for them to be well versed in every area. Bureaucracies like NASA have an important role in the provision of information about what is a problem, what is possible, and what it will take to make any changes. NASA's expertise in space exploration is perhaps one of the most important tools that NASA has at its disposal. Outside of a small number of former astronauts that have gone on to careers in politics, very few elected officials truly understand what it takes not just to get to space but what it takes to operate there. As such, policymakers must rely on the expertise NASA provides for information on which to base their decisions. This creates what political scientists call an information asymmetry. Asymmetry means something that is not balanced and when it comes to information and expertise about specific policy areas, bureaucracies often have far more of it than policymakers.[10] For example, suppose that Congress and the president are considering new policy directions for human space exploration and ask NASA for its thoughts. What NASA, the experts, report back can be a significant influence. Elected leaders might not know what is possible and so the options that NASA provides might either limit or expand potential policy choices. NASA, understanding this, might not provide *all* possible options but only the options that it, as an organization, might prefer. Information asymmetries can be even more significant in a niche and technical policy area like space since not many people are likely to be knowledgeable in it.

In terms of utilizing this informational power specifically for agenda setting, the rather low visibility of space policy generally and space exploration in particular, tends to work against NASA. As the Apollo program was reaching its climax in the late 1960s, NASA began to consider what its post-Apollo space program might look like. With grand visions of space stations and trips to Mars, NASA initially advanced an aggressive program that capitalized on its lunar experience. However, it soon became clear that space exploration did not have the cachet that it once had, the space race was over, and other priorities including the war in Vietnam were taking precedence. As a result, it became clear that a larger, and more expensive, space policy would not be pursued. Instead, NASA internally lobbied the Nixon administration and its Office of Management and Budget for what would eventually become the space shuttle program.

While NASA's move to set the agenda failed in the case of the space shuttle with the dwindling importance of space to the larger Cold War, by the 1980s with the space shuttle preparing for its first launch, NASA now set its sights on what to do with it. At the same time, the Reagan administration was preparing to reengage more aggressively with the Cold War including an emphasis on space-based weapons and missile defense. Seeing Reagan's interest in space and space technology, NASA began to lay out a case for a space station with the administration. Although there were still some concerns with the cost of the project, Reagan was far more supportive of the program and endorsed a policy that directed NASA to begin actively planning and designing a space station.

While these examples illustrate how NASA can utilize its position and information to influence political agendas and advance policy that it prefers, sometimes, NASA must react to other events that bring space policy to the fore. These events can include accidents like *Challenger* and *Columbia* or the actions of other policy entrepreneurs hoping to advance new directions in space policy. One interesting case of how NASA influences policy in the formulation stage comes from 1989 and George H. W. Bush's SEI. When Bush made the proposal to return the United States to the Moon and eventually to Mars, NASA was just recovering from the 1986 *Challenger* accident and was working on getting a space station off the ground. As a result, support within the agency for Bush's SEI was rather low, which contributed to NASA's early estimates of how much such a program would cost: upward of $500 billion.[11] The high price tag significantly limited political support for SEI and while the administration worked to salvage some of it, the proposal eventually died as Bill Clinton entered office.

NASA's use of its information asymmetry is interesting in these cases for several reasons. For one, while NASA used its advantage to set a high price tag to weaken SEI, it could have just as easily used its expertise to make the proposal more appealing to policymakers (as it did in the case of the space shuttle). How do bureaucracies decide when to support or not support policies? Or more broadly, How do organizations decide which policies they want to advance? Bureaucratic organizations do change and adapt at times—the creation of the US Air Force from the US Army and the USSF from the Air Force are just two examples where an organization has given up a role or mission. In the case of the space shuttle and space station, the emphasis on human spaceflight fits not just with NASA's organizational culture but its defining mission. With the example of SEI, while it, too, advanced human spaceflight, it would have threatened the still recovering space shuttle program and the emerging space station. As such, NASA preferred to protect its original mission rather than shift to new policy priorities. NASA might have gotten some additional resources to support SEI (which would have gone to another organizational priority), but NASA felt more threatened by the policy change and instead sought out stability by using its information asymmetry to its advantage.

In terms of policy adoption, NASA must rely on the acquiescence of its political principals in the White House and Congress. Even at that, NASA does have the ability to lobby presidents, members of Congress, other members of the executive branch, and even the public to whip up support for its preferred position. One statistic that NASA often touts in support of its policy goals is the extent to which NASA has contracts in most, if not all, congressional districts. In touting such facts, NASA hopes to convince members of Congress and even the people who vote for them that support for NASA is a positive thing because it brings funds and economic development to their local areas. Even still, NASA's ability to significantly influence policymakers can be somewhat circumscribed given the somewhat lower priority it is given as well as the public's lack of awareness of or support for greater spending on space exploration.

Policy Implementation

While information asymmetries contribute to bureaucratic power in the early stages of policymaking, perhaps the area in which these organizations hold the most power is in implementing and carrying out policy once it is adopted. To do this, bureaucracies often take very vague policy statements or legislative language and decide what it means and what they must do (with the appropriations given to them) to put it into effect. In regulatory agencies like the Federal Communications Commission (FCC) or FAA, this includes writing rules that the public must follow, publishing them, and enforcing them. Implementing space policy is a bit more difficult. For NASA it often involves determining what a space system, whether it is a telescope, satellite, or space vehicle, will need to do, asking its industry partners to help design and build it, and choosing the final design and contractor to build it. In the case of programs that operate continuously like the ISS or crewed spaceflight programs, internal rules and procedures must be developed to ensure safe operation and common operating parameters.

In implementing policy, one of the first things that NASA often considers is its internal organization. While the field center model that NASA utilizes imposes certain constraints and introduces certain behaviors, how NASA headquarters organizes itself for tasks can be rather telling. One way to think about this is to consider how NASA's placement within the federal government sends signals. There are four general types of bureaucracies in the federal government: cabinet departments, independent agencies, regulatory agencies, and government corporations. NASA is an independent agency, meaning that, in theory, its administrator reports directly to the president. If NASA had been placed as a subsidiary agency within one of the larger cabinet departments (for instance, the FAA falls under the Department of Transportation), there would have been another layer of bureaucracy between NASA and the president. The fact that NASA was broken out as its own agency is significant in the signals it sends about the relative importance of the organization. The same is true within bureaucratic organizations—the more layers there are between a subsidiary office and the agency's leadership, one can assume they have less influence and/or power. Thus, how NASA organizes itself internally with respect to the number of offices within it and their powers and abilities can signal the agency's priorities. For just one example, NASA's reluctance to support George H. W. Bush's SEI in 1989 was further signaled by its reluctance to engage in any significant restricting or reorganization. NASA declined to set up an office to oversee and engage with the SEI. On the other hand, following George W. Bush's announcement of the VSE in 2004, NASA immediately reorganized its front office to support both the adoption and implementation of the policy. In 2021, NASA split up its human exploration office into two different ones: the Exploration Systems Development Mission Directorate and the Space Operations Mission Direction. The first is designed to oversee NASA programs under development while the second would focus on operational, routine programs.[12]

While these types of reorganizations can be beneficial in terms of putting the right people in charge of the right programs and the further dividing up of responsibility, at the same time it highlights what is often a major critique of bureaucracy—that there is too much of it and the proliferation of subsidiary offices only makes getting things done harder. Certainly, any organization that oversees something like human spaceflight will have many tasks to coordinate among different field centers spread throughout the country. However, for a bureaucracy, this type of internal organization can be important in terms of signaling policy directions as well as in ensuring that policies are being given the appropriate amount of attention and power.

Like any bureaucracy, in the course of implementing policy and policy changes, NASA develops particular procedures and rules that contribute to the agencies' SOPs. These sorts of rules matter significantly because they essentially determine the way in which the agency "thinks" about problems and situations. One way to think about them is as if they were rules of a game; when the rules are written in a way to limit gameplay to a certain area or set of actors, they exclude others. All of this is incredibly important because of the high technology and dangerous activity that space exploration involves.

While somewhat granular, one example of the influence of SOPs in policy implementation comes from both the *Challenger* and *Columbia* incidents. Given the complicated nature of the space shuttle vehicles (and all space vehicles), NASA developed certain procedures for classifying potential problems—while some problems encountered are minor and do not affect the vehicle's operation, others might be more problematic. In the case of *Challenger*, examination of the solid rocket boosters after launch had shown some degradation. No major problems had been caused in those launches, however. Thus, despite evidence that there was a problem, the risk was deemed acceptable based on the procedures that had been developed.[13] Similarly for *Columbia*, foam loss off the external fuel tank was also a known issue but not considered potentially dangerous. In other words, the way in which NASA structured its rules and procedures for documenting and evaluating potential problems influenced decisions on whether shuttles could or should be launched. Crucially, though, NASA's organizational culture and attitude toward risk was also cited in later investigations as contributing to these problems. In other words, NASA not only had insufficient rules for dealing with problems but that was compounded by the agency's approach to risk.

While SOPs for crewed spaceflight might seem incidental to larger policy implementation, they are also influential for scientific programs. As noted previously, because human spaceflight takes up much of NASA's attention and resources, far less is left for scientific studies and exploration. With more potential projects to fund than funds available, how NASA makes decisions on which projects to pursue is quite important. If the SOPs dictate that science missions should be chosen based on one metric—for instance, scientific value—it may overlook other missions entirely. One means by which they do this is through a more formal process undertaken every 10 years by NASA's Science Mission Directorate with the involvement of other scientific agencies including the National Research Council. These decadal surveys represent the scientific community's consensus about what types of projects NASA pursues over the next decade providing NASA not only with expertise and information about what scientists are looking for but buy-in from a wide variety of stakeholders.

The reliance by NASA or any other bureaucracy on SOPs does highlight an additional critique: that of "red tape" or an unnecessary amount of bureaucratic paperwork. Anybody who has had to fill out what seems like an endless number of forms that do not seem to serve a purpose will likely understand the sentiment. While it can be frustrating, there are reasons for it. In large organizations that are often spread out physically, SOPs can aid in ensuring consistency and making sure that every customer or client is served in the same manner. In a democracy, this is even more important because government organizations seek to treat people equally and fairly, regardless of any distinguishing characteristic. For an agency like NASA charged with implementing programs that are very complicated, difficult, and spread across the country, SOPs are necessary to coordinate and document programmatic design, changes, and problems. While SOPs are necessary, their use and seeming rigidity contribute to the idea that bureaucracies are too rule-bound, inflexible, and inefficient. While NASA isn't an organization like the Social Security Administration or DOD where many people are likely to encounter these frustrations, nevertheless, they also have been criticized for ineffective operating procedures. In the case of both *Challenger* and *Columbia*, SOPs that perhaps accommodated too *much* risk were identified as contributing to the accidents.

A final tool that NASA exercises in implementing space policy is contracting and acquisitions. Unlike its early experience with hands-on design of the Saturn V and early crewed spaceflight programs, NASA has since adopted an approach wherein private companies play a large and primary role in designing and building equipment bound for space. While the process through which this is done might seem straightforward, it is anything but. There are significant regulations that government agencies like NASA must follow when soliciting contract bids and then choosing who should get the award; sketching this out in brief is difficult. However, when NASA and other government agencies are looking to acquire a good or service from private industry, the process often begins with a request for information that describes the type of thing that the agency is looking for and asks what private industry thinks they may be able to provide or what may be technically feasible at the time. Based on this, NASA will release a request for proposal asking companies to submit a bid for whatever project is being contracted. Companies not only tell NASA what it is they can do but also provide a timeline and estimated cost. NASA then must choose which company the contract is awarded to.

Because NASA contracts often involve significant amounts of money over a longer period, this process can become quite contentious and political. Projects like the space shuttle or SLS cannot just be switched from one company to another if the first company does not work out. While an agency may seek to make a contracting decision based solely on what company can provide the best product for the best value, there are a variety of other factors that it must take into consideration. Federal regulations often require bureaucracies to give preference to certain types of companies like small businesses or women- or veteran-owned businesses. Members of Congress might also become interested in the process if they represent districts or states where the bulk of the work may occur. While the contracting process is highly regulated in part to avoid this type of undue political influence, it is hard to keep it out altogether. As we noted in earlier chapters, there is a reason that so many NASA facilities are in the South. Putting NASA facilities there helped to solidify congressional support around a brand new agency.

Because how contract bids are evaluated and chosen is so highly regulated, the way in which NASA writes a request for proposal and what it chooses to put in it has a major policy impact. For example, NASA may stipulate that companies must have prior experience in something like building a satellite or launching a rocket or a certain *type* of experience. When this is included, it necessarily excludes newer companies that may be able to complete the task but have not done it already. On the other hand, if NASA has an interest (which it does not necessarily *need* to have) in building up a larger space industry, it may *want* newer companies or companies with less experience to be included. While this increases the risk of the task being completed and successful, it may further a longer-term goal of building up a set of companies that can provide such services and increase competition in the future.

A relatively recent contracting dispute illustrates some of these issues. As part of NASA's attempt to return to the Moon in the 2020s through Project Artemis, in 2019, NASA requested proposals for a human landing system (HLS) that would transport astronauts to and from the lunar surface. Instead of a contractor building the HLS for NASA to operate—the typical contracting model—the agency wanted to give commercial companies the opportunity to bid and provide the HLS as a contracted service, much as it had for the COTS program. In 2020, NASA initially selected three companies to further refine their proposed concepts: Blue Origin, Dynetics, and SpaceX. The next year, NASA selected SpaceX to receive the final contract to build and test its HLS concept. While NASA originally wanted to select at least two companies to build an HLS, it received only enough money from Congress to support one contract. Blue Origin immediately protested the award, contending that NASA had unfairly evaluated the proposals and made an improper selection. Additionally, Blue Origin claimed that NASA should have awarded more than the single contract; if it could not, it should have revised the request for proposals.[14] As part of the federal regulations regarding contracting, Blue Origin and Dynetics first protested the award with the GAO, which, in addition to assisting Congress with investigations serves as a referee of sorts in government contract disputes. The GAO denied Blue Origin's protest, and Blue Origin sued NASA in federal court, which agreed with the GAO that NASA had evaluated the proposals properly.

While this seems relatively straightforward, the issue also became a hot topic within Congress. After the SpaceX award and Blue Origin's protest, members of Congress (including Senator Maria Cantwell from Washington—Blue Origin's headquarters), became interested not only in the policy aspect of the HLS but the contracting and award process. Supported by Blue Origin's intense lobbying, the Senate inserted an amendment in a bill ostensibly about China to provide an additional $10 billion to NASA to award another contract, presumably to Blue Origin.[15] While that effort failed, in a draft appropriations bill, the Senate included a provision that would require NASA to present a plan to Congress on how it would include a second company in the award. Additionally, while NASA states that it only made one award because of its limited funding, the Senate report claims that the HLS program is not underfunded.[16] While it is unclear at this time whether the House will agree to the provision and it will be included in the final spending bill, this dispute is illustrative of several points about the role that contracting plays in enacting space policy. Not only is it a direct expression of how NASA chooses to implement policy, but it comes with a significant political dimension because of the time and money involved.

The final stage of the policy process is policy evaluation. Ideally, once a policy has been implemented, it is evaluated to ascertain several things including whether the actual process through which the policy is being implemented is effective and whether the policy is achieving the results it was originally designed to meet. For NASA, policies are often investigated by the GAO as well as its own internal Office of Evaluation. However, because of the long-term, high-cost nature of space exploration, whether a policy is successful or not is difficult to determine while a rocket or telescope is being built. It can often take years to determine whether any scientific discoveries or findings result from projects. Even if difficulties are discovered, it may be difficult to shift course as equipment is being built or missions are being carried out. While the lessons from these experiences in policy implementation may be useful for future projects, they can be difficult to ascertain in the short term in such a way that yields useful current changes.

The Significance of NASA's Role

Given this rather large role for NASA in setting space policy, how significant is it when compared to other actors in the policymaking process? In previous chapters, we have discussed the importance of presidents particularly in setting the agenda and the role of Congress as it involves policy formulation and adoption. And while NASA may clearly play the biggest role in the policy implementation phase as its actions can directly lead to how well the policy performs, their actions are still contextualized and predicated by the number of resources and requirements given to it by its political principals. Perhaps, then, the better question is, When may NASA prove to be more significant in policymaking and why?

The first factor that might influence this is NASA's recent performance—if it has done well in the recent past, one might assume that its amount of influence rises as well. Studies of other agencies like the FDA have found this type of pattern; Daniel Carpenter argues that the FDA, realizing the power of its reputation and performance, zealously guards its understanding of the power it gives them.[17] This does not appear to always be the case for NASA, however. NASA continues to rank high in public reputation according to Pew and continues to be named as one of the best places to work in the federal government.[18] Just as it was coming off its Apollo high, successfully sending humans to the Moon in less than a decade, NASA's ambitions for more expansive space exploration were quickly denied by a president more concerned with shifting resources toward other areas. On the flip side of this, a NASA failure often leads to policy change as was the case with the VSE the year after *Columbia* disintegrated as it reentered the atmosphere. Though *Columbia* did not demonstrate that space exploration was a failure per se, it did highlight how old and fragile the space shuttles were and how dangerous it had become to continue to operate them.

Both the post-Apollo and post-*Columbia* incidents suggest that other factors influence just how significant NASA is in the policymaking process. In the former, economic and geopolitical factors were contributors to Nixon's decision to have a less active civilian space program, while in the latter, policy failure was an essential cause. While we will cover economic and political factors in greater depth in later chapters, suffice it to say here that they contribute strongly to the larger political context in which NASA and space policy is considered in the United States. Policy failure, however, while it seems paradoxical, appears to be an important trigger that can allow changes to space policy that in turn give NASA an opportunity to influence its outcome. Key to understanding this relationship is to remember that space policy is an expensive and long-term undertaking and is often path dependent. Once a particular policy path has been adopted, it is essentially locked in for the foreseeable future. While some changes might be made on the margins, it can be hard to change quickly.

When policies exhibit path dependency, they can be very hard to change. Recall from chapter 1 that Baumgartner and Jones's theory of PE reinforces this. If the policy image remains stable, it undergoes only minor changes. However, events like policy failures can represent significant junctures. They open windows of opportunity where long set policies may undergo serious revision if not outright change. When looked at from this perspective, space policy changes in the wake of *Columbia* become more understandable. Of course, this sentiment begs the question of why policy changed in 2004 when it did not following *Challenger* in the 1980s. As we have discussed previously, NASA's actions following George H. W. Bush's SEI proposal contributed to its ultimate failure. In 2004, NASA was far more supportive. Thus, it appears that policy failures, quite counterintuitively, open windows for policy change, and it is in that window where NASA holds significant sway.

Another factor that can contribute to NASA's ability to influence space policy is its administrator. The NASA administrator, like many executive branch officials, is appointed by the president and confirmed by the Senate. There are no other requirements such as experience in government or in the space industry to become the administrator. While the way in which NASA has been organized and its organizational culture is already set, some administrators have been able to influence the agency and space policy more than others. While we will not delve into a long history of NASA leadership, some administrators are worth noting. NASA's second administrator, James Webb, played an important role in the 1960s. A former undersecretary of state, Webb had many years' experience in Washington, DC, having also served as director of the Bureau of the Budget under Harry Truman. At the beginning of his time at NASA, it was flush with cash as Kennedy's call to land on the Moon became accepted policy. It was not long, however, until the blank check began to be revoked and NASA had to continue the Apollo program with declining funds. Webb also had to maneuver NASA through the aftermath of the Apollo 1 fire in which three astronauts died during a ground test of their capsule. Henry Lambright argues that it was Webb's political acumen and skill that helped NASA not only achieve an incredibly difficult feat in a short amount of time but do so in a political environment that was turning hostile toward the agency.[19]

Another administrator who also proved quite influential was Dan Goldin, who was originally appointed by George H. W. Bush in 1992 and served until 2001. While not a politician like Webb, he was able to deftly shift NASA into the post–Cold War world. While Soviet-American antagonism helped drive Apollo in Webb's era, with the Cold War over, civilian space policy and exploration appeared to have lost a driving rationale. With the burgeoning space station program finding itself on the chopping block both in Congress and with the newly inaugurated Bill Clinton, Goldin managed to convince the new administration that the space station could serve another set of foreign policy goals. With Russia finding itself with a mounting set of economic challenges and highly trained scientists and engineers fleeing the country (perhaps into the arms of more hostile regimes), Goldin argued that bringing Russia into the project and providing some funding for them to do so would help keep these scientists in Russia and assist Russia economically. Thus, what had originally been part of a Cold War competition became part of post–Cold War cooperation between the US and Russia.

In addition to securing the future of the ISS, which continues to serve as a major point of cooperation among multiple countries, Goldin also initiated major reforms within NASA. By this point, NASA had long been criticized as moving too slowly with budgets that seemed to grow with every year. Recall from chapter 3 that Goldin initiated reforms aimed at NASA doing things "faster, better, and cheaper." McCurdy writes, "Through this initiative, they [NASA] sought to cut costs, take greater risks, and dispatch spacecraft that actually flew."[20] Despite some notable failures, these types of reforms helped contribute to some successful missions including the Mars Pathfinder that led to greater public and political support for the agency. As Dan Ward argues in a later analysis of NASA's faster, better, cheaper initiatives, despite losing 6 out of 16 FBC missions, cost, schedule, and performance actually improved across NASA and that, because of the reduced costs, NASA actually achieved more success per dollar spent.[21]

The ability of Webb and Goldin to wield power at NASA and significantly influence policy leads to a good question: Why are some administrators more successful than others? This question could obviously be asked about any leader in any position. A variety of circumstantial factors influence this including geopolitical context, the current state of policy, the domestic political situation of the day, and the ability and skill of the leader in question. What is clear is that there are some people who appear to be better able to use those skills to the agency's advantage to advance policy interests. Lambright argues that while the skills of individual administrators may vary, they do "matter in contentious, long-term human spaceflight programs that are subject to political instability."[22]

While the skill of administrators may vary and directly influence NASA's ability to influence space policy even if they do not directly dictate space policy, NASA plays a significant role at almost every stage of the process. The information and advice they provide can open or foreclose certain policy avenues, and their organizational culture and imperatives help guide long-term NASA interests. The way in which policy is implemented by NASA and ultimately its success or failure further contextualizes the amount of power that NASA has politically.

Conclusion

NASA is like a traditional bureaucracy and also quite set apart. It has all the traditional functions and responsibilities of a bureaucracy—it is hierarchical, specialized, and relies on a comprehensive set of SOPs. It is populated by experts in their fields who not only understand the intricacies involved in space exploration but can carry it out in an apolitical manner. As an organization, it seeks its own survival operating under a shared set of principles and ideas about its mission. Along with other government bureaucracies, it carries out the policies that are passed legislatively with the resources that are appropriated to it. And yet, the actual implementation of the policy is quite public in nature. People can witness a rocket launch, people can see stunning pictures from telescopes like Hubble, people can watch video from a helicopter flying around on Mars. All these activities are far more publicly available and shared especially when compared to something more mundane like renewing a passport or inspecting manufacturing facilities.

This public visibility highlights several things about NASA. First, what it does is quite unlike other bureaucracies, which can lend it more power to influence policymaking at various stages. The type of information asymmetry that NASA has obtained gives it perhaps a larger voice in policymaking. Second, despite the public visibility of much of what it does, public support for NASA may be said to be a mile wide and an inch deep. In other words, what NASA does is interesting and cool, but when compared to other priorities like climate change, health care, or national defense, it may pale in comparison. While NASA's activities make it more publicly available and approachable than an agency like the FDA or FCC, it still must behave and play politics like a bureaucracy to make such events happen. While political leaders must and do make decisions about the policy paths to take, NASA—shaped as it is by its history, organization, and strong culture—plays a significant role throughout the policymaking process until finally, those intricate pieces of technology that started as words in a bill and money in an account, cross the threshold of space.

While being a government bureaucracy, NASA's unique mission and knowledge set give it power and influence that other agencies may not. Even still, this power is not unlimited or sufficient. It is conditional and contextual, influenced not just by what political leaders dictate, but by society itself and the conditions in which it finds itself in at any given moment. The following chapters discuss these non-institutional influences.

Summary Points

- Though often criticized, bureaucracies are vital to the functioning of a modern government. They serve to implement and execute public policy and, while staffed by experts and career officials, often have a significant impact on the shape of policy.
- Created in 1958, NASA formed from earlier organizations including NACA, JPL, and the ABMA. These core groups and NASA's early experiences helped create an organizational culture that prioritized human spaceflight and in-house expertise. Their geographic spread also introduced some difficulties in how NASA headquarters might control the various centers.
- Bureaucracies in general and NASA specifically have several tools for influencing policy including access to information, the ability to lobby the White House and Congress for their preferred policy option, designing contracts and choosing companies to carry out various programs, and other choices in how policies and legislation are implemented.
- Several bureaucratic pathologies have influenced NASA's approach to space policy including information asymmetries, inflexible organizational culture, and SOPs that promote risk tolerance and can gloss over potential problems.

Discussion Questions

1. What are the pros and cons of bureaucracy in general? What are some examples of those pros and cons in NASA's history?
2. How did NASA's early history influence the organization's character and culture?
3. What are the tools by which NASA influences space policy? In your opinion, which is most important or influential? Why?

4. Is NASA's organizational culture a help or a hindrance to American space policy more generally?
5. How important is agency leadership in influencing space policy and the direction of NASA?

For Further Reading

Asner, Glen R., and Stephen J. Garber, *Origins of 21st Century Space Travel: A History of NASA's Decadal Planning Team and the Vision for Space Exploration, 1999–2004* (Washington, DC: NASA, 2019).

Kay, W. D., *Defining NASA: The Historical Debate Over the Agency's Mission* (Albany: State University of New York Press, 2005).

McCurdy, Howard E., *Inside NASA: High Technology and Organizational Change in the US Space Program* (Baltimore, MD: Johns Hopkins University Press, 1993).

Vaughn, Diane, "History as Cause: Columbia and Challenger," in *Organizational Collaboration* (London: Routledge, 2011).

Vaughn, Diane, "Regulating Risk: Implications of the Challenger Accident," *Law and Policy*, vol. 11, no. 3 (1989): 330–49.

7

Military Space and the Intelligence Community

Hollywood is quite good at depicting military and intelligence activities in space. By good, we mean that activities are often staged in terms of their excitement. Computer-generated images portray epic battles in deep space, with large starships shooting massive direct energy weapons at each other. Other scenes zoom in on small crews scrambling in a rover on the Moon or some distant planet, fighting with laser guns in efforts to establish and defend remote bases from a hostile enemy. And of course, in so many movies the military is called on to protect humanity from aliens. Closer to home, televised scenes often depict some sort of special operation in which the NRO or National Security Agency (NSA) satellites are tasked to provide timely intel that helps thwart the bad guys before they accomplish their nefarious deeds. It seems that Hollywood imagination for military space and space intelligence is only limited by production budgets! Unfortunately, while these depictions represent the first impressions of military and intelligence activities in space to so many people, they are far from what is currently possible and/or typical, for better or worse.

The reality of military space and space intelligence today is much less exciting, but no less important. It is a reality of satellites taking pictures of Earth's surface hundreds, or thousands, of miles below. It is one of military professionals hidden away in sensitive compartmented information facilities (SCIFs) doing data analysis on computers. It is one of research for the next optical sensor, or more efficient fuel source. It is one of strategy, planning, and preparation for the use of space-based capabilities in the next conflict. It is one of constant data gathering. In sum, the reality of military space and space intelligence is a lot of work being done by people on the ground, not by women and men in space suits and not with epic space battles.

Nonetheless, the rapid growth of activities in the space domain and more importantly, the increased threats to the domain and the assets in it, make the military and Intelligence Community (IC) vital components of US space policy in ways that they would not have been even a decade ago. As noted previously in the discussion of various space policy documents, the domain is congested, contested, and competitive. Dealing with this type of environment requires the capabilities to defend US interests as well as provide accurate intelligence to inform space policy. This chapter examines military space and the IC and their roles in this regard.

The Organization of Military Space and Space Policy

We begin by exploring the current organization of US military space, noting that this is always subject to change given the most recent dynamics associated with the establishment of the USSF. Additionally, this chapter explores in greater detail how this organization reflects military space policy. It is in this way that one can get a better sense of the complexity of military space policy to identify how the DOD and the military try to address space policy problems of which there are no easy solutions. As discussed further in chapter 11, the increased number of international players in the space domain make it more difficult for the DOD to implement and execute policy across a range of space issue areas, despite this being of the utmost importance for the United States if the country is to continue its utilization of space in ways that maintain American leadership and military superiority in the years to come.

It is the role of the military to provide for the security and protection of the nation's interests through the employment of military power. With respect to space, policymakers have found it challenging to effectively organize the use of space power. As a result, roles and responsibilities for space-based capabilities have varied over time, with efforts at consolidation and expansion in terms of the number of actors involved or the scope of mission sets for any individual organization. This has made the task of understanding US military space difficult for most people, including those in charge of it! Some of this complexity can be unraveled by laying out the current organization of the military space enterprise.

Military space legally and figuratively begins with the president. As commander in chief, the president is responsible for leading the way on national security and defense policy. As mentioned in chapter 4, the president and Congress tend to share the power of policymaking, but the president tends to have more power and influence over foreign and defense policy. Considering military space specifically, one should expect similar power and influence for the president and the executive branch over Congress and other actors. That stated, with space policy, power between the executive and legislative branches is shared, and one should expect this to continue to be the case as commercial and civil space share more of the policy stage with the military. For now, however, and certainly for the focus of this chapter, military space is dominated by the executive branch, with a healthy dose of congressional influence.

Within the executive branch, the NSC and the NSpC sit alongside the DOD in terms of influence on military space policy. While there is no hard and fast rule, meaning that it is administration dependent, the NSC tends to focus on national security space, leaving the NSpC to work on civilian and commercial space policy. As seen in the previous section, this division of labor is becoming less effective as the three sectors are increasingly interdependent. With the current Biden administration, there appears to be an active engagement between the two councils with the director of national intelligence (DNI), secretary of defense, and chair of the Joint Chiefs of Staff (JCS) each sitting on both. Thus, at least at the policy and strategy levels, there is a recognition of the need for integration and interagency coordination on military space policy.

This recognition is exemplified most recently in two of the current administration's space policy documents, the *United States Space Priorities Framework*, published in 2021, and the *National Cislunar Science and Technology Strategy*, published in November 2022. The *Space Priorities Framework* lays out the importance of space-based activities and US space infrastructure, associating space with US leadership and overall power. It also identifies the key areas for government involvement, to include supporting the maintenance of a "robust and responsible" space enterprise and keeping the space domain preserved for future generations.[1] The *National Cislunar Science and Technology Strategy* focuses on the coordination of science and technology sectors within the US government to establish an interagency approach to exploration, economic development, and scientific research.[2] What is important is the argument put forth by the administration that the efforts identified in these documents will require cooperation and commitment from the whole of government, regardless of which sector is being emphasized.

Having already discussed the role of the president and the two councils in space policy, our attention in this chapter turns to the DOD and the armed forces. There is a lot of guidance as it relates to military space. One might even argue there is too much. When considering the publication of national space policies, national space strategies, national security space strategies, or Defense Space Strategies, there are plenty of overarching documents to assist the DOD in carrying out policy consistent with national priorities. Additionally, there are Presidential Policy Directives (PPDs) on space as discussed in chapter 4. With a range of parameters set for the DOD based on these guidance documents, the DOD can work within them to achieve its own priorities as well.

The DOD

The DOD is the primary actor in the execution of military space policy as directed by the president. Its portfolio of responsibilities is extensive as it relates to the space enterprise, from organizing, training, and equipping (OTE) the USSF and the other services' space components to providing inputs shaping national security priorities in the space domain. While it is often discussed in terms of a single organization, with a civilian secretary of defense at its head, its composition suggests that there also is utility in thinking about it as a group of smaller units. This group is united by a common set of priorities identified in the *National Security and National Defense Strategies*, the DOD's own space policy directive and the National Military Strategy (NMS) of the JCS. As discussed previously, the NSS is fairly limited when it comes to space policy, thus it is helpful to elaborate upon these latter three documents in terms of the DOD's role in their creation, focusing specifically on how they shape space policy.

The NDS sets the parameters for funding and personnel for the military while also identifying the types of capabilities that will be required to carry out the DOD's role in support of the NSS. It is also responsible for setting the agenda for the development of the NMS. Thus, it serves as a policy bridging document by which the DOD can interpret and carry out its responsibilities in achieving broader US strategic goals. The NDS has been produced four times since 2005 and was mandated officially by Congress in the 2017 National Defense Authorization Act (NDAA), to replace the Quadrennial Defense Review that Congress previously established with the 1997 NDAA for publication every four years. The DOD published the most recent version in March 2022 for Congress. At present, the only version available is classified, so our discussion of the document relates to the unclassified fact sheet that the DOD released with it.

Before continuing, one might note a policy discrepancy that exists in terms of the publications of the NSS and the NDS. Most often the NSS is published after the NDS. This should beg the question, How can the NDS be developed prior to the NSS if the latter should be setting the agenda for the former? While it is often the case that the DOD moves a bit faster on its required NDS than the White House does with the NSS, coordination on the NDS takes place to make sure it is consistent with what will finally come out in the NSS. Both documents represent a longer process that requires considerable interagency coordination. As a result, the NDS always has top-down guidance from the president, even if the broader NSS is not ready for the public. In the most recent case, the Biden administration put forth interim guidance in 2021 that helped set the agenda for the NDS, prior to its release of the NSS in 2022.

The Office of the Secretary of Defense (OSD) is responsible for the preparation of the NDS. As mentioned previously, there is much that goes into policy development for defense—acquisition and sustainment, budgeting and financial management, intelligence and security, personnel and readiness, reform, research, and policy are all components of the DOD's mission. The current defense priorities focus on defense of the homeland based on the challenges presented specifically by China, deterring strategic attacks against the US and its allies and partners, deterring aggression, particularly in the Indo-Pacific and Europe, and improving the resiliency of the Joint Force.[3] Military space influences and impacts all of these priorities.

How space factors into these priorities is a job for the assistant secretary of defense for space policy (ASD/SP). The ASD/SP has the primary role in leading the space warfighting effort within the DOD. This role includes interagency coordination (i.e., engagement with the other actors from chapter 8) as well as engagement with international allies and partners. It also includes participating in budget discussions in terms of assessing DOD space priorities. The ASD/SP is a new position established in the 2020 NDAA, with the idea behind it to allow for a dedicated space policy professional to help advise the undersecretary of defense for policy. Even though the role is new, the ASD/SP responsibilities already have increased to include weapons of mass destruction (WMD), missile defense, and cyberspace operations. In terms of policy formulation, the ASD/SP provides policy on the employment of space forces, DOD policy on international agreements related to space, DOD cooperation on space with other nations' defense organizations, and policy related to the integration of space capabilities into operational plans.[4]

Beyond the roles and responsibilities identified for the ASD/SP, the under secretary of defense for policy has a space policy directive that sets the agenda for the formulation of policy across its office. DOD Directive 3100.10 is the most recent policy directive, published in 2022 and is instructive with respect to those actors identified in terms of their roles. Beyond the under secretary and the ASD/SP, there are 15 other actors specified: the directors of the Defense Technology Security Administration, Defense Intelligence Agency (DIA), National Geospatial-Intelligence Agency (NGA), NRO, and NSA; the under secretaries of Defense for Acquisition and Sustainment, Research and Engineering, Personnel and Readiness, and Intelligence and Security; the secretaries of the military departments and the Air Force; the DOD chief information officer; the chair of the JCS; the combatant commanders and the commander, USSSPACEOM.[5] Some of these actors are covered later, but first we continue with a discussion of what we consider to be the most important to military space in terms of the challenges associated with organization and threats. As one can already discern, however, the number of actors within the DOD alone that have some responsibility or role to play in the space domain makes policymaking incredibly complex when it comes to understanding who does what and who is immediately responsible for a particular issue. This is even with the knowledge of assigned responsibilities given that many of them overlap.

In terms of those actors most important in dealing with the challenges of organization and threat, our attention is directed first to the armed forces of the DOD, specifically the Army, Navy, Air Force, and Space Force. As separate branches, their roles highlight the importance of the organization, training, and equipping of the military to best achieve operational and strategic effects. Together, in the concept of a Joint Force, they integrate to provide policymakers the ability to employ force worldwide. For space policy, both the individual branches and the Joint Force provide good opportunities to uncover the role of military space in greater detail.

The Armed Forces

For the DOD, organizing for space has proved challenging. Since the early days of space and the focus on security, it has been difficult for a lead agency to take charge of organizing, training, and equipping for operating in the space domain. While the creation of the new USSF as a separate military branch is a policy move that recognizes the need for a dedicated focus on warfighting in the space domain, previous efforts to organize around the space domain failed.

As noted in chapter 2, none of the military branches were prepared to take on space fully as a new mission, nor were policymakers comfortable with that idea. From the responsibilities given to NACA to the establishment of the Advanced Research Projects Agency (ARPA), policymakers were keen to provide an overarching authority for space projects that would limit individual military branches' space activities. The Army was limited in its efforts to develop space launch vehicles because it could not gain permission to build long-range ballistic missiles; the Army's and Navy's pursuit of a joint space command was rejected early in the days of the space race, with the Air Force getting the role of development, production, and launching of space vehicles through a 1961 policy directive.[6] This role commitment followed on the heels of the National Space Act that helped separate civilian from military space. Yet, despite its role, the Air Force was unable to expand its military space program in the early days, at least publicly, with so much attention on space exploration being directed toward NASA and a US public approach to space best characterized as one devoted toward the idea of sanctuary. Eventually and beyond the public eye, however, there was a shift toward military space as the Cold War continued to heighten and security became a priority over prestige. The Air Force then found itself in a primary position for military space, becoming the space organization for the armed forces.

While the evolution of the military and its relationship to the space domain is of interest, we leave the details of that story for others to explore. Instead, we focus on the current organization of the armed forces given that policymakers are both responsible for and shaped by it in terms of the range of policy options available for the DOD when it comes to developing and executing space mission sets. One of the biggest problems when discussing current organization is its dynamic nature. While the Air and Space Forces dominate the military space conversation, the Army and Navy possess their own space-based capabilities and interests when it comes to military space. To that end, a brief overview of Army and Navy space follows, continued by a more in-depth examination of the Air Force. The military section concludes with a discussion of the USSF and the Joint Force construct in terms of understanding the wide range of organizational challenges associated with the armed forces.

Army Space

Army Regulation 900-1, last updated in 2017, provides the overarching space policy for the Army. Similar to DOD Directive 3100.10, it specifies the space responsibilities for Army leadership. What is important from our perspective is not the details associated with the myriad duties identified in the policy, but rather that it is a means to implement the DOD directives establishing the importance of space to the armed forces and the threats posed by the changing space environment. For the Army specifically, this means taking on space roles that allow it to support or enhance the service's main missions: mission command, intelligence, movement and maneuver, fires, protection, and sustainment. These roles include the integration of space capabilities, space control, space situational awareness, and satellite operations.

Current Army space operations run through its Space and Missile Defense Command (SMDC). Established in 1997, SMDC is responsible for dealing with operational challenges such as ASAT and ballistic missile threats as well as protecting Army battlefield communications through satellite communications. There are about 2,800 individuals that make up the organization, spread out over five subordinate organizations, several satellite communications centers, missile defense and early warning locations, and missile defense radar sites.[7] The SMDC also is charged with the development and maintenance of Army space doctrine, as exemplified in Field Manual 3–14, Army Space Operations. Finally, SMDC is committed to the education and development of Army space professionals. This tasking even includes sending Army astronauts to support NASA while helping to enhance its own use of space capabilities.

Navy Space

Given the shift of responsibilities to the Air Force for the majority of space, the Navy has made sure that it takes an active part in space-based capabilities important to its maritime mission. Today this is presented through Navy Space Command. As the primary actor for maintaining the freedom of navigation on the high seas, the US Navy relies on space-based capabilities as much as, if not more than, the other services. Most importantly is the Navy's need for satellite communications, precision navigation and timing (PNT), and weather information that drives its strong interest in military space. From its home in the Fleet Cyber Command/10th Fleet, Navy Space Command works to present forces for USSPACECOM, the joint force combatant command, and to represent the Navy's interest in meeting the service's own requirements.

Currently the Navy is working through five strategic initiatives for a US Space Command.[8] First are efforts to integrate space and maritime domains. This is important to ensure that naval officers are aware of the utility of space for the naval warfighter. This awareness ties directly to the second initiative, which is to identify the Navy's requirements and capability gaps to better position itself with USSPACECOM. Its third initiative is to achieve integration such that space capabilities can be used most effectively in operations and tactics, techniques, and procedures (TTP) development. The fourth initiative involves coordination and integration with the Department of the Navy in such a way that the Marine Corps can also utilize space capabilities across the spectrum of conflict and competition. The last initiative calls on the Navy to leverage commercial and industrial technologies and concepts as they develop, particularly those that have military applications.

All these initiatives focus on increasing the strength of space services provided for the Navy or by Navy Space Command. Most importantly for the Navy this focuses on connectivity and communications across the fleet through military satellites. Also worth noting is that the Marine Corps Forces Space Command stood up in 2020 to provide Marine support to Navy space operations as well, reinforcing the idea of increasing strength. To this end, the Navy has been able to provide capabilities for the service and the Joint Force.

Air Force

To this point, we have documented the historical legacy of the Air Force in the space domain in previous chapters. The status of the Air Force remains tied to USSF through a shared department in the DOD (Department of the Air Force), but also through the range of administrative functions that remain integrated as a practical matter. So, while each service is responsible for its own OTE, they continue to share resources to achieve these missions. Having stated this, the Air Force is discussed in terms of where it is likely headed as a service with the departure of most of its space components.

To begin, most, if not all functions assigned to the Air Force through DOD directives rely on space in some way. Nuclear operations, air superiority, global strike, ISR, mobility, command and control, agile combat support, and recovery operations would be severely limited without a supporting space component. This is important to note as it suggests the Air Force will retain particularly strong interests in the space domain and the use of space-based assets regardless of the mission sets taken upon by the USSF.

There has always been a tension (healthy or otherwise) between the air and space components in the Air Force. One of the central arguments for a separation of the space component was the sentiment by many within and outside of the Air Force that the organization could not adequately resource and think doctrinally about both air and space domains. As an example of this, the first Air Force space doctrine was the Air Force Manual 1–6, Military Space Doctrine released in 1982, three decades into the use of space assets by the DOD![9] Even then this doctrine was organizational in nature and as argued by some, made to fit into airpower doctrine versus being placed into the context of space as its own, unique environment.[10]

This tension between air and space made a great deal of sense. In terms of organizational culture, it often proved challenging to develop airmen who would put space ahead of piloted aircraft. Thus, it would be difficult to envision space in a primary position within an organization committed first and foremost to airpower. The nature of the US Air Force's commitment to space seemed lacking throughout the 1980s, leading insiders and outsiders to question the Air Force's role in military space. By the 1990s, the Air Force had once again tried to figure out how best to make space an integrated component of the Air Force. With the end of the Persian Gulf War, there were thoughts that air and space could coexist within one institution functioning to provide global power projection. There were even calls for the end of Army and Navy space commands in the early 1990s as they were thought to simply duplicate what the Air Force was providing for the DOD.[11]

By the end of the 1990s and into the 2000s, the Air Force was able to do much in the way of integration and operationalizing the role of space; however, no military organization outside of the Air Force felt as comfortable with the Air Force's overall role in space given its inability to treat the two domains of air and space equally. It is from this historical context that the calls for an independent Space Force continued to increase in the 2010s until its establishment in 2019.

While the next section discusses the Space Force, what remains of space for the Air Force today? In terms of DOD directives, the only essential space functions left for the Air Force involve the use of integrated command and control for air and space operations. Beyond this, previous Air Force missions are now supported fully by the USSF. This includes the transfer of missions such as missile warning, space launch and space intelligence, along with the operation of between 70 and 80 satellites. The one area where the Air Force will still be supporting the USSF is in terms of its infrastructure and personnel management since the end state numbers for the USSF will remain small at around 18,000 members.

Space Force

The discussion of the Space Force begins with an understanding of its relationship to the Air Force. First, the USSF is an independent branch of the armed forces. It has a chief of space operations that sits as a fully fledged member of the Joint Chiefs of Staff. Second, the USSF sits under the authority of the secretary of the Air Force as part of the Department of the Air Force. Third, the USSF has now acquired most of the space mission sets from the USAF, making it the primary military space organization among the armed forces.

Once the subject of fierce debate, many still ask, Why was the USSF seen as necessary to establish in terms of policy? One answer is that the need for the USSF reflected both domestic and international concerns. Domestically, President Trump drove the move toward the USSF by arguing its importance for national security. Historically, this had been brought up before as a possibility. The timing now, however, coincided well with the pressure from the rise in challenges from China, Russia, and others in the space domain. Advocates argued that a better organized approach to space security, for which the USSF would provide, would help identify and deter threats in space. Also, the change in organization would reflect what other space powers like China and Russia had already done. In this case, the United States was behind in its space organization and was simply catching up with other space powers rather than creating something new.

From a policy perspective, the change that comes along with the establishment of a new military branch is not an easy one. In part, it is a political decision. In this case, the desire of President Trump to create a Space Force created the momentum necessary to push a review and subsequent authorization by Congress in the 2020 NDAA. The change is also a reflection of the perceived lack of organization that existed among the space mission holders at the time. While noted earlier that this was directed primarily at the Air Force, the subsequent stand up of USSPACECOM as a combatant command suggests that the concerns regarding organization went beyond those of the Air Force. As discussed in the next section, space also was behind the curve when it came to the Joint Force concept. Thus, a rearrangement of the OTE roles also facilitated a rearrangement of the operational warfighting roles for military space.

Organizationally, the USSF is charged with five core functions.[12] First, it is the lead agent for providing freedom of operation in, from, and to space. Second, it is tasked with the sustainment of space operations. Third, it is obligated to protect US space interests. Fourth, it is charged with the task of deterring space aggression. This includes deterrence in, from, and to space. Finally, the USSF is tasked with the conduct of space operations.

It is not just about organization and function that drives the development of the USSF. Its creation is also about establishing an organizational culture that moves away from Air Force culture into something that is unique to the USSF. In other words, the creation of the USSF is meant to develop guardians instead of airmen. As it relates to military space policy, one argument is that a shift in organizational identity will put guardians in a better position to advocate for space in ways that airmen, soldiers, and sailors are not able to do. The result of this better advocacy should be the creation of efficiencies for the DOD in how space is approached. In essence, having space professionals that think about space power in ways akin to how pilots think about air power, or sailors think about sea power, should facilitate better space policies more aligned with US national space policy and strategic goals. Whether this is accurate remains an empirical question, however. In the future it will be interesting to see if these conditions occur as the USSF continues to develop. As a first start, the foundational doctrine for the USSF has been established to build this sort of organizational identity.

Projecting a little into the future, one might envision a more stable USSF that is fully capable of executing all the missions given to it from the Air Force, while expanding its mission sets to consider its role as the primary space agent for the US armed forces. The incorporation of the Space Rapid Capabilities Office, the Space Development Agency, and Space Systems Command all exemplify this type of expansion. These three organizations represent changes to the bureaucratic approach to acquisition and technology development that currently exist in the other military branches, suggesting a mandate for the USSF to alter its approaches to better achieve the mission sets it has been tasked with executing. One might also envision that the USSF will be better prepared to contribute to the Joint Force as it develops its capabilities and doctrine in a manner that reinforces its warfighting dimension in a domain that also has been characterized as such.

The Joint Force

In 1986, Congress passed the Goldwater-Nichols Act to address the chronic problem of interservice rivalry among the armed forces. In doing so, the armed forces were restructured to create a stronger JCS to advise the president, NSC and the secretary of defense, and a system of combatant commands to separate out the warfighting component from the service component focused on organizing, training, and equipping. The combatant commands were set up to operate under a joint concept, where each of the services provide personnel to create an integrated warfighting component that exerts operational control over these forces. The Joint Force would come to represent the way in which the US military executes its warfighting role, to include the training and doctrine development that goes along with it.

While the Joint Force concept has been successful, although not without its critics, the integration of space has always been much slower. Given the discussion of the difficulties associated with consolidating space operations, and the lack of a dedicated military branch to present forces to the combatant commands, this should come as little surprise. Add to this a period where space was not contested, and one can understand the lack of urgency associated with the meaningful integration of space.

Currently, this is beginning to change. In 2019, the previously defunct USSPACECOM was reestablished after a 17-year hiatus, following a merger with US Strategic Command (USSTRATCOM) in 2002. There are two important things to note here with its reestablishment. First, with the creation of the USSF, it was logical to consider a joint warfighting command so that space would mirror the other domains in terms of the armed forces' preparation for the employment of force in that domain. Second, policymakers and military leaders were communicating that space was now a warfighting domain. If true, then integration into the Joint Force was paramount.

Nonetheless, the decision to create a joint warfighting command focused on space raises questions. Is this combatant command necessary given that it is not focused on a particular area of the world like European Command (EUCOM) or Indo-Pacific Command (INDOPACOM)? All combatant commands need the integration of space capabilities, so would they not be better off utilizing personnel from the USSF and the other branches' space components? Does the establishment of USSPACECOM simply create another military bureaucratic element and not necessarily a warfighting one? While these questions have been settled in terms of policy for now, it remains to be seen how the Joint Force will utilize USSPACECOM in future conflicts. One clue about this use will be the next iteration of the NMS, a JCS publication that lays out military policy for the Joint Force so that it aligns with DOD directives, the NDS and the NSS. The last one was published in 2018 and it is likely that by the time of this publication, the next one will be out. It will be in the development of the future NMS and future conflicts that help validate these Joint Force policy approaches to space warfare in the years ahead.

The Intelligence Community

The challenges of organization and threat also extend to the IC. The IC overlaps with military space and comprises the other significant component of the national security space enterprise, providing the information required of leadership to craft both space policy and strategy, to include how best to use military capabilities in pursuit of political objectives. The collection of information from space-based assets goes back to the early days of the space race with the United States seeking to establish the norm of overflight for satellites, providing for the rationale and legal use of satellites for information gathering.

The IC has evolved significantly since the early days of space, with the most recent reorganization coming after the events of September 11, 2001, when Congress identified a failure with the various intelligence agencies to coordinate and identify the threat posed by al-Qaeda to the US homeland. The Intelligence Reform and Terrorism Prevention Act of 2004 led to the establishment of a DNI whose office is responsible for the coordination and integration of all US national security intelligence, which consists of engaging 18 separate agencies. When combined with the authorities provided for the secretary of defense, these two principals oversee and coordinate intelligence activities among several different IC members.[13] This section will focus on the agencies that serve under this joint oversight. Before discussing specific agencies, a broader discussion about the connection to space and intelligence is warranted. This discussion is meant to identify the relationship between the actors that utilize space for intelligence collection and analysis, separating out these behaviors from the warfighting and direct military applications discussed earlier in this chapter.

As mentioned in chapter 2, the Eisenhower administration was very keen on establishing the principle of overflight, which would allow for the legal use of reconnaissance satellites to gather data on targets deep in the interior of the USSR. After the launch of Sputnik, the race toward developing a satellite reconnaissance program began in earnest, leading to the birth of Corona, the first remote imagine satellite system. Despite its fits and starts, the success of the program allowed for the collection of information much faster and in greater detail than previous U-2 reconnaissance flights operated by the CIA. The importance of this mission cannot be overstated; it was the decision to engage in satellite reconnaissance that directly ended the missile gap myth for the United States regarding Soviet capabilities.

One of the historical legacies originating from the Corona period is who would lead and operate space-based intelligence platforms. The Air Force wanted to claim exclusivity, yet early on did not want to support satellite development; this was despite the close connection between space and politics throughout the early days of the space race. We will return to the Air Force's relationship to space in a later chapter, but what is important here is that its reluctance to embrace space completely opened the door for other actors to become involved in the reconnaissance satellite mission. The CIA and the USAF would jointly run Corona, for example, with the CIA managing Air Force assets from launch to operation and recovery of data. As also mentioned in chapter 2, President Eisenhower's creation of ARPA was meant to keep any one agency from monopolizing space operations. This legacy would shape the coming complexity of space policy organization through today.

The road to complexity in the IC begins with the CIA and USAF sharing of responsibilities for Corona but continued with the establishment of the NRO in 1961. Its existence remained classified until 1992, even though the agency would be assigned with initially consolidating the intelligence gathered from space for the DOD. This included establishing programs to design and coordinate systems with the Air Force, Navy, and CIA. Today, the NRO's mission has expanded to working in coordination with an increased set of mission partners such as the NGA, NSA, CIA, DIA, USSF, USSPACECOM, and the other military service intelligence branches to provide reliable and timely space-based reconnaissance and surveillance data to decision- and policymakers. The NRO is also engaged in developing and partnering with the commercial space industry, as well as foreign partners. To this end, the NRO has awarded contracts for commercial imagery data and purchased reusable rockets from SpaceX for some of its missions. Through its rapid acquisition program, the NRO has also increased its launch options regarding small satellites, increasing overall launch efficiency. It has worked with Australia and the UK on launches of satellites from remote sites.

So where does the NRO sit in the organizational conglomerate of relevant space actors? Historically, there has always been a back and forth between the intelligence community, at first the CIA and now the DNI, and the DOD with respect to who controlled the organization. Since 2010, the DNI and the secretary of defense work under a memorandum of agreement that charters the NRO to both parties stipulating agreement on the expenditure of funds and program approval.

In terms of the policy process, the NRO specifically provides information that helps with the formulation of both national security space policy and national security policy more broadly, while engaging foreign commercial partners as a part of national space policy goals. The agency also has an evaluative element as well. Through its Center for the Study of National Reconnaissance, the NRO has an established research unit that provides useful analysis of the discipline, practice, and history of national reconnaissance. Included in this is a process to incorporate lessons learned from the NRO's reconnaissance practices.

The importance of the NRO for national security space cannot be understated. Even in its early days, this importance was recognized, with President Johnson once noting how invaluable Corona and the intelligence it provided on missile threats was to the United States. "We've spent $35 billion or $40 billion on the space program," he claimed, "and if nothing else had come out of it except the knowledge we've gained from space photography, it would be worth 10 times what the whole program has cost."[14] Today, the NRO is an organization that maintains independent control over its space assets and launch, despite the creation of the USSF. While its budget is classified, the scope of its ground stations and launch offices, spread across the globe, suggests the US government allocates significant resources to the organization, consistent with the idea that it has sizable influence and remains a key national security space actor within the broader space security enterprise.

Keeping with a theme of complexity, however, the intelligence community's involvement in space goes beyond coordination with the NRO. Most members of the IC are directly involved in the operation of space assets to conduct their missions, in essence, implementing broader national space policy that includes not just maintaining space leadership, but also engaging in activities that exercise the freedom of space operations and utilizing space for national security purposes. The DIA, for example, provides reports on space security threats as a combat support agency for the DOD; the organization also supports other military missions outside of combat. Space is just one part of its portfolio, as it also provides intelligence on a broad range of foreign military capabilities. The data provided by the DIA offers the information needed to plan strategy for the implementation of policy at the national level all the way down to tactical execution in support of policy goals. Another producer of space-based intelligence is the NGA. Like the previous organizations, it has a broad customer base to provide geospatial intelligence (GEOINT), ranging from national leadership and the DOD to nongovernmental organizations. Its history also shows a confluence of organizational interests, as it initially evolved from a joint effort of the CIA and DIA to provide imagery analysis for policymakers.

These examples only skim the surface of the intelligence community's space-based efforts. As discussed earlier in this chapter, while all the military branches operate and utilize space assets for more direct warfighting needs, they also utilize space for intelligence production. When added to the mix of actors involved in intelligence, one quickly develops a picture of a very dense interconnected network of IC interactions in support of national space policy and broader national security efforts.

Returning to policymaking impacts directly, how does the IC make its contributions beyond the provision of information? Already mentioned are the efforts by IC agencies to coordinate among the other members; this of course means the establishment of SOPs and understandings, like the one between the DNI and DOD for the NRO, to achieve the IC's mission most efficiently. With respect to national space policy, there is a clear line tied to its implementation. What should also be identified, however, is the IC's role in agenda setting and the formulation of policy. As discussed in the previous chapter, policymakers tend to suffer from information asymmetry, a condition where information is not uniformly possessed among actors. This asymmetry provides the IC a great deal of influence in helping set the stage for what is important or not as it relates to the development of policy. The DIA's publication of space threats, for example, provides a starting point for policymakers to lay out a plan for dealing with those threats, which in turn will be implemented by the broader national space enterprise. Diving deeper into how this type of publication matters, consider the problem of increased counterspace capabilities by Russia and China. Former DNI Director Daniel Coats, in a 2017 hearing by the Senate Select Committee on Intelligence put it clearly: "Russia and China perceive a need to offset any U.S. military advantage derived from military, civil, or commercial space systems and are increasingly considering attacks against satellite systems as part of their future warfare doctrine."[15]

With the provision of this type of information by the IC, policymakers have begun to develop policies that focus on the development of a resilient space architecture. This includes the establishment of policies and legislation that allows organizations like the NRO and the DOD to incorporate rapid acquisition processes into their operations, potentially purchasing technology that can be replaced much faster than more exquisite space capabilities typically used for mission sets. One specific initiative includes the creation of the IC Commercial Space Council; its five voting members (NRO, NSA, NGA, DIA, and CIA) are coordinating and working through the best way to incorporate commercial capabilities into IC strategies and policies.

The influence of the IC also comes with its share of challenges. Given the various missions and roles, there is a coordination challenge that exists in the collection and dissemination of space-generated intelligence. The IC Commercial Space Council is one example; notably there are only five IC members that have a voting interest in the council, leaving others on the sidelines. Another recent example of this challenge includes the NRO, USSF, and USSPACECOM's efforts to resolve overlaps and gaps in their space operations with the joint publication of their "Protect and Defend Strategic Framework." While the document is classified, the organizations have described it as a guiding document that helps to deconflict roles and spells out how cooperation will take place among them.

Additionally, with a range of agencies both needing and wanting operational control over their own satellites, there is significant competition for resources and attention, both common characteristics of bureaucratic competition. The increasing use of space for national security has only made these challenges more problematic. The number of space-related mission sets across organizations continues to increase, with little interest in consolidation by policymakers or the organizations. One can observe an expansion of related mission sets taking place. The DOD, for example, has been trying to take on responsibility for acquiring commercial satellite imagery right alongside current IC efforts. In some ways, this supports the goal of resiliency in the space domain. In other ways, it also creates a host of inefficiencies that potentially lead to bottlenecks in information sharing and an overemphasis on threats relevant to a particular organization's mission. Even within organizations, sometimes space functions are relegated to various offices that are further removed from the organization's leadership, potentially limiting an individual office's impact on policy.

The IC is a complicated and dense web of organizational relationships; its space-based components spread across a host of actors but share a common goal of providing timely and accurate space-based information to a range of consumers, but most importantly to policymakers. The sheer number of organizations that have an influence on space policy requires a significant effort at coordination to most effectively provide what policymakers need to create policy and what other organizations need to implement it. The question that lingers given the complexity of the IC is whether the coordination is working. An additional question is, How effective can a conglomerate of agencies be in synchronizing efforts toward a set of common national policy goals? While these questions are left unanswered here, these are important to consider in our continued discussion of governmental actors in the broader space enterprise, the subject of the next chapter.

Conclusion

As the space environment has changed, the United States has crafted policies to better shape the organization of military space and the IC to deal with these changes. Consolidation and integration stand out as two attributes of these efforts. Consolidation has taken place in the establishment of the USSF and USSPACECOM, representing a service and combatant command component. Integration is occurring across the DOD, within individual military branches and throughout the IC.

Currently, the organizational effort in military space has been toward more consolidation, although with the growing importance of space, this has been difficult to do. Just within the context of military space, we observe a wide range of roles and responsibilities that exist in the space domain focused on security and defense. The establishment of the USSF and USSPACECOM seem to have addressed DOD organizational concerns of overlapping responsibilities, space advocacy, and integrating space into warfighting strategy and operations.

The changes to military space organization also represent a more current view of how the United States thinks about space. There is a clear understanding of the importance of space and the need to defend both the domain and US interests in it. For military space, the challenge will be how to expand its portfolio to consider the growing interests of the United States that now encompass an important commercial dimension.

The IC is also a significant player in the national security space enterprise, working alongside the military in defense of US interests. Coordination instead of consolidation is a key goal among the IC to get information to policymakers in a timely fashion. As the activities and number of actors in space continue to multiply, the IC will continue to be challenged in keeping up with the flow of information and data. Also important is the ability to maintain resiliency in the face of growing threats to the IC's mission, including its partnership with military space to provide options for leaders in the defense of US interests and national policy goals.

Summary Points

- Military space and the IC are organized beginning with the president and ending respectively with the specific branches and the joint force, and the various intelligence agencies that create intelligence products.
- The USSF was established to provide a separate branch to organize, train, and equip space cadre solely focused on the space defense and security missions.
- The IC works in tandem with military space to provide information and data to policymakers in support of national security and defense.
- Both military space and the IC are organized in ways that create tension between consolidation and expansion in terms of roles, missions, and number of actors. The result is a complex network of relationships that make it challenging to coordinate effectively.

Discussion Questions

1. The Air Force was viewed by some as not being able to advocate equally well for airpower and space power. Based on your understanding of organizational culture, do you think that it would be possible to do so? Why or why not?
2. What role exists for the military in protecting commercial space? In thinking from an organizational perspective, which branch would you task with this role once it goes beyond LEO?
3. Of the USSF functions, which is most important and why?
4. The IC's space roles and responsibilities seem to overlap significantly. What sorts of arguments might make this a compelling approach to organization? Why might it be problematic?

For Further Reading

Berkowitz, Bruce, and Michael Suk. *The NRO at 50 Years—A Brief History,* 2nd ed. (Chantilly, VA: NRO, Center for the Study of National Reconnaissance, 2018).

Coletta, Damon, and Frances T. Pilch, eds. 2009. *Space and Defense Policy* (New York: Routledge, 2009).
Defense Intelligence Agency, "Challenges to Security in Space: Space Reliance in an Era of Competition and Expansion," 2022, https://www.dia.mil/Portals/110/DOCuments/News/Military_Power_Publications/Challenges_Security_Space_2022.pdf.
Johnson-Freese, Joan, *Space as a Strategic Asset* (New York: Columbia University Press, 2007).

8

The "Other" Actors

The story of US space for much of its history has been one centered on the exploits of NASA, the DOD, and the IC. As such, they are discussed in separate chapters, as their outsized roles require detailed examination. As mentioned in chapter 2, often their exploits intertwined to help the United States achieve gains in prestige and security. Late in the Cold War, however, these actors carved out their respective niches in the space enterprise. This is not to say collaboration and cooperation did not continue, it does through the present day, only that the evolution of US space policy has allowed for the conceptual development of civil, national security, and commercial space sectors in ways that offer clearer divisions of labor when it comes to space missions and roles.

The number of space missions and roles has increased both quantitatively and qualitatively, however, resulting in an increasing number of government actors involved in space policy, well beyond NASA, the DOD, and the IC. While satellite technology is useful for DOD communication and missile warning, it is also useful for studying crop yields and climate. Policy is thus required for both sets of uses making space policy relevant not just for NASA and the DOD, but for organizations like the Department of Energy. For policymakers, the explosion of actors in the domain has increased the complexity of the entire policy process. As a result, policymakers have made efforts to better organize and identify the bureaucratic organizations across the government necessary for achieving success at all stages of the policy process. It is in this way that any efforts at achieving a coherent national policy is likely to succeed.

Unfortunately, this organization and identification, a policy process itself, has not been easy. Identifying and understanding the roles and missions of the numerous actors involved in some stage of the policy process is far from straightforward, even for those inside the government. As considered in this chapter, the "other" governmental bureaucracies involved in space policy create a dense network of relationships that often overlap. Additionally, the relationships with respect to space policy vary in their stage of emphasis. While all the organizations are involved in policy implementation, some of them also are involved in agenda setting, formulation, and adoption, much like NASA, the DOD, and the IC. A few even pay some attention to evaluation.

This chapter identifies and discusses some of the "other" governmental bureaucracies involved in the space policy process. Given the number of actors, the chapter briefly provides a sense of their contributions to the policy process, highlighting the most relevant aspects and working through a discussion of a few key themes. One overarching theme, and perhaps what is most remarkable about the following discussion, is the uncovering of just how complex and interwoven US space is into the fabric of the US government. In 2024 it is fair to say that space is ubiquitous across the government in terms of its influence and policy touch points. Another theme, building off of this ubiquity, is the importance of bureaucratic alignment of organizational policy and goals with those of national policy. Achieving this alignment is more challenging for some organizations than others; regardless, efforts to seek alignment require a great deal of coordination within and across organizations. To this end, a third theme arises: in the aggregate, the success of national policy implementation requires a complicated interplay between organizations that have overlapping interests and varied responsibilities. In the end, the cumulative effect of organizational behavior and relationships lead to questions regarding the nature of how the space enterprise is organized. What are the strengths and weaknesses of the current organizational framework? Where might there be places to gain efficiencies as it relates to the stages of the space policy process?

Consider the simple question of where to begin. One way to work through the challenge of identifying and discussing the other relevant organizational actors in the government is to begin with the NSC membership. As discussed in chapter 4, the NSpC has developed since the Eisenhower administration, waxing and waning with respect to influence based on a given presidential administration. Despite its lack of formal authority, NSpC membership continues to grow, with the Biden administration most recently increasing the number by 5 to 20. What is important for us in this chapter is who's included beyond the traditional major actors. With the heavy influence of the executive branch in space policy, its identification of those actors deemed most relevant provides a useful point of departure for exploring the importance of these other space enterprise actors. To this end, however, we focus on how organizations are involved in the policy process. While even this categorization is not perfect, it provides some conceptual clarity versus something more chaotic.

Department of State

As the diplomatic arm of the US government, the Department of State (DOS) is the lead agency in international engagement for space-related issues. Its primary goal as it relates to space is the promotion of US interests abroad and within the United Nations. This includes fostering international cooperation in the space domain, advancing US leadership in space, facilitating agreements with other countries in the exploitation of the space domain, and articulating US positions as they relate to international space policy and international space legal matters. Space diplomacy then covers a wide range of activities, but above all is symbolized by international interactions on all things space on behalf of the US government. While often this is not done alone (for example, both NASA and the DOS were engaged in the development of the Artemis Accords), the DOS remains a vital, statutory link between the United States and the international community when it comes to space. As you will see in a later chapter, its efforts are sometimes overshadowed by those of the DOD, creating interagency tension as it relates to roles in US space policy. Nonetheless, much of the day-to-day and routine relations between the US and other countries on international space issues runs through or is sanctioned by the DOS.

In some ways, the DOS's role in space represents a microcosm of the problems associated with the complexity and ubiquity of space. It is tasked with involvement in a wide range of space activities that influence other aspects of its primary diplomatic mission, in addition to those of other organizational missions. Its space diplomacy role is spread out across different parts of the department. The Office of Space Affairs, located within the Bureau of Oceans and International Environment (OES) is the designated agency within the DOS to manage the complexity associated with developing and maintaining a coherent national space policy that can be articulated globally. In doing so, the office has prioritized five key areas of focus that help ensure the implementation of US national space policy: space diplomacy, exploration and commercialization, the preservation of outer space, global satellite navigation, and satellite-based Earth observation.[1]

Space diplomacy is really an umbrella term that includes the other key areas. Specifically, space diplomacy involves leading and taking part in negotiations of new agreements, articulating US space policy positions, and shaping international policy agendas in the space sector. These actions are consistent with agenda setting, formulation, and adoption of policy in the space domain. The aforementioned Artemis Accords, agreements securing international partnerships for the United States in the conduct of cislunar space exploration and an eventual Mars mission, provide an example of this type of diplomacy. Exploration and commercialization include outreach to increase commercial and private partnerships in space in a manner consistent with US goals and the broader framework of the OST. One US goal is the maintenance of the space domain for peaceful purposes; as such, the DOS is a primary actor in reinforcing norms and the adoption of standards of behavior in space operations. The focus on global satellite navigation for the DOS includes collaboration with international partners on the utilization and adoption of the American GPS system as a primary means of navigation and seeking compatibility with other navigation systems such as Russia's GLONASS, China's Beidou, and the European Union's Galileo.

The final focus includes helping to facilitate agreements on the use of Earth observation satellites for such things like disaster relief and weather processes. The goal in this latter case is to provide resources for the United States and partners such that SOPs are in place should the need arise to utilize another's capabilities in the event of an emergency or simply to advance the well-being of civilian communities. Combined with US domestic space policy, particularly presidential space policy directives, these five areas help advance policy across civil, military, and commercial space sectors, and in doing so, seek to facilitate whole of government approaches to maintaining leadership and stability in the space domain. The challenge lies in the need for the DOS to coordinate between US agencies in addition to between states.

The foundation for the DOS's international engagements is the OST and the US commitment to the peaceful use of outer space. As noted in chapters 2 and 3, the relationship between space and security is rather tightly connected. To this end, the department is also active in the diplomacy of space security, having responsibilities for the negotiation and monitoring of agreements that involve international security cooperation in the space domain. The Bureau of Arms Control, Verification, and Compliance (AVC) is a key actor in this regard, seeking to develop transparent processes within the international community on the deployment and use of space-based capabilities. This function continues to increase in importance given the role of space assets in international security, ranging from missile warning to space-to-space threats. Missile defense cooperation among US allies, for example, is an active area of engagement for the AVC.

Another important role for the DOS with respect to space security, is the administration of the International Traffic in Arms Regulations (ITAR) and US Munitions List (USML), executed by the Directorate of Defense Trade Controls (DDTC).[2] Due to an overlap in technologies that create dual-use problems, in addition to the increasing militarization of space, ITAR strictly regulates the export and import of materials viewed as advantageous militarily or with respect to US intelligence capabilities. These materials are identified on the USML by category. Category XV, for example, includes spacecraft and related articles that may be used for the detection or mitigating the effects of a nuclear detonation. Still, there are other categories relevant for space, such as Category XVIII, which regulates technology related to directed energy weapons. In short, ITAR and the USML are in place to keep defense-related materials and services out of the hands of foreign nationals that may put them to use against US interests.

While keeping technology out of the hands of potential US adversaries is an important goal, there are many that think the current ITAR and USML process is too cumbersome when it comes to facilitating the other aspects of the DOS mission. Commercial partnerships internationally, for example, can be difficult due to the costs associated with maintaining ITAR compliance of US companies. International partnerships for exploration can be challenging for NASA in its working relationships with others as it must seek to obtain licenses to transfer USML items to its partners. There are also second-order effects of this process—transfer to one country potentially means a subsequent transfer, thus there is additional scrutiny required to make sure that one legal transfer now is not illegal later. One policy challenge then becomes how best to balance US commercial and civilian interests with the security concerns that arise by the exchange of technology and goods in the space sector. While this is not a problem the DOS solves, it is a problem for which the DOS is directly responsible in terms of its role in executing current ITAR policy.

In sum, the DOS is an organization that is involved across the entire space policy process. Through its international engagement, the department is able to help with agenda setting and policy formulation. Additionally, the department plays a role in coordinating between domestic and international audiences on policy, in many ways a critical cog in policy adoption at the global level. And of course, with respect to implementation, the department is the key player in keeping other actors in adherence to US foreign policy goals as they relate to space, to include the prevention of secure materials and technologies from being transferred to hostile foreign actors.

Department of Commerce

In highlighting the role of the DOS in controlling the transfer of sensitive technologies and services through ITAR and the USML, such concerns extend into the dual-use domain and commercial regulation, areas managed by the Department of Commerce. The commercial equivalent of ITAR are the

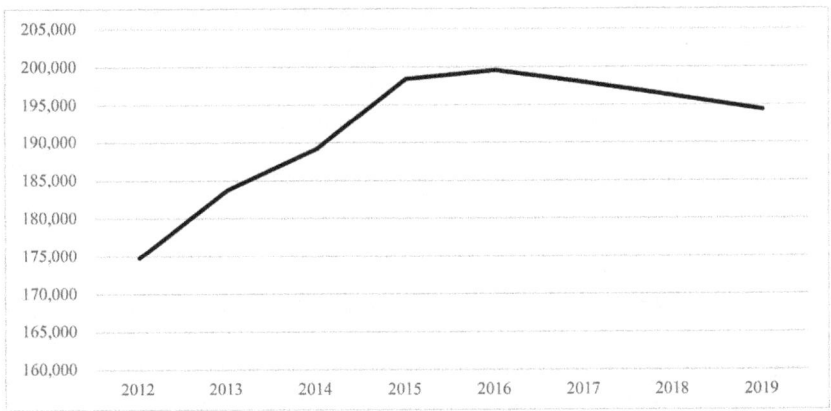

Figure 8.1. Space economy output in millions of current US dollars, 2012–2019. Figure compiled by the authors with data from Bureau of Economic Analysis.

Export Administration Regulations (EAR) that govern the manufacture, distribution, and sales of goods and services with dual defense and commercial uses. Similar to the USML, the Department of Commerce publishes a Commerce Control List (CCL) to identify things that fall under its purview. There can be overlap as well, so that such technologies like satellite and spacecraft thrusters fall under both USML and CCL categories since they are dual use. Additional items include such things as global navigation, computing, communications, electronics, and sensors.

Broadening out beyond EAR regulations, the Commerce Department is responsible for improving the conditions that strengthen commercial and economic growth for the United States. In the space domain, this means taking on the responsibility of helping to regulate and facilitate the expansion of an increasingly growing market. Space commerce has risen significantly over the past decade. The US Bureau of Economic Analysis (BEA) reported that the American space economy accounted for about $194 billion in gross output and 345,000 private sector jobs in 2019 alone.[3]

As figure 8.1 illustrates, the trend suggests that space will continue to grow in its importance to the overall economy. This creates the need to support the market with specific attention devoted toward it; the Office of Space Commerce was created with this idea in mind. Established in 1988, the Office of Space Commerce was not formally recognized by Congress with legal authorities until 1998, when it was codified into US law.

According to its congressional mandate, the Office of Space Commerce is designated as the primary actor for the coordination of space issues with the Department of Commerce to include: fostering the conditions for the economic growth and technological advancement of American space commerce; coordinating issues and actions related to space within the Department of Commerce; representing the Department of Commerce in negotiations with other countries to promote American space commerce; promoting space relevant US geospatial technologies through interagency working groups; and providing support for the US government on position, navigation, and timing (PNT) systems.[4] This coordination involves a range of other actors within the department that includes the National Oceanographic and Aeronautical Administration (NOAA), the Bureau of Industry and Security (BIS), the National Telecommunications and Information Administration (NTIA), the National Institute of Standards and Technology (NIST), and the Bureau of Economic Analysis (BEA). In general, these actors are responsible for the broad array of coordinating functions related to meteorological systems, communication spectrum management, aerospace, licensing, satellite calibration standards, and space economy analysis. All of these adoption and implementation efforts in turn feed into a broader strategic goal of advancing US leadership in space commerce.

One of the more high-profile space responsibilities of the Department of Commerce is the assumption of civil space traffic management from the DOD. This process began in 2018 under a Trump administration Space Policy Directive, SPD-3, and was directed to reduce the DOD's burden while also recognizing the need for a civilian entity to manage a growing civilian space sector. This increased the budget of the Office of Space Commerce from an annual allocation of $1.8 million to $10 million in the fiscal year 2021 budget.[5] Despite this recognition of growing importance, the Office of Space Commerce remains small, and questions abound relating to its ability to carry out the roles and missions it has been developing over the last three decades. To this end, the fiscal year 2023 budget proposed by President Biden increases the Office of Space Commerce's budget to $87.7 million, signaling a clear recognition of the need for more resources.[6]

Finally, while weather satellites are explored more fully in chapter 10, one additional highlight worth discussing in greater detail in the space mission of the Department of Commerce is that related to the satellite operations of the NOAA. The current operations, in many ways, links to the past, going back to the early space race and the International Geophysical Year discussed in chapter 2. Similar to the motivations to study Earth from space, current NOAA satellites do so to better prepare the United States and others for extreme weather, climate change, and even space weather threats. In its current inventory, NOAA utilizes five geostationary satellites (the GOES network) that provide imagery and atmospheric measurements of the Western Hemisphere, four polar orbiting satellites (Joint Polar Satellite System [JPSS]) for severe weather forecasting, and a Deep Space Climate Observatory (DSCOVR) that monitors solar activity from about one million miles away from Earth. It also partners with NASA, the Air Force, and the DOD to operate some of their weather satellites.

In sum, the Department of Commerce is an important actor in the implementation of commercial space policy. Given its wide range of roles and responsibilities, it is also in a position to help shape agendas in ways that facilitate the growth of the commercial space sector, while also helping to maintain US leadership in areas that go beyond the traditional bounds of commerce. With such a varied set of responsibilities, the Department of Commerce will be challenged to balance its duties effectively as the expectations for it continue to grow. Like the DOS, the Department of Commerce also will be hampered by the difficulty of managing and coordinating space policy within the organization, as well as across other organizations with which it has overlapping interest.

Department of Energy

Building off the legacy of the Atomic Energy Commission's role in the early days of space, including participation in the Apollo mission, the Department of Energy (DOE) remains a valuable actor in space policy. Its current vision is to support space policy through its science, technology, and engineering efforts. In its most recent space strategy, the DOE highlighted four contributions it makes to the national space enterprise: the development of space-capable energy technologies, research and scientific discovery for and in space, supporting the peaceful use of space and enabling space development.[7] Similar to the Departments of State and Commerce, these are all done to effectively support the broader policy goal of space leadership for the United States.

While the State and Commerce Departments have very clear congressional mandates for their roles in the US space enterprise, the DOE has a bit more flexibility for how it defines its contribution. Nonetheless, its core mission involving research on energy, environment, and nuclear issues places it in a very important support role for the advancement of US space policy goals across the civil, military, and commercial sectors. The Artemis program, for example, will require DOE expertise to assist in the development of a more robust space infrastructure, including energy generation and sustainment for lunar and Mars missions. One particular aspect of energy research along these lines is the development of more efficient and more powerful fuel sources for exploration that can be generated beyond Earth. Given its responsibilities through the Atomic Energy Act, the DOE is the source for nuclear materials production and acquisition for other governmental actors. Nuclear fuel remains a critical component for space power systems across the US space enterprise, dating back to its first use for satellite power in the 1960s.

Another important role for the DOE is supporting fundamental science research in the public and private sectors. Particularly relevant is collaboration with other institutions such as universities and private industry through its national laboratories where ideas and information can be shared and developed through multilateral efforts. This again reinforces a broader method to achieving policy goals through a whole of government and in this case, a whole of national approach.

While the DOE and NASA work very closely together to achieve gains in civilian space exploration and research, the DOE also works with the DOD to support national security space missions. The area most significant in this regard is the monitoring, modeling, and evaluation of nuclear explosion testing through space-based sensors. This area includes helping provide nuclear treaty monitoring and atmospheric testing data. Also important is the development of capabilities to better protect space-based technology. Beyond the DOD, the DOE has worked with the USSF and the US IC to ensure resilience for US space assets in the event of natural- or human-caused disruptions.

Finally, while much of the DOE space mission reflects efforts to develop technology, energy, and science, there is another role that perhaps has cataclysmic implications. In collaboration with its National Nuclear Security Administration (NNSA), the DOE assists with Earth defense from space objects that could cause great harm upon their collision. This role includes computer modeling and simulation of events such as asteroid impacts and the development of redirection efforts. In this, the DOE plays a rather important function in agenda setting, formulation, and adoption of policy directed toward reducing the risk of such impacts.

Department of Transportation

The mission of the Department of Transportation (DOT) is clear and concise: "To deliver the world's leading transportation system, serving the American people and economy through the safe, efficient, sustainable, and equitable movement of people and goods.[8]" Over the last three decades, this mission has included commercial space transportation. While the United States is not where many people in July 1969 envisioned the country would be, space transportation has increased in a steep upward trajectory and is likely to remain this way for the foreseeable future. Primarily, space transportation involves the movement of a lot of equipment and technology and relatively few people; however, the number of people likely to travel into space makes it important to lay the foundation for the application of the DOT mission to space in a meaningful way.

The FAA is the operating agency within the DOT that is responsible for space transportation. Given the dangers and risks associated with space travel, the main focus for the FAA is on safety. As a result, the FAA, and more specifically, the Office of Commercial Space Transportation (AST) is responsible for implementing policy, both at the national level and from within the DOT. Relevant national policy beyond the National Space Policy includes legislation regulating commercial space launch and space transportation infrastructure, as well as a broader national space transportation policy. Also relevant for the FAA are international treaties, including the OST but also more specifically, the Convention on Registration of Objects Launched into Outer Space and the Convention on International Liability for Damage Caused by Space Objects. These two conventions are responsible for establishing a mechanism for identifying space objects, creating a register of space objects, and for establishing state liability for any of its space objects that cause damage to others, respectively.

In carrying out its mission, the FAA, like the other actors discussed thus far, must work with a range of partners. This is particularly the case given that while the FAA is responsible for safety through regulation and licensing, it is not actually the principal actor engaged in the transport of people or goods into space. With respect to regulations, there are many to which the FAA/AST attends, including such things as lightning safety for launch, establishing risk limits associated with areas affected by the reentry of space objects, providing guidance on assessing the reliability of reusable launch and reentry vehicles, and human spaceflight protocols. The regulation process is one that allows inputs from the commercial, military, and civilian space sectors, creating challenging dynamics with respect to coordination and support for overall outcomes that benefit all parties. In the new space era, there is a heightened importance for these matters as innovation and the speed of technology growth continue to create new problems and obstacles for the FAA and broader DOT mission.

One future policy area that will require a great deal of FAA input, for example, is that of space tourism. As companies like Blue Origin carry out their space tourism projects, there will be pressure on the FAA to maintain safety levels like those found in civilian air travel. This means additional considerations for how spaceports are operated and maintained, in addition to implementing policies geared toward maintaining safe, routinized travel for space passengers in an environment considerably more dangerous than that of other domains of passenger travel. There will be competing interests between those that want to push farther and faster when it comes to the space tourism industry and those that want to ensure the infrastructure and safety processes are ready for an explosion of growth (discussed further in chapter 12). This tension is something also observed in chapter 6 between NASA and commercial space more broadly. While these interests will be adjudicated in the higher echelons of policymaking, the FAA will have a key role to play in providing the expertise and analytical resources helpful for decision-makers to shape space tourism's future.

For the moment, human spaceflight remains more of a novelty by comparison to the frequency of satellite launches. The launch industry is the foundation for all space-related issues, and certainly a primary activity for the implementation of policy from the vantage point of the DOT, FAA, and AST. For a long time, the ULA, a partnership formed between Boeing and Lockheed Martin, was the only actor able to provide launch services for the US government. ULA historically has been known for its exceptional track record in launch success, in some ways mitigating the FAA's mission of ensuring safe launches and enforcing regulations. The problem with ULA, however, has been its overall cost, creating opportunities for companies like SpaceX to provide a competing model. As discussed in more detail in the commercial space chapter, these efforts have created a host of challenges for the United States in terms of policymaking. The increase in the number of launches, the different levels of risk acceptance, and differences in the willingness to push the boundaries of regulations and policy will make it difficult for the FAA and its fairly small Office of Space Transportation to keep pace. This is more significant when considering the inability of both the president and Congress to keep pace as well. Thus, one problem with policy development is the speed at which it can be delivered by comparison to the speed at which it is needed.

Finally, when thinking about the future for the DOT (FAA and AST) as it relates to their roles in space, it is worth thinking about the future of policy as it relates to space infrastructure, specifically with respect to spaceports. Further down the organizational tree is the Office of Spaceports and the National Spaceport Intergovernmental Working Group run by the AST. Their charge is to create a network of spaceports that will serve as the backbone of space transport for both the United States and the international community. As mentioned in chapter 2, the early efforts of southern states to become active players in the space enterprise has led to their continued importance, which is reflected in the location of today's spaceports. Florida and Texas dominate spaceport and infrastructure locations, with others in Alaska, California, Colorado, New Mexico, Oklahoma, Virginia, Alabama, and Georgia. In the long term, supporting national policy will mean continuing to develop spaceports throughout the US both qualitatively and quantitatively to ensure continued and sustainable access to space.

Also important in maintaining consistency with the broader policy goal of maintaining US leadership in space, the DOT, FAA, and AST point to being an exemplar for other countries looking to establish a more robust space infrastructure. This includes a willingness to assist others with respect to general regulations, launch safety, and spaceport development. Through the establishment of the National Spaceport Intergovernmental Working Group, intergovernmental, academic, and industrial cooperation continues to strengthen the promotion of spaceport growth through innovation, resourcing, and standard setting, both increasing the robustness of the US space infrastructure and serving as a model to emulate.

The Federal Communications Commission

As a part of maintaining the broader space infrastructure, the FCC is responsible for coordinating and allocating the use of the electromagnetic spectrum for nongovernmental, domestic uses of space telecommunications. This set of responsibilities means that the FCC is involved in all commercial space activity, active in licensing and regulatory functions. Its regulatory authority has increased and evolved; a growing number of satellites means more coordination, but it also means the need for greater regulations in its operations. For the FCC this requirement has led to the provision of regulatory policy regarding satellite debris, an increasingly relevant problem for the international space enterprise as we mentioned in chapter 1.

Through its International Bureau, the FCC has been active recently on a number of initiatives that help shape both domestic and international policy. One way it has sought to align its policies with national space policy, for example, is to focus on supporting White House strategy on the promotion of in-space servicing, assembly, and manufacturing (ISAM). In August 2022, FCC commissioners voted in favor of creating an inquiry that would identify challenges and problems associated with the current state of ISAM as well as how best to facilitate it from a policy perspective moving forward. This inquiry begins a formal process for the FCC in the areas of innovation and transformation in the space economy, and according to the FCC, serves as an effort to promote US leadership in the space economy.[9]

Highlighting the interplay of coordination and engagement with other space actors, the FCC also works with actors like the DOS and the Department of Commerce in representing US interests internationally in its interactions with the ITU. International spectrum management is a key part of this coordination, as well as the registering of US satellites with the ITU and taking part in meetings to discuss and review international policy on the use of spectrum and satellite orbits. One problem that remains significant in this interplay is the lack of clarity with respect to agency responsibilities. The last several administrations have not settled which authorities should ultimately be granted to a particular agency when it comes to current and the need to develop new regulations for a quickly evolving commercial space enterprise. For the ISAM initiative, for example, it will be helpful to know which actor is responsible for satellite servicing regulations, something that could easily fit into the Department of Commerce's portfolio just as well as the FCC's. Short of congressional or consistent presidential guidance, it is likely we will continue to see space regulatory actors like the FCC try to fill the gaps, potentially creating more redundancy and confusion for commercial space actors as they seek to gain the permission necessary to improve the very area that national policy is seeking to develop. Such issues reinforce the problems associated with complexity in the national space enterprise.

OMB and GAO

Follow the money. Like everything else in the federal government, resource trails can tell one a lot about the priority a particular policy has across an organization and for the country as a whole. As seen in the early days of the space race, the competition for resources can lead organizations down certain paths with long-term impacts. The Air Force could not capitalize early to achieve a monopoly on space in part due to a lack of internal resourcing for satellites, with Air Force leadership lobbying for money to go primarily to crewed aircraft. This in turn helped facilitate efforts to spread resources for space across other agencies. This process has persisted such that space resources continue to be shared across a wide range of actors, creating concerns about funding at all levels of government involved in space policy as well as concerns regarding the effectiveness and efficiency of space functions across the government.

When it comes to the management of funding and assessing effectiveness and efficiency of organizations, the OMB and the GAO are the two agencies that have primary responsibilities in these areas. The OMB is a White House office and thus handles these duties for the president and cabinet. The GAO is the agency that provides these services for Congress. While they perform these duties separately, both utilize the work of the other in their efforts to bring accountability and efficiency to the policy process. Space policy in particular, given the increase in funding requirements and overall resourcing, has been increasingly a part of the oversight of these agencies as both the president and Congress evaluate where best to allocate taxpayer dollars to achieve maximum effect in the development and execution of policy.

In chapter 4 we discussed just how much the president's agenda drives space policy. As the executive agency responsible for preparing the president's budget proposal, the OMB has significant influence on the space budget. While the details regarding the OMB's broader budgeting function is beyond the scope of our concerns here, it is important to note that the preparation of the budget involves touch points across the entire policy process. These touch points include setting the stage by providing the president with the current economic conditions of the country, disseminating guidance to the executive's agencies about the SOPs for their budget requests, coordinating with agencies on budgeting issues, issuing guidelines for budget submissions, and assessing whether proposals are in line with the president's policy guidance and overall priorities. Support for space programs works through this process, where agencies' budget requests essentially compete for dollars with others' requests. This competition helps explain why in this discussion of these "other" actors, there is a clear attempt to align policy and priorities with those of the president's; to do otherwise risks a negative OMB assessment that may downgrade an agency's priority for funding.

The OMB also is tasked with oversight of agency performance across the executive branch, and through its Office of Information and Regulatory Affairs (OIRA) weighs in on issues involving proposed changes to regulations or policies involving the roles and missions of executive agencies. Given its tasking, the OMB is a very powerful actor in policymaking in its own right. Its influence on space policy can be made clear with a few examples. In June 2022, the OMB recommended against a DOD initiative to create a Space National Guard based on concerns regarding its start-up cost. This cost was calculated by the Congressional Budget Office (CBO) at approximately $20 million for the start-up plus an additional $100 million per year. Despite National Guard Bureau, DOD, and some degree of bipartisan support in the Senate, the establishment of the Space National Guard did not receive presidential support, nor did the initiative make it in the most recent NDAA. In another illustration of the OMB's impacts, in the 2017 NDAA, the OMB was tasked to assess the leadership, management, and organization of DOD space and to provide recommendations on how to strengthen these areas. These recommendations included the need for change, the establishment of a new sub-unified command that would eventually elevate to a new unified combatant command, and the consolidation of space authorities within the DOD. These recommendations have all been acted upon since the publication of the OMB's report, with a range of changes culminating in the creation of the USSF, but also including the reconstitution of USSPACECOM as the 11th combatant command.

On the congressional side, the GAO provides hundreds of nonpartisan reports for its members to provide them with information to make the most informed decisions about the effectiveness and cost of ongoing or proposed policies. Alas, the GAO does much more, including auditing, adjudication of bid protests for federal contracts, and the issuing of decisions regarding appropriations and possible violations of appropriations law. Similar to the OMB, the GAO offers both assessments and recommendations based on requests and mandates from Congress. It was a 2016 GAO study, for example, that identified the problem of too many space authorities within the DOD. It was this study that subsequently helped develop the aforementioned 2017 OMB report on DOD space.

NASA often is the target of GAO activities and helps illustrate GAO's policy role. NASA is responsible for billions of investment dollars for space exploration and research. Since its early days, NASA has been known to overspend on its programs to meet policy goals. Acquisitions particularly were known to suffer consistent overruns, something that is fairly typical across the entire space acquisition enterprise. Nonetheless, given its high profile, NASA has been very scrutinized by Congress over the decades, with the GAO being the mechanism by which congressional members can gain information in their preparation of making arguments for or against various funding levels in a given budget year. The GAO consistently identifies NASA as a high risk with respect to acquisition management. In its most recent assessment of major projects at NASA, the GAO pointed out that while the agency managed to complete six major projects in 2021, including the launch of the James Webb Telescope, it also exceeded expected costs by $3 billion, and schedule timelines by 10 years![10]

Figure 8.2 presents the GAO report's depiction of the cost and schedule overruns for NASA from 2013 to 2022. What the GAO concluded is that these overruns threatened the ability to fund new missions or continue to fund some of its other current missions. As one of 14 assessments of NASA, the GAO also pointed to its suggested recommendations for NASA over the years, many of which NASA accepted, despite not implementing many of them fully or at all.

In terms of policy, the GAO's reports matter because they provide the type of information required by members of Congress in their decision of how to

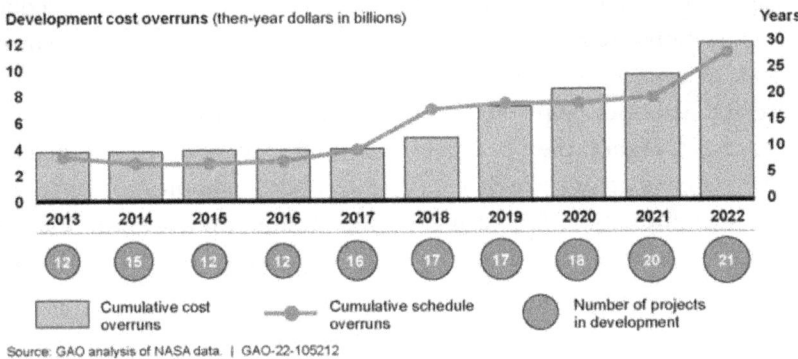

Figure 8.2. NASA cost and schedule overruns, 2013–2022.

allocate funding across NASA's range of programs. More than this, however, they provide guidance on where the organization can improve on its budgetary performance and how it might do so. While these reports are not the final say on the allocation of money, nor are they the only inputs on policy decisions regarding NASA's space projects, they do play a significant role in the process.

The bid protest process is also an important GAO function. Companies making bids for government contracts can protest a lost bid that is adjudicated by the GAO to determine that agencies awarding contracts did so legally. This is not always so straightforward as legal processes can be complicated, with protests taking place over seemingly trivial legal interpretation differences. For commercial space, this process has significant implications given the importance of government contracts for commercial enterprise at this stage of its development. As discussed in chapter 6, in one recent protest, Blue Origin claimed that NASA improperly awarded a procurement contract for the human landing system for the Artemis program to SpaceX over its own bid. The value of the contract totaled $2.9 billion. After a thorough review, the GAO denied the protest, concluding NASA did not engage in any legal violations, nor did the agency conduct an unfair bidding process. With such a large sum of money involved, the GAO was directly responsible for a decision of sizable impact for the commercial space industry.

For both the OMB and GAO, the aforementioned themes on the ubiquity of space, the need for bureaucratic alignment, and the complicated interplay between organizations in the policy process should be clear. First, space policy is inherently an increasingly larger part of the budget process, affecting the resourcing of so many agencies across the government, resourcing for which both organizations have an important advisory and evaluator role. Second, both organizations are responsible for alignment of policy between the principals and the agencies serving them. The extensive nature of their analytical and research activities helps provide the means by which this is done on a consistent basis. Finally, the two organizations often serve as an intermediary between a range of actors, whether it's with respect to adjudication or simply in terms of coordination. One can even see the interplay between the two organizations as the executive branch and Congress try to agree on a final budget and policy priorities for space.

Conclusion

When accounting for the "other" actors in the space enterprise, those not typically thought of when one considers space policy, it should be clear just how ubiquitous the influence of space currently is for US policy. While we have made a significant effort to be thorough in our discussion of the many actors involved, we undoubtedly are forced to leave others out, others that even in small ways play a relevant role in helping the United States achieve its policy goals in the space domain. What should also be clear is the importance of space policy for the United States. With so many roles, responsibilities, and missions to consider, the need for well-developed policy is more important than it has ever been. The challenges, however, loom large in the quest to do so. How effective can coordination across so many departments and offices be given the differences in interests and variation in authorities across the different space actors? How can commercial actors reliably plan on developing their business models given the complexity of the regulatory process? What do foreign partners and allies make of this complexity when it comes to new agreements and initiatives to foster cooperation? These are some of the questions that arise when considering how actors maneuver through the cumbersome policymaking and implementation processes brought about by the complexity of the US space enterprise.

This complexity cannot be understated. While it is in part driven by the ubiquitous nature of the space enterprise across a wide range of government functions, there is some complexity that seems to be driven by bureaucratic inertia. Space is involved in so much that the government does, policy has no choice but to follow. In many cases, policy has followed too slowly, leaving departments and agencies trying to decipher the direction of national space policy, leaving them to cling on to ideas such as maintaining space leadership and protecting the freedom of access to space. These concepts are nebulous enough to allow some interpretation on the part of agencies with respect to their own initiatives in the space domain. It is up to national policymakers, often through the use of the OMB and GAO, to make sure these bureaucracies are properly aligned in ways that are best likely to achieve their identified policy goals. Despite all of these efforts, however, the complicated structure of the current system leaves an observer wondering if any national policy can be fully implemented successfully across the bureaucratic morass of space actors. As the demand for more policy increases with respect to the increasing importance of space, is the US space enterprise subject to an endless expansion and creation of authorities to keep up? Since the early days of the space race between the United States and USSR, the trajectory continues in this direction, fairly unabated. Time will tell if this approach is the most effective in achieving space policy goals.

Summary Points

- Space is ubiquitous across the US government in terms of its importance. This has created a complexity that makes carrying out space policy particularly difficult.
- Bureaucratic organizations both seek and are pushed to align their organizational level policies and priorities with those of the principal policymakers to gain support and remain relevant.
- Organizations across the space enterprise interact with numerous other organizations on space issues. This interplay creates a dynamic that makes it difficult to single out individual contributions to national space policy, since it is the aggregate effect of these interactions that determine whether policy is successful or not.

Discussion Questions

1. Why are there so many space actors involved in the space policy enterprise? Would policymakers be better to consolidate the number of actors? Why or why not?
2. What does it mean to achieve leadership in space?
3. What types of policies are necessary to shape the development of spaceports in the United States?
4. Of these "other" organizations and actors involved in space policy, which is most influential? Why?

For Further Reading

Department of Energy, *Energy for Space: Department of Energy's Strategy to Advance American Space Leadership, FY 2021–FY 2031*, January 2021, https://www.energy.gov/sites/prod/files/2021/01/f82/Energy%20for%20Space-DOE%20Space%20Strategy%20Paper%2001-06-2021.pdf.

Government Accountability Office, *NASA: Assessments of Major Projects*, 2022, https://www.gao.gov/assets/gao-22-105212.pdf.

Lambakis, Steven, *On the Edge of Earth: The Future of American Space Power* (Lexington: University Press of Kentucky, 2001).

Pappalardo, Joe, *Spaceport Earth. The Reinvention of Spaceflight* (New York: Overlook Press, 2017).

9

Nongovernmental Forces

On May 30, 2020, America ended a nearly 10-year drought in human spaceflight missions launched from American soil. With the retirement of the space shuttle after the last flight of *Atlantis* on July 8, 2011, Russia was the only country participating in the ISS able to launch humans into space. Since then, the United States relied on the Russian Soyuz to transport American and other international crew members to and from the ISS. Even though the two had worked cooperatively on the ISS for almost 30 years, the relationship was still somewhat uncomfortable—the Russians charged the United States an average of $56.3 million per seat with one costing over $90 million and came amid a period of increasing tensions between the two countries.[1] By the time of the shuttle's retirement, it was clear that a replacement, which NASA had been working on since 2004, was nowhere near ready. As such, in 2010, the Obama administration ordered NASA to turn to the emerging commercial space industry to provide crew transportation to and from LEO.

This policy shift was significant. To that point, no private company had succeeded in launching anything to the ISS, let alone a crew of astronauts. SpaceX was only just beginning to operate its Falcon 9 with a demonstration flight to the ISS still ahead. Many NASA officials and politicians were skeptical—then Senator Bill Nelson helped lead the fight in the Senate to have NASA develop a new generation rocket (what would become the Space Launch System) as well as cut NASA funding devoted for commercial crew.[2] In 2021, as Nelson became NASA administrator, however, it was clear that shifting responsibility to the commercial sector for crew flight was not only possible but possibly preferable. The lack of public support for a more expansive space agenda and the growing technological capability of the private space sector meant that NASA might be able to do more in LEO with the same or less resources. To date, SpaceX has launched four NASA crews to the ISS as well as one private trip, Inspiration 4, to orbit and another, Axiom-1, to the ISS itself, and other companies including Boeing and Blue Origin are advancing programs for private-crewed orbital flight.

To this point, this book has largely focused on the institutions of government, both national and international, that influence space policy in the United States. However, it is becoming increasingly clear that there are nongovernmental forces that also are important in setting the direction. The American public, in either supporting or not supporting a wider space policy, can enable the general direction either positively or negatively. Private corporations, as they become increasingly capable and economically more valuable, can independently influence space policy through the capabilities they offer and their political influence. The growing economic value of space as well as the ability of space to aid economic development and advances help to condition what is done and how. And finally, technology and its availability can either hinder space advancements by limiting what is possible or enable it. This chapter explores each of these factors in brief beginning with public opinion.

The Public

Previous chapters have introduced you to the different institutions of American government that have a significant role in setting space policy. In a democracy, however, those institutions are populated by people who have been chosen to represent the wider public. Recall from chapter 5 that the assumption is typically that elected officials will behave in a way that their constituents want them to (thereby providing representation), but that assumption isn't always correct. The public may not always have complete information on which to decide, or they might not be interested. Even if they did, some representatives choose to act in a different way than their voters prefer simply because they believe it is the right thing to do. In any case, to understand and explain the behavior of elected leaders when making space policy, it is necessary to understand the contours of public opinion and perhaps more importantly, its limits.

Before starting, it's important to note that there are some limitations on how well we can understand how the public feels about space. First, as this book has shown, there are many different aspects to space policy. In general, there is a lack of public awareness and understanding of how complicated space is and the variety of different efforts that fall under the general heading of "space" or "space exploration." Second, because space is, at best, a secondary policy concern, there is a lack of consistent public opinion data about space. While one such source will be discussed later, it offers only a more general question about space exploration spending. Relatedly, while people may express a certain opinion about space exploration on its own, people are rarely asked about space in conjunction with other policy priorities. In other words, an individual may support the idea of space exploration but when asked about it compared to something like health care or climate change, it likely holds a lower priority.

Before exploring the extent to which public opinion can or has influenced space policy, we must first try to understand patterns of public opinion about it. Public shock at the Soviet Union's launch of Sputnik in 1957 helped push the Eisenhower administration to support the creation of a new space agency, NASA, and public support remained somewhat high in the agency's early years. However, as early as the mid-1960s, the initial wave of space enthusiasm began to wane as the public, and as a result Congress, became more concerned with its growing cost. Notably, Roger Launius finds only one instance in the 1960s where more than 50 percent of the public believed that the Apollo program was worth the cost: July 1969 immediately following the Apollo 11 landing.[3] This suggests that public opinion mattered very little in deciding whether to go to the Moon or not, once again reinforcing the Cold War motivation for the project.

Unfortunately, public support for space exploration does not appear to get much higher. One of the only sources of long-term data on public opinion about space comes from the General Social Survey (GSS), a representative survey performed roughly every two years. Since the 1970s, the GSS has asked respondents whether they feel that money being spent on space exploration is too little, about right, or too much. Figure 9.1 shows this data over time. The post-Apollo hangover is evident in the early 1970s as 63.4 percent of respondents in 1974 believed that too much was being spent on space exploration. Over time, those numbers have come down as the percentage of respondents who feel that space spending is about right or too little has increased. Despite that promising trend, in 2018, the percentage of respondents who felt there was too little spending was only 22.4 percent. The GSS also asks respondents how interested they are in space (figure 9.2). Between 2008 and 2018, an average of 22.8 percent of respondents said they were very interested in space, 45.3 percent moderately interested, and 31.3 percent not at all interested. Taken together, this data suggests that the American public is only somewhat interested and supportive of space exploration but that it is not a primary concern.

Though the GSS data gives us a rather consistent view of data on the public's opinion of civilian space exploration policy, we have a more limited view of how the public feels about military uses of outer space. Several polls from 2018 show a split in support for the creation of the USSF—one poll found that 57 percent of respondents supported it (the highest among this group of

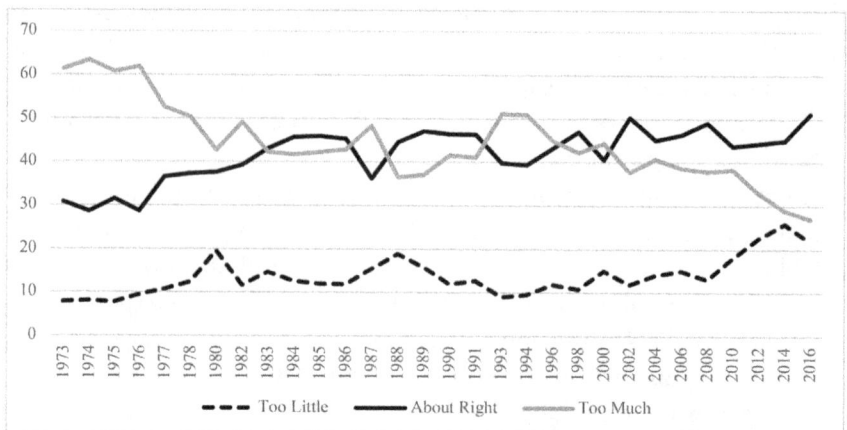

Figure 9.1. General Social Survey attitudes on space exploration spending, 1973–2018. Figure compiled by the authors with data from the General Social Survey, https://gss.norc.org/.

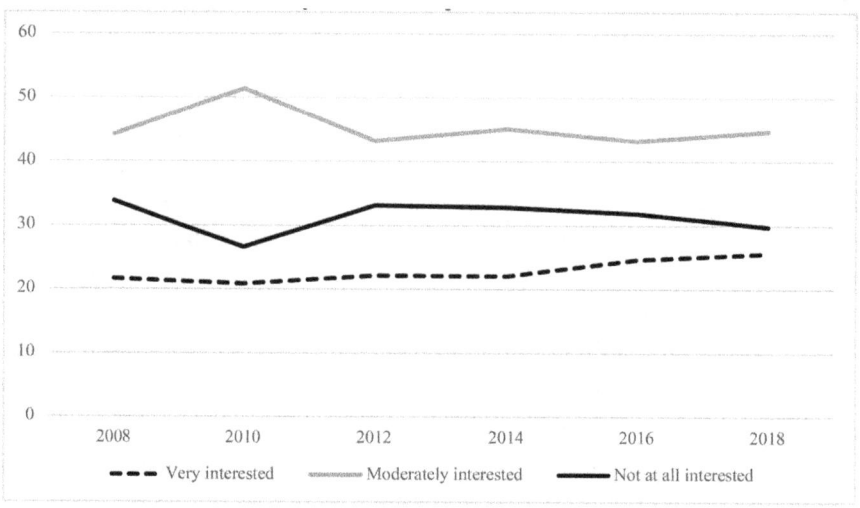

Figure 9.2. General Social Survey interest in space, 2008–2018. Figure compiled by the authors with data from the General Social Survey, https://gss.norc.org/.

polls) while in another, that figure was just 33 percent.[4] A third poll showed a 2:1 split.[5] Overall, these polls showed a deep partisan split with Republicans almost overwhelmingly supporting the USSF's creation, another poll showed that only half of Republican respondents were supportive.[6] Though there is little further public opinion data, the public does seem to continue to be skeptical. Several writers have highlighted the USSF's public relations battles and the service continues to receive a fair amount of criticism in social media.[7] Reactions on social media and the internet are by no means a representative method of gathering public opinion data, but taken as a whole, the public seems to have a similar skepticism for expanded military space policy as it does for more civilian activities.

Data like that provided by the GSS can indicate general trends in how the public feels about space, but it alone cannot tell us why they feel as they do or how influential public opinion is in influencing space policy. To begin with the first question, there are several different factors that influence how people feel about space including age, gender, political party, education, and socioeconomic status. In terms of age, François Nadeau finds older generations have a higher level of support for space spending as a sort of Apollo-era nostalgia. In other words, those who were able to experience the Apollo program firsthand tend to support more space spending.[8] Education is also an important influence as Nadeau and Wendy Whitman Cobb both find that those with higher levels of education and knowledge about space also support more spending.[9] Partially related to different levels of science education, women are two-thirds less likely to support more space spending, confirming a significant gender gap in space support.[10] The findings about political party are more mixed— Whitman Cobb found that more Republicans support greater space spending, while Nadeau showed no significant political difference. Though both Democrats and Republicans support space spending, they do so for different reasons—David Burbach finds that Democrats support space because of its ability to monitor and tackle climate change and other environmental issues while Republicans support it for national security reasons.[11]

The notion that people support space activities for deeper underlying reasons underscores the secondary nature of space policy in the United States. Space and space exploration appear to be part of a postmaterialist policy agenda in that people tend to support it once other needs such as food, security, and education are filled.[12] This is also reflected in the varying degree to which public opinion influences space policy directly. Using the GSS data, Alan Steinberg explored the extent to which it was correlated with NASA's budget. He finds that NASA's budget is both responsive and not responsive to public opinion: while NASA's funding does increase as tolerance for space spending increases, this only applies to the actual NASA budget. If NASA's budget is examined as a percentage of the overall federal budget, however, NASA's budget has decreased considerably even if tolerance for spending rises.[13]

It is worth taking a moment to discuss the tools that the general public have through which they can influence space policy. This section has implicitly assumed that elected officials, interested in election and reelection, take into consideration how their constituents feel on issues such as space. While the percentage of the public that likely votes solely based on a candidate's views on space is quite low, leaders cannot move too far away from the median position of their voters. In addition to this mode of influence, there are other ways that the public can influence policy including writing letters, spreading awareness through grassroots campaigns, and getting involved with different space-focused interest groups. However, if the number of people who are likely to decide who to vote for based on space policy, the number of people who will go above and beyond this standard is also quite small.

Taking public opinion on space as a whole, there are several findings important for our purposes. First, public awareness of and interest in space is relatively low. Further, the percentage of the public who might be predisposed to supporting space activities is quite small, thus limiting the amount of public support. The result is that rather than supporting more space activities, public opinion more often represents a limit on what can be done. Second and related, while elected officials may be able to claim that they are following the public's lead in giving NASA more absolute funding, they are also responding to this secondary importance as NASA's budget as a percentage of the overall budget declines. Finally, public support for space policy appears to be more contingent on what it is related to—space activities that can claim to have a national security or larger environmental purpose may be likely to receive more support from the public in the abstract. This also highlights an irony in support for NASA: human spaceflight, a mission central to NASA's organizational identity and culture, tends to receive less support from the public. While this may suggest that NASA pursue other activities or applied missions, NASA's continued focus on human spaceflight highlights just how powerful organizational culture can be for bureaucratic agencies.

The Space Industry

If the public at large is not entirely supportive of space activities, other actors more intimately involved with space policy certainly are. It is perhaps a bit ironic that as Dwight Eisenhower left the presidency in 1960, just as the space race was getting underway, he warned of the growing power of what he termed the military-industrial complex. Concerned about the growing power and influence of a permanent arms industry, he worried that it may overwhelm "our liberties and democratic processes in the future." While Eisenhower was surely right to be concerned, the size of the military industry has only grown since, and, because of the deep connection between military technologies and space, it is many of those companies who have become intimately involved with space policy and activities in that domain.

While the government fully designed and built many of the earliest rockets and satellites, private industry has been involved almost from the beginning. For example, while built and operated in conjunction with the US government, TIROS-1, the world's first weather satellite, was built by RCA, and Echo 1, the first passive communications satellite was built by Bell Labs.[14] Rockets like Delta (itself a variant of the USAF's Thor ICBM) were built by the Douglas Aircraft Company. At NASA, engineers helped design spacecraft and rockets, but they were ultimately built by private contractors spread throughout the United States. Several companies and even universities were involved: MIT's Instrumentation Laboratory worked on guidance and navigation systems, North American Aviation was the prime contractor for the Apollo spacecraft, and Grumman built the lunar module. Numerous other subcontractors were used including Honeywell, Northrop, and Lockheed.[15] For the Saturn V rocket, a collection of companies provided various components including North American (F-1 engines), Boeing (first stage), Douglas (third stage), and IBM (guidance).[16] While this approach made coordination across these companies more difficult, it also helped to spread government largesse to multiple states where the work would actually be done. This in turn helped to garner greater political support for the effort.

While these and most other early space efforts were undertaken by the US government and built by private industry under contract, companies also considered how they might exploit the commercial potential of outer space. One of the earliest efforts to do so was in the field of communications. Operated by AT&T under an agreement with several international companies, Telstar 1 and 2, launched in 1962 and 1963 respectively, successfully relayed a variety of information and communications including television, radio, phone calls, and even computer data around the world.[17] Development of a corporation, Comsat, initiated by the US government soon followed. Communications satellites were one of the first areas in which purely commercial space activities were undertaken followed later by things like remote sensing and satellite television.

While still difficult, satellites are relatively easier to design and build than rockets. It's in this area where industrial development has been more difficult and slower. In the United States, early launch services for commercial satellites were done under contract with NASA who in turn contracted for rockets like the Thor-Delta. Following Apollo, the development of a commercial launch industry was stifled by the development of the space shuttle, which was intended to launch every government satellite.[18] In Europe, however, a group of countries led by France developed the Ariane rocket in the 1970s and in 1980, Arianespace became the first commercial launch provider in the world. The failure of the space shuttle *Challenger* in 1986 led to reinvestment in expendable launch technology in the United States; companies like Lockheed Martin and Boeing stepped in to provide these services, which would in turn be available to commercial customers. Proactively, the Reagan administration also put in place a legal and regulatory framework to support the establishment and growth of a launch services market and make launch (and landing) a government-regulated activity.

Despite the obvious benefits of space in multiple arenas, the growth of a space industry was still hobbled. For most companies, space served only as a secondary line of effort; their primary expertise and sources of funding often lay elsewhere including general aviation, military equipment, and terrestrial communications. This combined with the relatively low value of space commercially at the time meant that the burgeoning space industry did not have as much influence as the military-industrial community already did. The high cost and difficulty of access to space exacerbated this because to do anything for a commercial profit in space would require huge amounts of money up front to develop and deploy space-based assets.

The growth of the internet and the dot-com boom of the late 1990s changed this trajectory. Newly made millionaires and ambitious companies seeing potential profit in space proposed new satellite constellations that would provide both internet and satellite telephone services. Both Boeing and Lockheed Martin worked to update their expendable launch vehicles to accommodate what they saw as growing commercial and military demand. However, as highlighted in chapter 3, much like the dot-com bubble itself, the space industry was also hit hard by reality. Teledesic, a constellation of 840 satellites designed to provide broadband satellite internet backed by Bill Gates, closed in 2002 after only four years of satellite design. Even relative space success stories like Iridium, a satellite telephone system, were affected. After spending $5 billion to develop and launch a set of 66 satellites, it declared bankruptcy in 2000.[19] After a reorganization, it continued and still operates a large communications constellation. XM and Sirius, two separate satellite radio companies both launched satellites in the early 2000s only to find the need to merge in 2008. Boeing and Lockheed Martin, seeing their potential launch market dry up, were forced by the US government to create a partnership, the United Launch Alliance, to provide assured launch services to the US government.

And yet, while the space market seemed bleakest, a new generation of tech millionaires began to invest not in satellite technology but one of the more difficult aspects of space activity: launch. As discussed in chapter 3 and more fully in chapter 12, in 2008, SpaceX became the first to privately develop and launch an orbital rocket with its Falcon 1. As the Falcon 1 did not allow for any reusability, SpaceX quickly moved on to developing the more advanced Falcon 9, which incrementally developed reusable capability. This early success also coincided with significant changes at NASA as they implemented George W. Bush's VSE. As discussed in earlier chapters, SpaceX and other space companies participated in NASA's COTS program, which aimed to stimulate commercial developments. In 2012, the first Dragon was sent to the ISS and SpaceX began work on a crewed version. Since then, they have launched and recovered Falcon 9's first stage booster dozens of times, continued to deliver cargo to the ISS, and developed and launched a crewed version of Dragon allowing SpaceX to become the first private company to send humans into space. Other companies like Blue Origin are quickly following.

In addition to new technological approaches to rocket recovery and reusability, other developments are also assisting in growing the space industry. In previous decades, satellites were purpose-built, unique, and large, making their costs quite high. New generations of satellites have taken advantage of standardized and smaller approaches in the form of cubesats and smallsats. These types of satellite technologies not only make manufacturing easier but take advantage of both miniaturized electronics and off-the-shelf technology. Their reduced size allows them to be launched at a lower cost, often in bunches of several satellites on a single launch vehicle. The reduced cost of satellites and launches means that more people can access space including schools and universities and countries that have been previously unable to afford such activities. Additionally, like the satellite boom of the 1990s, several companies and countries have proposed massive satellite constellations in LEO to provide services ranging from internet (SpaceX's Starlink for example), remote sensing and tracking (Planet), and to support military activities (the United States).

The sum of these developments is that the value of the space industry has grown enormously in recent years. In 2020, the space industry was worth $447 billion with some estimating that it could reach $1 trillion in the coming years.[20] The economic impact of space activities is impressive as space-based infrastructure contributes enormously to life on Earth. The global economy is predicated on many of the services that space-based assets provide including precise timing measurements that enable banking and financial transactions, weather monitoring that allows industries like aviation and shipping to adjust to changing conditions, and communications, which have given rise to immense flows of information and data. As the value of space and space companies has grown, the power of the space industry has as well. While traditional defense contractors like Lockheed Martin, Boeing, and Northrop Grumman used to see space as a small portion of their overall defense portfolio, space is an increasingly important concern for them and the various new space companies that have arisen. And beyond the ideological motivations that lie behind Musk's SpaceX and Bezos's Blue Origin, it is this financial motivation that is a significant driver of behavior in the space industry. Even for both SpaceX and Blue Origin, the pursuit of profit is an important factor as it enables the broader ambitions that both companies have. Given this, the space industry has an enormous financial motivation to influence US space policy in ways that support further growth. In addition to angling for contracts from the DOD and NASA, the space industry is also concerned with laws and regulations pertaining to launch, technology transfer and import/exports, and operations.

First, the dual-use nature of space technology again plays a significant role in terms of what types of technologies US companies are allowed to take overseas as well as what they can launch. Pivotal to this is ITAR, which puts in place a framework to control the manufacture, sale, and distribution of space- and defense-related technologies outside the United States. This process ensures that sensitive technologies are not being sold or distributed to potentially harmful or dangerous actors and begins with an official list, the US Munitions List, of which technologies or products must be controlled. In the late 1990s because of concerns over technology leakage from the United States to China via several satellite manufacturers, "satellite technology" was added to the Munitions List as something to be tightly controlled and regulated. The result was that American companies found it more difficult to work with suppliers or customers who were not in the United States as it was not at all clear what "satellite technology" entailed. Companies suffered significant market losses as international companies began marketing their technologies as "ITAR-free." As the effects of this regulatory change became clearer, American space companies lobbied for regulatory change, which finally came in 2014.[21] While the control of sensitive technology is important, space companies are interested in balancing that with their ability to compete on the global economic market.

The same has been true for operations of space technology like remote sensing. As more advanced cameras and sensors have been developed and become commercially available, space companies focused on providing remote sensing imagery and data have become more plentiful. For instance, Planet operates a constellation of more than 200 satellites, many of which are small cubesats, that image the whole Earth daily. While this is a potentially valuable commercial service for everything from businesses monitoring daily traffic to agriculture, it can also serve an intelligence and defense purpose particularly as these off-the-shelf technologies grow in terms of their capabilities. The result is that their deployment and operations have been heavily regulated in the United States. Beginning in the 1980s, limits were placed on commercial remote sensing satellites that not only restricted who the data could be sold to but also their capabilities. This placed American companies at a disadvantage compared to international firms who either had no such limits or less restrictive ones. Concerns about economic competition led the US to further reduce restrictions, but the extent to which remote sensing is regulated is a highly contested affair. While the regulatory regime today is far more amenable to companies like Planet, how to balance national security and intelligence concerns with economic incentives continues to be a pressing issue for commercial companies.

As this section has detailed, commercial space companies obviously have many motivations to be interested in US space policy. They also have some significant tools at their disposal. While they may not be able to vote for political leaders, businesses and their associated industry groups can lobby the government for desirable policy outcomes, donate to political campaigns, and even lead public influence campaigns to try to build a groundswell of public support. Musk and Bezos even have very powerful media tools at their disposal—Musk completed his purchase of Twitter in 2022 and Bezos has owned *The Washington Post* since 2013. While neither has explicitly used either platform (to date) to advocate for their space companies or plans, these outlets provide them with powerful ways to do so.

To begin with lobbying, companies seeking to influence policy in general often employ the services of lobbyists, people who advocate on their behalf to elected officials. Lobbying is a multimillion-dollar industry with scores of individual lobbyists who can monitor ongoing policy debates and attempt to influence policymakers. In 2020, the defense aerospace industry spent almost $47 million on lobbying, with Lockheed Martin and Northrop Grumman coming in at one and two in spending. ULA, the Boeing and Lockheed Martin joint venture, ranked number three, spending $1.8 million.[22] New space companies have also gotten in on the act—SpaceX's lobbying totals have reached as high as $2.38 million in 2019 with Blue Origin spending $1.93 million in 2020.[23] Congressional lobbying has played a significant role in the debate over NASA's HLS contract as discussed in chapter 6. Both SpaceX and Blue Origin undertook significant lobbying campaigns on Capitol Hill to demonstrate to lawmakers the benefits of their respective approaches.[24]

These figures, however, speak only to lobbying of Congress—the reality is that corporations can also lobby the executive branch. This is particularly the case when it comes to the formulation of federal regulations. As has been highlighted in previous chapters, regulatory agencies like the FAA and the FCC have a significant impact on space policy via rules that oversee everything from launches, operations, safety, and landings. Like the contracting process, the way in which regulations are made are quite detailed but provide multiple opportunities for public comment and company lobbying. Since legislation is often quite vague, regulators wield significant power in how to interpret that legislation to put it into action. This is yet another arena where lobbyists can influence regulators to interpret the law in ways that are favorable to their clients and industry.

In addition to lobbying efforts, commercial companies can do what all sorts of other organizations and individuals do that may help influence how politicians view them: donate funds to candidates for offices, political action committees, or political parties. While there are mixed findings on the efficacy of political donations, that does not stop companies from donating, thus justifying some attention to them. Traditional aerospace companies have a long history of political donations and, while new space companies have sought to differentiate themselves from their competitors, they also have taken up the practice. Consider the 2020 election cycle. Boeing contributed $7.6 million to political candidates, Lockheed Martin $6.9 million, and Northrop Grumman $6.6 million.[25] Given that these companies also have a larger stake in the general defense industry, these numbers are significant but not necessarily surprising. In terms of new space companies, SpaceX donated $1.2 million and Blue Origin $461,524.[26] Although these numbers are far smaller, they are part of a pattern of increasing political donations by both SpaceX and Blue Origin over the past several years and indicates both companies' interest in securing more favorable government action across the space policy spectrum.

A final means by which companies seek to impact policy is through coordinated public campaigns. Because companies realize that most elected officials are concerned with reelection and are thus very aware of how their constituents might feel on various issues, persuading the public on an issue can be a valuable means toward political support. While lobbying is very much an insider activity with little involvement by the public or even awareness to it, companies and interest groups can try to raise consciousness of major issues and move the public to concern and hopefully action. Company leadership can make statements or give media interviews. Advertisements in various media including print, television, or social media can be utilized. All of these efforts, however, are typically geared toward public persuasion and the development of support for their business efforts.

One recent example can help to illustrate. As part of its efforts to develop its Starship system, SpaceX has sought permission from the FAA to launch prototypes to orbit from its facility in Boca Chica, Texas. Among the things the FAA must consider to grant a launch license are the facility, the impact of such activity on the surrounding community and environment, and the safety procedures being used. In the fall of 2021 as the FAA scheduled public hearings on the SpaceX application, Elon Musk took to his social media feeds to urge the public not only to submit favorable comments to the FAA but also to speak favorably at the forthcoming public hearings. The result was that a large portion of the people who commented during the hearings were not only pro-SpaceX but resided outside the immediate Boca Chica area and, presumably, would not be as deeply impacted by space launches as local residents.[27]

The growth of the commercial space industry along with increasing government reliance on it means that space companies are becoming more influential economically and politically. These companies, while sometimes possessing higher motivations, are also propelled to seek out new revenue streams and opportunities for profit, activities that inevitably depend on favorable government policies. While traditional aerospace companies have long used their resources to influence public policy, new space companies are also rising to the challenge as indicated by the increasing amount of money spent on candidates, campaigns, and lobbying. Because of the public's acquiescence to space policy and its place as a secondary policy area, the space industry has both the means, motive, and opportunity to significantly influence the course of space policy in the United States. Their economic success, while partially predicated both on government policy and their own behavior, is also dependent on the final two subjects in this chapter, economics and the availability of technology.

Economics

Economics writ large has been found to be both something that supports greater attention to space as well as a beneficiary of increased space investment. Though "the economy" doesn't have any specific motivation, a good economy often leads to greater space spending, which in turn can support greater economic benefits. This section will examine both sides of this equation. We begin first with how space policy and space spending contribute to economic growth.

Outside the original space race, advocates for a more expansive space policy have had trouble finding a compelling justification for increased spending. To counter the implicit argument that money would be better spent elsewhere, advocates have variously touted species survival, the urge to explore, education, and the achievement of new frontiers as rationales to undertake space activities.[28] Perhaps the most frequent reason given, however, is economically oriented. Not only does investing in space lead to the development of new technologies and technological spin-offs that then create their own economic returns, but investing in high-education, high-paying space-related jobs is a form of economic development of its own.

NASA often touts such factors in public to help justify further spending. For example, NASA has an entire web page devoted to highlighting technologies that have come about because of their efforts including things like air purifiers, water conservation systems, biometric sensors, software, and microelectronics.[29] New technologies discovered and developed by NASA are made available for public use and licensing contributing further to economic development. In terms of larger economic impact, one report shows that in 2019, NASA generated $65 billion in economic returns with a budget of $21.5 billion and supported more than 312,000 jobs across the United States.[30] At a local level, NASA's impact is also substantial: Kennedy Space Center in Florida, for example, had a local impact of almost $4 billion with every 10 direct jobs at the center leading to the creation of 13 more across the state.[31] This doesn't account for defense spending on space, which likely has similar impacts.

The potential for a significant economic impact from space spending means that members of Congress often take this into consideration when making policy. In one study, Martin Machay and Alan Steinberg found that economic variables including presence of the space industry, NASA centers in a state, and NASA contracting dollars all had some influence at one time or another in votes on NASA authorization bills.[32] Their findings suggest that as the number of people involved in the space industry has grown as compared to the larger aviation industry, the influence of space jobs in a member's state or district have as well. Given the rapid increase in the space industry in recent years, the economic motivation may only continue to grow in importance.

Even though there is likely a strong economic argument for greater spending in space policy, economics also serves as a limiter of sorts, particularly for civilian and scientific aspects of it. While military- and national security–related space spending is often considered a mandatory element of the federal budget, NASA and the scope of its civilian activities is another matter entirely.

Space exploration and scientific study is very much a discretionary element of the federal budget, something nice to have but optional. As such, it makes sense that when the US economy is performing well and the federal government is taking in more money, spending for NASA tends to increase. On the other hand, as economic indicators such as debt and inflation flash warning signs for the American economy, NASA's budget is one of the first ones to suffer the blows of a budget cutting axe.[33] This dynamic is also supported by Steinberg's findings on public opinion that, even though NASA's absolute dollars increase each year, its percentage of the federal budget generally decreases.

Even though there is a definite economic benefit that comes from space spending, these findings show that the state of the US economy is more often a factor that limits space policy. While the influence of a bad economy might be overcome if there was wider public support for space spending or even greater public awareness of the benefits that come from it, that base of support is not something that is a regular occurrence in this field. Even the influence of the space industry is somewhat stunted given the rhetoric around spending on Earth versus spending on space. As both Musk and Bezos have come to play more prominent roles in space policy, some politicians have become more openly critical. Senator Bernie Sanders, for example, has taken to the Senate floor several times to criticize the billionaires and their influence on space policy while Representative Earl Blumenauer has proposed a tax on space tourism.[34] Thus, even as the industry's economic power grows, it must still compete for scarce budget dollars against policy areas like education, social services, and health care that many people see as having a more direct influence on their everyday lives.

Technology

Unlike the concepts discussed previously in this chapter and in this book, "technology" does not necessarily have any intrinsic motivations ascribed to it, nor any particular behavior to understand. There are no tools that are necessarily at its disposal by which it can influence space policy. Instead, technology *is* a tool that other actors can use to influence the potential course of space policy. Technology is also a goal or result often sought by many actors that in turn enables future developments. As such, it can exert an independent force on the possible paths that space policy may take in the United States. To that end, this section focuses on the ways in which technology in general impacts space policy with a particular eye toward both its enabling and limiting functions.

Figure 9.3. Technology spectrum. Figure created by the authors.

Before delving into this further, we must define "technology." While we might think this is relatively easy, definitions of technology may be quite varied. It can mean the various tools that we use to perform tasks, or it may mean the process through which humans come to discover new knowledge. Technology can also encompass certain sociological and anthropological dimensions including how knowledge is applied and the ways in which it is employed.[35] Since our purpose is not to provide a definitive meaning of technology, here, we take it to mean primarily the *means* or equipment through which or with which people act. Defining technology in this way allows us to focus on what is possible to undertake—scientists and engineers may have ideas about what is possible, but it is not those ideas that necessarily enable or limit space policy. Rather, it is the *ability* to make those ideas a reality that does. Someone may think that a particular form of rocket propulsion is possible but the means through which that can happen may not be or may only be available via intense effort and high cost. Viewing technology in this way, we can place it on a spectrum (figure 9.3) that goes from technological means that are highly unlikely or impossible to achieve on one end and technological means that are prevalent and widely used on the other. Where a particular piece of technology falls on this spectrum can impact the content of space policy in the United States.

Perhaps the easiest influence to identify is from those technologies that are far furthest to the left and are therefore incredibly costly, difficult, or impossible to achieve. The unavailability of these technologies serves as upper boundaries for what may be carried out in space or via space. One such example comes from the tyranny of the rocket equation. Recall from chapter 1 that to reach orbit, rockets must achieve a certain velocity and to do so requires a certain amount of fuel. But as the rocket becomes heavier due to its fuel load, it in turn requires ever more fuel to lift what it needs. This explains why such large rockets are needed to launch what appear to be rather small payloads or spacecraft. To be sure, this problem can be overcome—multiple rocket launches may be used to loft more complicated pieces of machinery that can later rendezvous in orbit or larger rockets may be built. Each of these approaches, however, comes with high costs. Given the lack of political willingness to devote too much in the way of resources to space, the limits of this technology and the expense associated with going further is a limit on what the state can do in space.

In a very real way, this technological limitation played a role in the development of the space shuttle. Early concepts for the shuttle in the late 1960s and 1970s proposed a system that was not only reusable but was a single-stage-to-orbit (SSTO). SSTO rockets use only a single booster stage to loft their payload to orbit rather than the multiple stages. However, to get by on a single stage, the rocket engines would need to be immensely powerful and capable of lifting not just the spacecraft but its entire supply of fuel to orbit. As NASA came to understand how expensive this would be and the political limits in terms of funding that they would be facing moving out of the Apollo program, the concept was quickly discarded. What NASA still tried to achieve, however, was a shuttle system that was at least partially reusable in the hopes of reducing the cost of going to space.

The limits presented by conventional rocket technology have continued to impact the types of rockets and systems that NASA and other space actors have developed. For several decades, the cost of launching a payload to space was steady at just over $18,000 per kilogram, a cost that was partially determined by the technological capabilities available.[36] The shuttle, though partially reusable, still came with a high price tag. The high cost in turn limited how many actors could access space and the types of activities that were being undertaken. In recent years, however, we have witnessed what changes in launch technology are capable of inducing. As companies like SpaceX have pursued reusable rocket technology, the costs have been significantly lower to just over $2,000 per kilogram, thereby enabling the enormous growth in the space sector.

While technology limits what is possible in space policy, it is worth discussing what limits the availability of technology in the first place. Most commonly, it is physics. Faster than light travel, something that is quite common in science fiction tales of space exploration, isn't possible. While it would make space exploration more feasible, the physics of the universe we live in don't allow it. Other technologies that might make space travel somewhat easier like cryogenic hibernation systems, artificial gravity outside of centrifugal systems, or instantaneous communication, while potentially possible in the future, will likely not be available soon if ever. Even if physics and the physical laws of nature were not a significant impediment to these types of technologies, cost is. If cost were not a consideration, there is likely a wide range of technologies that could be pursued, not just in space and space exploration and exploitation, but in many fields. Unfortunately, as much of this book has detailed, space policy in the United States typically operates within significant resource constraints. Thus, the cost of advanced technologies and the willingness, or more often, unwillingness, to appropriate them to areas relating to space compounds the impact of technological limitations in general.

On the other end of the technological spectrum are technologies that are widely available. While this might not seem like an influence on space policy per se, there are some who argue that it is. Theories of technological determinism deal with the relationship between society, societal development, and technology, and scholars have posited a variety of specific ideas concerning them. The debate includes arguments that technology does not influence societal development whatsoever to the perspective that it is meaningless to talk about the state of a society without considering how technology has influenced it.[37] One version of technological determinism argues that if a technology is available, it is inevitable that it will be used. Consider, for instance, nuclear technologies. Theorists who adopt this version of technological determinism argue that nuclear weaponry was essentially inevitable following the discovery of basic nuclear physics. In the same way, James Clay Moltz notes that one of the significant influences on space policy generally and military space policy specifically has been the availability of technology and the idea that if available, it will be used.[38] This is most clearly seen in the development and usage of dual-use technologies of which nuclear technologies are a part. Like nuclear technologies, which can be used for peaceful purposes, space technologies, even if they also serve peaceful purposes, will inevitably be used for more nefarious reasons. The availability of technologies, then, can be a significant determinant of space policy.

The pursuit of human spaceflight by the Soviet Union and the United States during the Cold War can be viewed through the lens of technological determinism. The key technological development to focus on are creating rockets powerful enough to launch a nuclear payload halfway around the world and not necessarily to space. Using the same (or similar) missiles as rockets to go into outer space enabled the United States and the Soviet Union to demonstrate their military capabilities in a relatively peaceful manner. This competition could easily have stopped at that point but quickly progressed to the dimension of human spaceflight. As discussed in previous chapters, there has been a fierce debate since the 1950s about the scientific value of human spaceflight when more—and perhaps better—science might be done in the absence of people. The key question to consider, then, is why would a Cold War competition that was primarily driven by the need to demonstrate the power and reach of ICBMs almost automatically invoked the notion of humans going to space. While there was a long history of science fiction stories and dreams of people going to and living in space, it is not automatically clear that these desires alone would foster the investment of such large sums of money. Technological determinists would argue that given the availability of a certain technology (powerful missiles and rockets) and their ability to launch a payload capable of carrying a human, it is natural that the competition would continue into this dimension.

Simply because a technology and application exist does not mean that the other factors that we have discussed in this section, and throughout this book, do not affect space policy. Given the demonstrated ability of humans to go to the Moon, why not continue to Mars or other destinations? While the development of technology needed for long-term, deep-space exploration is ongoing, it is not far-fetched to think that with the appropriate resources that it could not be achieved. Rather, the dynamics of Cold War politics and resource constraints served to limit both the development of needed technology and its application in new ways. Despite this, theories of technological determinism and the sheer availability of technology contribute to the ways in which space policy is developed in the United States.

The topic of technology has a relationship to each of the topics discussed in this chapter. First, technological capabilities are now often discovered, developed, and deployed by private industry and others including universities, though it is often done at the government's behest. The result is that governments become deeply involved in what has been traditionally assumed to be a nonpolitical, purely scientific endeavor. By devoting scarce resources to particular lines of technological development, the pursuit of knowledge and technology becomes inherently political. This is a very real critique that has been leveled at various policy areas including climate change, health care, and space policy. Since World War II, governments have increasingly provided greater support for basic research and development.[39] It is difficult to know in what directions an entirely privatized space industry would have taken the exploitation of space, but it is equally difficult to see it happening at all without the involvement of the state.

Second, and relatedly, the growth of technology and the government's use and increasing dependence on it, has led to charges of government as a technocracy with scientists leading the way rather than the more democratic involvement of the people.[40] The backlash against scientists, doctors, and other public health experts during the COVID-19 pandemic is indicative of this. With respect to space policy, this has not been a significant problem, but it is easy to see how it may become one given the lack of attention that the public gives to space issues in general. Finally, just as economics appears to serve as a ceiling for what is possible in space policy, the availability of technology does as well. If the technology is available, according to technological determinism, it is inevitable that it will be used. With respect to the interaction between economic concerns and technology, the wider availability of commercial, off-the-shelf technology for space has been a significant enabler of the space economy. Remote imaging services allow businesses to monitor traffic coming in and out of stores, farmers rely on it to plan and monitor crops, and government officials use it to monitor and respond to emergency situations. As more uses for space-based activities become clearer and the technology cheaper and more available, the general economy is likely to find ways to take advantage of it. Ideally, this creates a virtuous cycle of sorts in space-based improvements to the economy, which then enable more investments in advanced science and technologies that might be exploited in the future. To be sure, the unavailability of certain technologies will continue to be an impediment, but as governments, including the US government, face greater spending challenges, private industry's ability to fund not only research and development but operational space systems will supplement the state of space technology.

Conclusion

For most of the space era, governments have driven how the world views and utilizes space. Geopolitics, great power competition, and the race for international prestige motivated states to invest in space technology. Over the past several decades, this situation has changed dramatically as private space companies have not only been successful but become independent actors in and of themselves. This is not unlike the greater global situation and the rise of globalization. Partially enabled by space-based technologies, the global economy has become highly integrated with any given state's economy increasingly dependent on the health of the global one. This condition of globalization with its complexity and interdependence has made the world feel smaller and forced the global community to consider the increasing power of non-state actors including businesses, nongovernmental organizations, and international social movements. Individuals and organizations can communicate and spread their message almost instantaneously around the world, bringing together people globally for both good and bad purposes. Multinational corporations can leverage their power to move operations around the world depending on which countries can offer the best advantages.

As this chapter illustrates, while space-based assets have enabled globalization to a large extent, the effects of globalization are now affecting the space domain in turn. A single private company, SpaceX, has more space power than many of the world's countries—combined. Even the world's most powerful country, the United States, is arguably dependent on the services that SpaceX provides. Companies like Planet that provide remote imaging services are enabling people outside government offices to analyze the actions and operations of foreign actors, identifying potential missile silos in China or monitoring illegal fishing off the coast of Africa.[41] And yet, even while space-based assets and services are proving to be ever more valuable both in the services they provide and the economic value they generate, the public remains staggeringly unaware and unappreciative of the many roles that the space domain plays.

As the power of private actors in space grows, there are several questions that will need to be addressed. Foremost among these is the extent to which private actors may be given a formal seat at the international table along with states. While private companies have played informal roles in a variety of space-related forums at the United Nations, the responsibility to regulate the behavior of private actors in space remains with state governments under the OST. Another question specifically is the extent to which space policy is essentially outsourced to private companies like SpaceX. While government via a technocracy can present a threat to democracy, a space policy that is essentially determined by the capabilities of private actors is equally questionable. With respect to the public, a major question is, What will it take, if anything, for the general public to take an interest in space and raise their consciousness about the major issues? As has been the case for much of public opinion across a variety of policy areas, all too often it takes a major failure of public policy for the government and the public to come to grips with what needs to be done. If that is what it will take for space, the only question then is, How bad must the failure be and to what extent would that failure prove catastrophic? We will return to many of these questions in the remaining chapters.

Summary Points

- Public opinion data shows that the public has little knowledge about space activities and little enthusiasm for increased spending. This only reinforces the idea of space policy as a secondary policy concern.
- Though major private companies have been involved in American space activities from the start, the increasing power of new space companies has given the commercial space industry even more power especially to influence space policy via tools like lobbying and campaign donations.
- The larger economic situation can impact space policy as it affects the number of resources the United States (or any country) can devote to it.

At the same time, investment in space can also contribute to economic growth and technological spin-offs.
- The availability and feasibility of technology relevant to space not only enables space activities but limits it as well.

Discussion Questions

1. Of the factors discussed in this chapter, which do you think is most influential in US space policy?
2. Would increasing public support for space exploration necessarily increase its political support? Why or why not?
3. In your assessment, how strong is economic development as a reason for undertaking space activities?
4. To what extent should private companies be allowed to influence US space policy? How does this fit with notions of democratic governance?
5. Technological determinists argue that a technology, if it exists, will eventually be used for all the purposes that can be found for it. Do you agree with this argument? Is it inevitable, for instance, for weapons to be deployed in space?

For Further Reading

Bainbridge, William Sims, *The Meaning and Value of Spaceflight: Public Perceptions* (New York: Springer, 2015).

Launius, Roger D., "Public Opinion Polls and Perceptions of US Human Spaceflight," *Space Policy*, vol. 19 (2003): 163–75.

Nadeau, François, "Examining Public Support for Space Exploration Funding in America: A Multivariate Analysis," *Acta Astronautica*, vol. 86 (2013): 158–66.

Pyle, Rod, *Space 2.0: How Private Spaceflight, a Resurgent NASA, and International Partners Are Creating a New Space Age* (Dallas, TX: BenBella Books, 2019).

10

Civilian Space Policy

US space policy consists of a wide variety of activities, not just the major ones that tend to get all the media coverage. However, ask a random person what NASA does, and they might only identify things like flying people to the ISS or perhaps even operating major projects like the Hubble Space Telescope. We shouldn't be surprised. Even in politics and government, the place where space policy is made, these types of activities tend to receive the most attention, not just because they tend to cost the most, but because they are the most visible and easily understandable. That does not mean, though, that the other elements of space policy aren't worthy or important. NASA, and other government agencies, do a lot of things that are not readily apparent.

Recall from chapter 1 that civilian space policy includes those space activities that the government undertakes for nonmilitary purposes. Though NASA is the primary face of civilian space policy in the United States, there are several other organizations that also carry out civilian space policy. The Department of Commerce is increasingly involved in space traffic management efforts. The NOAA runs several weather-related satellite systems. The DOS is involved in diplomatic efforts such as the Artemis Accords. Previous chapters have already delved into the history and policy details of many of these initiatives; as such, this chapter does two things. First, it highlights some of the other kinds of civilian space policies including planetary defense, Landsat, and NOAA's weather imaging programs. Second, it highlights some of the major policy constraints and choices involved in civilian space policy. For instance, as previously mentioned, cost is a major constraint on what the United States does in outer space, especially for civilian space programs that might not have an overriding national security rationale. This sets up major choices about the type of space exploration pursued, where, and who to include in carrying it out. Additionally, the legacy of excluding women and minorities from NASA and other space activities (most prominently in the astronaut corps but in other ways as well), has increased the attention paid to equity, equality, and diversity across US space policy. Finally, the fact that space is a global commons with an increasing presence of multiple actors makes issues like space situational awareness and space traffic management increasingly significant. This chapter explores these choices, these questions, for civilian space policy.

Other Civilian Space Programs

While NASA and its human spaceflight programs might be the most well-known of the US civilian space activities, there are several other major civilian space programs. Many of these seek to benefit Earth directly. This section will outline three of them: planetary defense, Earth sensing, and weather.

Planetary Defense

Hear the term "planetary defense" and you might start thinking about how to protect Earth from alien invaders as in the plotline of many television shows and movies. You may also think that planetary defense has something to do with military-like defenses or organizations. However, the planetary defense mission rests with NASA, and it is fundamentally one that revolves around scientific discovery and understanding of the things that may threaten life on Earth as we know it.

Earth is almost constantly being bombarded by asteroids and meteors, rocky planetary objects that orbit the sun. While most of these burn up harmlessly in Earth's atmosphere, sometimes they are large enough to survive and impact Earth. Of course, the significance of the impact depends on the size of the rock—the larger it is, the more likely it is to survive its descent through the atmosphere, though it will be much smaller by the time it does. Occasionally, larger asteroids and meteors cross paths with Earth and result in spectacular explosions. For example, in 2013, an asteroid approximately 20 meters in size burned through Earth's atmosphere to explode over the Russian city of Chelyabinsk. The blast generated between 26 to 33 times the energy as that created by the Hiroshima atomic bombing. While events like these are rare, they do occur and leave potentially disastrous consequences in their wake—just ask the dinosaurs.

The first step in protecting Earth from such space dangers is knowing what threats exist in the first place. As such, one of the key missions for NASA's planetary defense office is the NEO Observations Program. The goal for this project is to "find, track, and characterize at least 90 percent of the predicted number of NEOs that are 140 meters and larger in size."[1] Even this can be quite difficult as often these objects are hard and difficult to spot, especially the farther out they are in space. And once spotted, it takes repeated observations to gather enough data to know which way the asteroid is going. Smaller meteors like the one that exploded over Chelyabinsk can be quite difficult to spot and might not be seen until they are in close proximity to Earth. Once any of the smaller programs that NASA supports spots an NEO, it is followed up on by another survey to get a better understanding of the object and its track. This is an incredibly important step—to undertake a mission to redirect an asteroid that might be a threat, we must have as much lead time as possible to put a mission together, launch it toward the asteroid, and influence its track.

But knowing about an NEO is only half the battle. Various means have been suggested to deal with potential incoming threats, but since very little is known about the asteroids, it is hard to know what it will take to deflect one that might be on a collision course with Earth. Movies like *Armageddon* and *Deep Impact* tried to deal with their fictional threats through sending a human mission to drill into the rocks and place bombs that would break up the incoming threats. For many reasons, missions like those would be nearly impossible to undertake in the real world—in both *Armageddon* and *Deep Impact*, things do not go as planned. Given that, much scientific thought has gone into other means such as using gravity to draw an asteroid into a different trajectory or even painting one side of an asteroid so that it attracts more heat from the sun thereby influencing its path.

In 2021, NASA launched its first mission to test out some of these methods with the Double Asteroid Redirection Test (DART). DART is being sent to a double asteroid system called Didymos, which features two asteroids—the larger Didymos and the smaller Dimorphos. DART is designed to crash into the smaller one, Dimorphos, in an attempt to change its orbit around the larger one, Didymos. Not only will this provide data on what it will take to physically crash into an asteroid, but it will provide scientists an idea of how much it will change an asteroid's orbit. The data generated from the test is an important starting point in assessing just what it will take to move an Earth-bound threat.

Even with DART, there is still a lot that needs to be done to protect Earth from threats in the solar system. For much of the early 2000s, little funding was provided for NASA's planetary defense efforts, generally in the range of $4 million per year. Several developments led to dramatic increases in recent years including the Obama administration's overhaul of human spaceflight policies. With Moon missions canceled in 2010, NASA struggled to articulate what its new SLS rocket would do—sending humans to study an asteroid was one suggestion. Increased funding for NEO identification followed, and in

2014, NASA proposed a more ambitious asteroid redirect mission (ARM), which would have sent a probe to rendezvous with a near-Earth object (but not land on it) and gently "tug" it toward Earth for further study. To carry out either one of these, much more needed to be known about the population of asteroids that might be appropriate targets.[2] While neither of these missions has ultimately been carried out, the increases in funding have led to the detection of a growing number of NEOs.

More recently, some have argued that the mission of planetary defense should belong with the USSF. While the US military does contribute some to NASA's ongoing efforts including providing data from their vast array of space monitoring systems, little support has been generated for the proposal. First, much of the planetary defense mission at this point is focused on identifying, tracking, and monitoring rather than active defense. Second, more scientific research is necessary to understand what will be needed to move near-Earth objects should they threaten Earth. Finally, because the NEO tracking effort necessarily involves multiple scientific, civilian, and academic organizations both in the United States and internationally, giving primary responsibility over planetary defense to a military organization might hamper needed cooperation and collaboration.

Landsat

From the beginning of crewed space missions, the photographs that have often captivated the public the most were those that looked back on Earth. One need only consider the striking Earthrise photo captured on Apollo 8 as it orbited the Moon to understand this fascination. As early as the mid-1960s, some scientists proposed a series of remote imaging satellites that could monitor Earth, specifically natural resources. As useful as such monitoring may seem, other government agencies were skeptical of the proposal. Not only would the cost be rather high, but there was a concern about photographing other countries, and the DOD feared that such remote imaging might compromise military secrets.[3] Despite this, the United States Geological Survey (USGS) convinced the Department of the Interior to begin an Earth remote sensing program. The USGS's progress finally convinced NASA to move ahead on construction of what would become the first Landsat satellite.[4]

In 1972, NASA launched the Earth Resources Technology Satellite-A (ERTS-A) with a second, ERTS-B, following in 1975. The mid-1970s led to a name change from ERTS to Landsat and the launching of three additional satellites. The satellites were placed into near polar orbits that allowed each to photograph Earth once every 16 days. Not only did the returning images allow scientists to monitor the availability of natural resources, but, perhaps more importantly, monitor change over time. With Landsat data going back to the 1970s, scientists now have a wealth of imagery to assess how things such as climate change and natural disasters have altered Earth's landscape.

Given these wide uses of Landsat data and its operational success, the Carter administration considered moving Landsat from NASA to NOAA, which it believed would allow the program to grow and perhaps eventually be privatized as more private companies became capable of launching and operating satellites. With Landsat's transfer to NOAA in 1979, the Reagan administration moved to fully privatize it in the early 1980s.[5] This move was fully in line with other administration efforts to expand the commercial space sector, which included the development of the first commercial space regulations and legal framework signed into law. Proponents of privatization had several arguments. First, some argued that the Landsat imagery was not a public good. In other words, Earth resources monitoring was a private good that benefited the people who were willing to pay for it. As such, the government should not provide it. Second, some feared that the Landsat program would be canceled due to its high cost—moving it to the private sector would not only preserve the program, but, ideally, allow it access to new developments in hardware and software. Third, if Landsat were to be moved to the private sector, the hope was that costs would drop thereby enabling greater use of the system.

Congress was initially supportive of the privatization of Landsat and the Earth Observation Satellite Company (EOSAT) was chosen to operate the program. To aid the transition, the government promised a $250 million subsidy to assist with new satellites and their launch costs.[6] However, several problems soon became apparent. The biggest one was the lack of a commercial market. While privatization advocates believed that it would reduce costs and bring in new users, the market was small and getting smaller as costs for the satellite imagery were continuously raised.[7] The government also went back on its funding agreement, putting EOSAT in an increasingly difficult budget situation. These problems and congressional concerns over privatizing Landsat and other weather satellites ultimately led to the passage of the Land Remote Sensing Policy Act of 1992, which transferred responsibility for Landsat to NASA and the USGS.

Landsat continues to operate today with three satellites in orbit, the most recent launched in 2021. However, it has not been without some continuing concerns. As Congress considered whether to support the development of the latest Landsat satellite, Landsat 9, in the mid-2010s, there was a reconsideration of whether privatization or some form of commercialization may be appropriate. This proposal reflects just how far the commercial space industry has come since the 1980s and 1990s, particularly as remote sensing companies were starting to expand their own satellite systems. At the same time, supporters of Landsat were concerned with the continuity of data should the system be privatized once again. Congress was also concerned with increasing costs.[8] Regardless, the development of Landsat 9 was initiated in 2015.

The partnership between NASA and the USGS also continues as NASA helps develop and launch the satellites while the USGS oversees operation and provides access to the images. In 2008, the USGS made all Landsat data publicly available at no charge, which has led to a significant increase in data usage. In 2017, researchers found that Landsat data resulted in $3.45 billion in economic benefits.[9] Landsat data has been used in a wide variety of contexts including "agriculture, forestry and range resources, land use and mapping, geology, hydrology, coastal resources, environmental monitoring, disaster response, and national security."[10] While commercial systems like Planet's are now contributing to the market for Earth imaging, NASA and the Department of the Interior are now looking ahead to the future of their Earth sensing programs. Both the USGS and NASA are currently investigating both user needs and potential new capabilities that a new Landsat satellite might be able to provide. At the same time, the availability of commercial alternatives or public-private partnerships may become more attractive given the decreasing costs of some of the commercially available systems.[11]

Weather

Even before the value of Earth sensing became obvious, utilizing space assets to monitor Earth's ever-changing weather patterns was a major priority. The importance that the United States placed on weather satellites at the dawn of the space age is reflected in the fact that in the same speech President John F. Kennedy proposed that the US go to the Moon, he also asked for additional funding for a "worldwide weather observation" system. NASA launched the world's first weather satellite, the Television Infrared Observation Satellite (TIROS-1), in 1960. Though it only operated for 78 days, it was the start of a series of weather satellites launched throughout the 1960s including additional TIROS satellites and seven Nimbus satellites. While the early TIROS satellites were mainly equipped with television cameras that beamed back images of weather patterns, later satellites were further equipped to measure things like atmospheric temperature, moisture, and water vapor, all of which are essential in not merely watching the weather but predicting it.

As the nation's weather satellite program rapidly expanded, the Environmental Science Services Administration (ESSA) was created in 1965 to oversee the operation of weather satellites. ESSA became NOAA in 1970, which continued to innovate and launch more powerful weather satellites including the Polar Operational Environmental Satellites (POES) and the Geostationary Operational Environmental Satellites (GOES) series. GOES satellites, as the name implies, were placed in geostationary orbits meaning that they could provide constant coverage of a section of Earth. The growing constellation meant that by the mid-1980s, American weather satellites could provide updated weather data to ground stations every five minutes.[12]

Much like Landsat, the gathering of weather data via satellites in these early decades remained the sole province of the government. Where an argument could be made that the Landsat data was a private good, there was widespread agreement that weather data was a public good. While some commercial weather prediction firms started to emerge, they still relied on the data generated by the government system. Thus, while the Reagan administration attempted to privatize Landsat, the same pressure to privatize US weather satellites did not exist.[13] There were also other reasons that mitigated against weather privatization including the need to preserve free and open data exchange and the requirement of government subsidies for the scheme to work. Further, legislation passed in 1984 implemented a prohibition on charging fees for use of the weather data.[14]

While weather satellites did not have to face the privatization challenge of the 1980s, they have been affected in recent decades by the increasing pace of space commercialization. Though advance imaging and sensing satellites remain quite expensive, a new means of acquiring weather data was developed in the early 2000s utilizing the already ubiquitous GPS satellites. As demonstrated by NASA's COSMIC mission, small satellites could be placed in orbit that measure the time it took for GPS signals to reach the COSMIC satellite as the signals moved through Earth's atmosphere. The ways in which the signals are slowed and refracted depend on the conditions in the atmosphere allowing meteorologists to determine weather conditions.[15] Later COSMIC demonstration satellites expanded their capabilities to pick up on other location services like Europe's Galileo, Russia's GLONASS, and China's Beidou, leading to this method being called global navigation satellite system radio occultation (GNSS-RO).

Since GNSS-RO requires rather small satellites equipped with radio transmitters and receivers, there is a growing commercial opportunity to supplement weather data in a commercial market. This option has become increasingly attractive given the expensive nature of NOAA's weather satellites and how effective the commercial space industry has become. However, the previous legislation prohibiting the privatization of weather satellites remained a hurdle until new legislation passed in 2017, the Weather Research and Innovation Act. In addition to allowing weather data to be sold, it also mandated that NOAA purchase commercial weather data to supplement its own databases and models. Concerns remain, however, over how government and commercial data will be collected, maintained, and disseminated to protect what most have come to consider a public good—weather prediction.[16]

These brief histories of planetary protection, Landsat, and the weather satellite programs demonstrate several important points about US space policy. First, it is far more than just what NASA does or the human spaceflight program. Civilian space programs are varied and provide significant benefits that most of the country, indeed the world, take for granted. Second, just because the programs are assumed to exist and contribute does not mean that they are cost or controversy free. They are still subject to many of the same political winds that the larger and more well-known programs must sail through. Cost is a major concern along with who should properly be developing and operating them. Third, the growth of the commercial space industry is also impacting these programs by providing less costly and perhaps more efficient options for data collection. These developments have led to serious policy reconsideration regarding whether the government should continue to provide such services or become a consumer of commercial services. While Landsat's experiment in privatization failed in the 1980s, conditions are much better today that it becomes a more feasible option. However, just because it is feasible, does not mean it is likely. What appears to be more likely, as in the case of NASA and human spaceflight, is supplementing governmental activities with commercial ones.

Inclusion, Cost, and Cooperation

Previous chapters have highlighted several trends in US space policy that continue to be important questions and considerations in moving forward. First, the lack of wide public support not only means that civilian space policy becomes, at best, a secondary policy issue, but that there are significant differences in support across different demographic groups. In addition to the gender gap noted in chapter 9, Black Americans also express less support for increased funding for space exploration. Black Americans, like women, were often excluded purposefully from early space activities—an exclusion intensified by the location of many NASA centers in southern states in the 1960s, which still practiced segregation. The resulting legacy is that women and Black Americans have had historically few role models through which to encourage greater participation in science, technology, engineering, and math (STEM) professions or the space industry specifically. This history is especially significant as the importance of space increases across society, highlighting the need to ensure that all populations not only see the benefits of heightened activity in the space domain but can participate in them.

A second trend that should be clear by now is that space is expensive. Though the civilian programs discussed previously are not nearly as costly as larger ticket items like human spaceflight, even they have been subjected to concerns about cost and budget cuts. While national security arguments tend to support higher budgets for military- and national security–related space activities, civilian space policy is often short on this resource. The lack of wide public support highlighted at the beginning of this chapter and in chapter 9 often means that civilian space policy becomes, at best, a secondary policy issue. Following from this, such activities and agencies then become major targets for budget cuts, particularly in difficult economic times. This becomes quite apparent when we look at NASA's budget history in chapter 1—following a high in the 1960s during the space race, NASA's budget has almost consistently fallen. While this decrease might not be so bad if the overall federal budget also fell, but that is not the case. NASA's 2020 budget of $22.63 billion represents just 0.48 percent of the federal budget.

While $22 billion seems like a lot of money, given the high costs associated with operating in space, it is anything but. Given this, NASA, and the United States, must make certain choices about how to spend those resources and how to make the budget do more. This section will detail these four trends: inclusion and diversity, science or exploration, humans or robots, and international partnerships.

Diversity and Inclusion

Looking at the STEM field as a whole, of which space is certainly a part, there is a long-recognized gender gap—women are less likely to pursue careers in STEM.[17] Scholars have identified several different reasons for this gap including harmful stereotypes that women and girls are bad at math and science, lack of visible role models, discrimination against women in STEM professions, and difficulty in achieving appropriate work-life balance.[18] In terms of space policy and politics, this phenomenon has been most closely studied in the diversity (or lack thereof) in the most visible space participants—astronauts. As detailed in chapter 2, women were excluded early in the US space program despite the existence of women who wanted to be included and demonstrated the physical ability to participate.[19] It wasn't until the 1970s when NASA, under pressure, finally chose the first woman, Sally Ride, and first Black American, Guion Bluford, for the astronaut corps. People with disabilities are still excluded from the astronaut corps.

Astronauts are not the only area affected by low levels of diversity—space leadership has also been largely male and white. This began to change in 2009 when President Barack Obama appointed the first Black American to lead NASA, former astronaut and retired Air Force general Charles Bolden. Alongside Bolden, Obama also appointed Lori Garver as deputy administer of NASA. Perhaps unsurprisingly, Garver writes in her 2022 memoir that even then, she still had trouble in her job, difficulties she believed were partially due to her gender and the organizational culture of NASA as a "boy's club."[20] This male-dominated culture is also found in many of the major commercial space companies including Blue Origin and SpaceX. Not only are those two companies led by white men (though notably, SpaceX's president is a woman, Gwynne Shotwell), but they have also been subjected to several legal complaints of sexual harassment.[21] Given the increasing importance, if not dominance, of commercial space companies, this may become problematic if similar cultures are exported to future space settlements.

The lack of diversity in space has not gone unnoticed. During the 1960s, NASA and its Apollo program were criticized for ignoring the plight of minorities and poor people in the country, with some leaders wondering whether the money being spent on going to the Moon might be better spent reducing poverty or improving health care. In response, NASA devoted some resources to find ways to use space technology to reduce pollution and energy consumption in low-income and Black communities and improve living conditions.[22] However, as Neil Maher observes in his study of NASA and the "age of Aquarius," most of these efforts were not done with the primary intention of serving underserved communities but to increase NASA resources in a time when budgets were being cut.[23] More recently, NASA has made concerted efforts to rectify the lack of diversity in its workforce and has increased its outreach to women and minorities. When the Artemis program was announced, NASA leaders made a point to emphasize that the project would not only return Americans to the Moon but would take the first woman to step on the Moon. With the election of Joe Biden in 2020, NASA now emphasizes that it will not only be the first woman going on Artemis but also the first person of color. Groups like AstroAccess are actively studying how people with disabilities might be able to participate in spaceflight, testing novel techniques on zero-gravity flights.[24]

Many hurdles remain for increasing diversity in the space field, however. Structural barriers remain in recruiting more women and minorities to STEM professions including the expectation in many commercial companies that their employees work more than 40 hours a week and be available at all times—something that disrupts work-life balance for many people who have family responsibilities. As we have seen throughout this book, organizational cultures are notoriously difficult to change. As Garver's experience at NASA just 10 years ago attests, NASA is still struggling to instill a welcoming environment for women and minorities. Finally, compared to how much attention has been given to encouraging women and minorities into the space and STEM fields, little thought has been given to how space systems might be used to advance solutions to problems women and minorities face not just in the US but around the world.[25] For example, one issue affecting women, minorities, and other disadvantaged groups around the world is structural or state-based violence. With significant improvements in commercial remote sensing, these resources might be used to better identify, track, and publicize mass violence against these groups. However, as the Satellite Sentinel Project discovered after doing just this, being able to identify violence does not help if people are not paying attention and willing to do anything to disrupt it.[26]

These trends mean that some of the major considerations for civilian space policy moving forward include examining the ways policy might be used to increase women and minorities in the space industry and professions as well as how space systems and resources might be used specifically to advance the specific circumstances women and disadvantaged groups around the world face every day.

Science or Exploration

In many ways, the choice between spending precious dollars carrying out major scientific research or space exploration is one of the most fundamental decisions that is made about civilian space policy. While the line is somewhat fuzzy, space exploration typically involves human spaceflight and missions whose primary function is not necessarily generating scientific data whereas the science category includes probes, missions, and satellites whose primary mission is to generate scientific findings. Historically, exploration, including human spaceflight, has received about half of NASA's budget while scientific programs receive about 30 percent.[27] While this doesn't mean that no science is being performed on human spaceflight missions, it is indicative of what the US government is prioritizing via its budgeting choices.

There are several reasons why funding for human spaceflight tends to be higher. For one, sending humans into outer space is expensive not just because of the life support systems that are necessary to keep people alive but because launch and vehicle systems need to be more reliable with more redundant systems in case something goes wrong. These redundancies cost more in terms of time, weight, and testing, leading to higher budgets. Second, human spaceflight missions tend to get the most public attention and thus, enthusiasm. Part of this is also a Cold War space race legacy where human spaceflight missions got the most attention and thus prestige. Finally, NASA as an organization values human spaceflight for both its sense of mission and the number of resources it guarantees to the agency.

Despite the attention that human spaceflight missions receive, proponents for more science funding argue that scientific findings of the type provided by missions such as the Hubble Space Telescope, Mars rovers, or even Earth sensing missions contribute more to our understanding of space and the universe. While they might not be as "sexy" as human missions, the benefits are far higher in terms of pure knowledge, particularly since we don't know what we don't know. In addition, dollars spent on scientific missions can go further and do more since redundant and life support systems are not necessary with probes and satellites meaning that the science budget can be stretched even further than the human spaceflight budget.

Humans and Robots

Questions over the proper balance between science and exploration necessarily flow into whether humans are necessary for scientific missions let alone space exploration. This argument has a long history dating back to the 1950s. While pressured into supporting a human spaceflight program, even President Dwight Eisenhower wasn't all that impressed with the *need* to send humans into outer space or even a more expansive crash program to put the United States ahead of the Soviets in space achievements.[28] James Van Allen, famous for discovering the Van Allen radiation belts, long argued that sending humans to space was both too costly and unnecessary for major scientific advancements. Reflecting on both the space shuttle and ISS programs, he argues that neither greatly contributed to scientific advancements despite their promises to do so.[29] Even many of the early rocket scientists who worked to construct crewed systems were skeptical of allowing too much human interaction, preferring instead fully automated systems that would reduce the chances of humans making mistakes.

Van Allen also points out that far from being less exciting than human spaceflight, uncrewed missions have often generated similarly high levels of enthusiasm from the public. For several decades now, the public has often enthused over the spectacular images sent back from the Hubble. When one last servicing mission to the Hubble was canceled following the *Columbia* accident, public and political outrage led to it being put back on the schedule. A series of Mars missions continues to receive high public attention as well. When the Mars rover *Opportunity* finally died in 2019, there was a great deal of public sadness expressed on social media with a later documentary showcasing its achievements.³⁰ At the same time, the newest rover on the red planet, *Perseverance*, has continued to generate public interest with its first-of-its-kind helicopter and the spectacular images that it has sent back.

Van Allen's argument is representative of many science proponents who believe that humans only make science more costly and more difficult to undertake in space. And yet, he acknowledges that one of the remaining rationales for sending humans to space (the only one in his opinion) "is the ideology of adventure."³¹ In addition to this, advocates for human spaceflight have also argued that humans can respond better and more quickly to emergency situations that might arise. For examples of such situations, proponents point to Apollo 11's landing on the Moon when Neil Armstrong needed to move away from the planned landing site because of a rock field and Apollo 13's survival following an explosion involving its oxygen tank. In both scenarios, individual people were able to take over from computers and act in ways to save the mission. In terms of science in particular, advocates also point out that humans on the Moon, Mars, or other planetary body would be better placed to pick and choose appropriate rock samples to bring back with them than robots who may not be able to move as deftly or to places that have intriguing characteristics. Robotic explorers would also be limited in the types of scientific experiments they could carry out whereas humans can move about more fully to deploy more types of experiments. Even the shuttle missions to service Hubble contribute to the idea that human spaceflight is beneficial for scientific exploration.

While both sides of the human and robot debate have excellent arguments, part of the reason that this debate is so important is the limited budget that NASA must have to support the two endeavors. The larger political importance of human spaceflight to national prestige and reputation lead policymakers in Washington, DC, to be more supportive of such endeavors. In turn, NASA also tends to emphasize human spaceflight because it knows it will receive more resources for carrying it out. Despite the merits of the argument over science and exploration and humans and robots, these very real political dynamics will likely continue to make human spaceflight a priority in US civilian space policy.

International Cooperation

One means of reducing the cost of space exploration and science is to spread the cost across multiple countries. Thus, one of the major incentives for the United States to pursue international cooperation is to reduce cost. While international cooperation might not have been possible in the 1950s and 1960s as the United States and the Soviet Union were the only ones with major space capabilities, the possibility started to arise as early as the 1970s with the space shuttle. At the same time NASA looked to spread costs across the US government, they also brought on board several international partners including Canada and the new European Space Research Organization, the forerunner to today's European Space Agency. Canada's contribution was in the form of a robotic arm to be placed in the shuttle's bay, the Canadarm, and ERSO built a laboratory module to also fit in the payload bay called Spacelab. A similar model of cooperation was utilized in the 1980s as the space station started to be developed with Europe, Canada, and Japan coming on board to contribute key elements.

While international cooperation and collaboration is most apparent with human spaceflight, it has also been extended to scientific missions as well. Major scientific missions like Hubble and the new James Webb Space Telescope have been partnerships among the United States and others like the ESA and Canada. The Cassini probe, launched to Saturn in 1997, carried a probe from the ESA called Huygens. Other countries have provided instruments for planetary probes that have gone to Mars while space communication is enabled via an international network of receiving stations.

Aside from cost, there are several other benefits to international cooperation. First, it creates and sustains relationships across countries, which can lead to better relationships overall. For example, the involvement of Russia in the ISS in the 1990s supported overall US foreign policy goals and has continued despite several periods of turbulent relations between the United States and Russia. Cooperating in an area such as space science, for example, can allow countries that might not otherwise have good relations to work together in something that is not as sensitive as national security. Second, it can help build up space capabilities in other countries, particularly those that might not be as developed in the United States. Canada's robotic arm technology, for example, has been intrinsic to the success of both the space shuttle and the ISS. Assisting countries in this way leads to economic benefits at home and allows the United States to take advantage of capabilities that we might not have.

On the other hand, working with international partners also has its problems. As in any collaborative endeavor, compromises must be made resulting in a project that might not fit all participants' goals equally well. Working together also requires trust that partner countries will follow through on their obligations. This is a factor that has often cut both ways for NASA. When the space station project was almost ended in the early 1990s due to cost overruns and concerns, international partners would have also suffered from the unilateral US decision. This has often happened during major international projects with NASA—budget cuts have often forced changes on projects that have in turn affected how international partners are contributing. As such, the United States is often seen as a less reliable partner making other countries more wary of working together.

Finally, choosing to work with partner countries could also impact calculations of international prestige. If the credit for a particular mission must be shared across multiple countries, the United States may not be seen as deserving of full adulation for the achievement. In other words, if the major goal the United States is trying to achieve via its space program is international prestige, international cooperation may not assist in such an endeavor. Consider if the United States had cooperated with other countries in carrying out the Apollo program—they would likely not have been able to reap the same reputational benefits. However, there is also prestige value in being seen as a country that is willing to cooperate with its partners. With the ongoing Artemis program to return to the Moon, for example, the United States has purposefully involved more international partners. This involvement, though, is very specific. While NASA remains fully in charge of the overall program, it has worked to negotiate the Artemis Accords with dozens of other countries to govern activities on the Moon. In this way, it can claim both international cooperation as well as full credit for whatever achievements Artemis makes.

The high cost of spaceflight, whether for scientific or exploration purposes, imposes major policy choices on US civilian space policy. The commitment of budgetary resources to particular activities is a statement about what an organization, or a country, values. While the US has made choices to this point to prioritize human spaceflight and exploration and international cooperation, that does not mean other options don't exist that can influence policy direction in the future.

Space Situational Awareness and Space Traffic Management

The issue of international cooperation and collaboration in space becomes even more important as we consider the nature of the space domain as a global commons. Global commons are natural resources that no one and no state can be excluded from benefiting. Examples include things like the sea and air in addition to space. The fundamental problem with global commons is that, since no one can be excluded from taking advantage of them, people are likely to take advantage of the resource, often harming it or depleting it for everyone else. The classic example of such a problem is the environment. It is easy for one state to pollute the environment, but the consequences of that pollution cannot be kept over just that one state. In turn, as more and more states increase their pollution, the entire environment is harmed and those states that may be behaving properly are in turn harmed.

Space, even though it is hard to reach for most states and more or less difficult to take advantage of directly, is also a global commons. The OST recognizes that no state can claim sovereignty in space and that every state has a right to access and use it. But like the global environment, it is beginning to suffer from problems of overuse. Stemming from the notion of space as a global commons, there are two issues in particular that are becoming more significant challenges for civilian space policy: space situational awareness and space traffic management and space debris.

As we have detailed throughout this book, there is a lot more going on in the space domain than has been the case in years past. Even though space seems like a limitless domain with plenty of room for all the things we want to use it for, this is not the case with the space near Earth. Even small satellites require significant amounts of distance between them so that the signals they emit do not interfere with other satellites nor other satellites with it. Additionally, space assets are traveling at such swift speeds that what seems like a monumental distance can be closed quite quickly when two objects are moving toward one another. The more companies, states, and individuals utilize the space domain, then, the less there is of it for others to use, the very definition of a global commons.

Given the increasing congestion, the world must find ways of managing it, particularly as the number of satellites and crewed missions increase. There are two aspects to doing this: knowing what is in space and where it is going, what is called space situational awareness (SSA) and an active means of managing that traffic, space traffic management (STM). As the pace of activity picks up, there is a higher chance that different platforms might come near to one another or even collide—in these instances, rules of the road are needed to determine how to behave, how close satellites can come to one another, and who needs to act and when.

Unfortunately, because this has not been a major problem until recently, no internationally agreed on rules exist. What is in place has been an ad hoc system of tracking assets in orbit for SSA purposes, mostly performed by the United States military. The US Air Force, and now the USSF, has had the responsibility of keeping a database of satellites and debris, tracking their orbits, and making most of it publicly available. The exception to this has been in the case of some military and national security satellites that the military does not want to provide full information on. When the database flags potential conjunctions between objects in orbit, the satellite operators are notified. They can then either take action to move their assets or not.

There are some obvious problems with the system as it currently stands. First, as a military product, some of the SSA data is necessarily downgraded. While this is generally understood, it also introduces elements of uncertainty in terms of space operations. Second, and connected to the military origins of SSA, some space actors may not fully trust the data that is being provided. Third, even though space operators may be alerted to potential conjunctions, there are, for the most part, no STM rules stipulating who must move what asset. What rules exist are often limited and ad hoc; SpaceX, for instance, has an agreement with NASA that SpaceX must move their Starlink satellites whenever they might potentially collide with any NASA mission. That is not the case for SpaceX and China, however. In 2021, China sent a protest to the United Nations noting that several Starlink satellites came close enough to their space station to threaten it, a claim that SpaceX subsequently denied. Finally, even though the USSF has substantial means of detecting and tracking space debris, not all of it is adequately tracked. Pieces smaller than 10 centimeters are very difficult to locate meaning that there is always the potential for debris impacts, even with the system that is now in place.

Recognizing some of the shortfalls with SSA and STM, in 2018, the Trump administration issued SPD-3, which began to shift the responsibility for both away from the military and toward a civilian organization, the Department of Commerce.[32] While a civilian approach to SSA and STM would still take advantage of the SSA data that the military provides, the goal of SPD-3 is to reduce some of the friction and distrust that comes with the military providing a civilian service, allow the USSF to focus on its core mission of defending US assets in space, and open more doors to collaboration with commercial efforts at SSA. In terms of STM, SPD-3 also calls for the United States, through the Department of Commerce, to develop best practices in terms of satellite operations. While the document does not specifically call for work toward rules of the road that might be accepted internationally, it does suggest that the policies and procedures adopted for STM might help establish them in the future.

While SPD-3 was highly praised for starting to shift SSA and STM responsibilities away from the military and toward the civilian sector, there have been some problems over the following years. First, the Department of Commerce has little in the way of expertise and experience in either of these areas. As a result of this shortcoming, Congress requested a study of the question, which concluded that Commerce was indeed the correct agency to carry out the task. The presence of a small organization was lauded in the report since it would force them to be creative and agile and cooperate with a variety of government and commercial organizations.[33] Even with that endorsement, however, Commerce's efforts to take on SSA and STM remained stalled for several reasons. First, the office that has been put in charge of SSA and STM, the Office of Space Commerce, falls under the authority of another Department of Commerce agency, NOAA. Considering this from a bureaucratic lens, this has very real policy consequences. Not only does it put several layers of bureaucracy between the Office of Space Commerce and department leadership, but it also suggests that the department does not see the effort as important. Second, the Office of Space Commerce, instead of proceeding directly to taking over the military database and system, has been working on means of not just using the military system but creating a system that commercial organizations can contribute to and directly participate in. Finally, there has been a significant lack of funding for the effort both on the part of Congress and the Department of Commerce.[34]

The Department of Commerce's assumption of responsibility for SSA and STM highlights several important issues about civilian space policy in the United States. First, it reflects widespread recognition that space is becoming far more congested, and policies are needed that can adequately respond to it. Second, even with that acknowledgment, there is still a lack of funding for the efforts. Similar to NASA's efforts, the lack of political and public visibility that a problem exists in which resources must be dedicated to solving also harms SSA and STM. Finally, even if the US can organize and create its own SSA and STM systems, it does not guarantee international acceptance. To the extent that it becomes the only viable space management system in the world, countries and space actors may be forced into using it, but it does not mean that there will also be agreement on the STM rules that the United States may decide to abide by.

Conclusion

There are several common ideas found among these different types of civil space policies and the challenges they face in the years ahead. While cost will almost always be a concern, civilian space policy in the United States is being confronted with the challenge of how to encourage, accommodate, and take advantage of the advances in the commercial space industry as well as how these advances might be used to encourage and amplify the concerns of diverse communities. We've already seen a number of these adaptations: NASA utilizing SpaceX to launch astronauts to and from the space station, utilizing lower-cost launch services to launch civilian satellites, and finding ways to make SSA systems more open to commercial systems and operators. Work must still be done in terms of how space systems and policies may be used to advance the goals of diversity and inclusion, but there is no doubt that this will continue to play a role in civil space policy for some time to come.

The more significant question as it regards civil and commercial space's relationship, however, will be the balance between them. In other words, how much should the US government take advantage of being able to contract out for services versus being the primary operator and owner of the systems? There have been some clear benefits to utilizing commercial space systems in terms of cost reductions and technological advances. In many ways, this has helped alleviate some of the larger cost constraints that are in play when it comes to space. However, a company's relationship with a government, even the US government, cannot and does not protect it from the larger dangers of market-based economies or other things that may harm its operational systems. Any number of situations may lead to a company going out of business or being sold, all of which could affect the state of a relationship between that company and a government.

Another major question is whether NASA is still necessary in a world where companies can provide the services that NASA once organized. Or, even if NASA is necessary, does it need to be as big as it historically has been? To this end, some have speculated that it may be time to transition NASA back to an agency like it had been when it was NACA. Recall from chapter 6 that NACA was established in the early 20th century as a means of researching advanced aviation technology; rather than develop and operate new technology and programs, NACA researched them with the goal of providing that information to the larger aviation community, which could then take commercial advantage of it. With the commercial space industry growing at such a swift pace and the US government in a position to take advantage of it, NASA might return to being an agency in charge of contracting for services and pursuing advanced research that could support further commercial developments. While this major organizational change is unlikely to happen barring major policy failure, it does beg the question, What should NASA's role be as the commercial industry continues to develop?

This chapter has shown that civilian space policy must still contend with all the typical policy challenges that face policymakers today: proper organization, cost, cooperation, extent to which services should be contracted out, how to get the appropriate people involved and even the proper focus of the policy. However, these challenges are often made worse by the fact that civilian space policy is not a major priority for most elected officials. And when it is, it is often for local, parochial reasons having to do with jobs or economic development in their home states or districts. This often inhibits larger consideration of the proper balance between crewed and uncrewed space exploration as well as the larger role for commercial companies. The extent to which civilian space policy is spread out across several different government organizations further complicates consideration of it as a whole when it necessarily implicates a number of congressional committees and a number of different activities.

Summary Points

- Though human spaceflight often dominates the headlines, there are other elements of civilian space policy in the US including planetary defense, Earth sensing, and weather.
- The legacy of excluding women and minorities from the space program and STEM professions continues to influence civilian space policy not just in who gets to go to space but in how space systems may be used to advance the goals of equity and equality.
- The high cost of space exploration and technology often forces policymakers and space organizations to make choices about the types of activities to undertake and with whom. These choices have included what elements of space exploration to focus on (science or exploration), whether to carry them out through robotic technology or humans, and the extent to which international partners are involved.
- Space is a global commons, a natural resource that no one can be excluded from. This has made cooperation on a large scale in space difficult particularly for issues such as space situational awareness and space traffic management.
- One of the most significant questions in civilian space policy moving forward is the extent to which commercial space companies should be utilized to carry out US space policy.

Discussion Questions

1. Even though we know that NEOs hit and threaten Earth, why do you think it has been difficult for this policy area to gain attention and funding?
2. What does the failed privatization of Landsat say about the state of space policy both then (the 1980s) and today?
3. Space policies having to do with weather are arguably very important to our lives and the global economy. Why don't we pay attention to activities like these that are perhaps more impactful than human spaceflight?

4. What sorts of policies might be possible to encourage greater involvement of women and minorities in civilian space programs?
5. What is the value of cooperation in civil space policy? Does it outweigh the negatives?

For Further Reading

Johnson-Freese, Joan, and Brian Weeden, "Application of Ostrom's Principles for Sustainable Governance of Common-Pool Resources to Near Earth Orbit," *Global Policy*, vol. 3, no. 1 (2012): 72–81.

Maher, Neil M., *Apollo in the Age of Aquarius* (Cambridge, MA: Harvard University Press, 2017).

Straub, Crista L., Stephen R. Koontz, and John B. Loomis, "Economic Valuation of Landsat Imagery," *US Geological Survey*, 2019. https://pubs.er.usgs.gov/publication/ofr20191112.

Van Allen, James A., "Is Human Spaceflight Obsolete?" *Issues in Science and Technology*, vol. 20, no. 4 (2004). https://issues.org/p_van_allen/.

11

International Politics and National Security Space Policy

The way the United States perceives and interacts with the world significantly influences its conduct in the space domain. Chapter 2 discussed the competition between the United States and USSR during the Cold War period, seeing how the back-and-forth efforts to gain prestige and security over the other led to a range of policy choices by both countries to press their relative advantages in each area. As the Cold War ended, the United States was in a clear position to assert dominance, not only globally but also in the heavens. This chapter looks at the nature of international politics, particularly theoretical explanations for how interactions between countries work, and how the nature of international politics affects contemporary US space policy. Conceptually, this can be accomplished by dividing the contemporary period into two parts. First, there is the immediate post–Cold War period when the United States faced no real threat to its security and maintained a free hand in the space domain. The second period encompasses the disappearance of that free hand, the rise of several threats, and the response by the United States to these new challenges. What becomes clear is that understanding the perception of policymakers' beliefs about how the world works has direct impacts on our understanding of space policy across the civilian, commercial, and national security dimensions. Each of these were examined in greater detail in their own chapters, however, for now it is sufficient to paint with broad strokes the picture of cause and effect, or at the very least, correlation between perspectives on international politics and space policy preferences and outcomes.

Prior to examining the two periods, it is necessary to discuss broadly the dominant theoretical explanations in international politics. This chapter will focus on three perspectives: realism, liberalism, and constructivism. There are other sets of explanations, but given the limited space available, these are most prevalent and most likely in the minds of policymakers on a day-to-day basis, consciously or not. The term "theoretical explanations" is used because each of these perspectives represents a way to understand the behavior of actors in the international system through the influence of one or more factors. Realism, for example, focuses on the utility of power and relative gains for understanding international political behavior. Liberalism, conversely, emphasizes the need for cooperation and the focus on absolute gains. Constructivism requires understanding the importance of social interactions, identity, and norms to better identify and comprehend behavior. To be clear, all these perspectives tend to abstractly relate to ideal types; explanations are always made up of different factors across the perspectives. It is this real-world complexity that makes the study and application of international politics difficult. In space policy, such difficulty is apparent in the range of international problems brought on by the increasing importance of space and the number of actors involved.

These theoretical perspectives do, however, allow one to match various schools of thought about how states might view the domain vis-à-vis other actors. Early in US history, the idea of space as a sanctuary stood out prominently in American rhetoric, if not reality. As the space domain has increased in its complexity and utility, scholars and policymakers have offered competing perspectives, mostly derived directly from the larger theories of international politics examined in this chapter. It is the connection between these perspectives and schools of thought that also helps shed light on the possible future of the space domain and the US role in this future.

Returning to theory, if one reflects on the Cold War chapter, the influence of realism on space policy was quite apparent. Both the United States and USSR were seeking to gain relative advantage over the other in the space domain and the broader context of the Cold War. If the USSR was successful in one area of space, launching Sputnik for example, then US policymakers viewed themselves as having lost something relative to its adversary. While prestige was important, the Cold War quickly focused on relative advantage in security—which state had the most ICBMs, which state could better spy on the other using its satellites, which state could defend itself better against the other in the event of a nuclear exchange? These were all questions of which state had the most power, and how best could they keep it.

The fall of the USSR allowed the United States to de-emphasize the role of power in international politics and space policy. The Persian Gulf War in 1991 demonstrated that the United States was unmatched in its global power capabilities. Consistent with this was its ability to harness space power to achieve its strategic ends. This meant developing space policy in ways consistent with an international political order of hegemony or primacy, where the United States found itself as the sole superpower with the opportunity to shape the international system in ways that best supported its interests in the development of a liberal order. This also meant maintaining a military capable of projecting power anywhere in the world with very short notice. Space technology enabled this in a manner unrivaled by any other country for the next two decades. Over this time, the United States placed a strong emphasis on the utilization of the space domain for military purposes. The one exception, and admittedly a large one, was the use of GPS for civilian purposes.

In the two decades after the Cold War, the use of GPS revolutionized a host of systems and institutions. Computing, banking, and telecommunications powered through space technology and the concomitant growth in cyber power fueled tremendous growth in the exchange of goods and services, as well as the flow of information. Driven in large part by security concerns, space technology eventually came to be meaningful well beyond military capabilities. As we'll discuss in chapter 12, this growth would fuel the beginning of the new space era and the rise of commercial actors. In the meantime, the United States was able to realize dominance both in military and civilian space endeavors.

What did this mean for policy beyond the focus on security and its subsequent spillover into the civilian world? One major effect involved how space projects were acquired and developed. Given a lack of rivals, a lack of threats, and the high cost of launch, space policies focused on big, long-term projects that were perceived as the most cost-effective. Large, exquisite systems from the major aerospace corporations became the status quo. This approach meant a long supply chain of domestic and international manufacturers and many opportunities for delays and overruns in the space enterprise. These systems also were not built with the need to defend them in mind. This feature has led to vulnerabilities today, as the international system has shifted away from a unipolar system (an international system with one dominant state) with the United States to one with more competition and challengers to the status quo.

Today, the question of power is one that remains relevant between the United States and Russia, but more importantly it has shifted to the power balance between the United States and China. The fall of the Soviet Union and the end of the Cold War reinforced realist ideas of power in international politics for China. If China was going to rise as a great power, it would require the capabilities to challenge the United States, or at least be able to deter the United States in the event of conflict. Given the lessons observed from the performance of the US military in the Persian Gulf War, this meant China would need to excel in space. In doing so, China's actions and policies have set the stage for a return to the classic security dilemma.

Chapter 2 also reflected some degree of cooperation between the United States and the Soviet Union. For some political scientists and practitioners, cooperation remains the way to achieve peace and security, since it is not possible for a state to unilaterally do so. These individuals are typically described as having a liberal perspective. This perspective should not be confused with liberal in the domestic politics sense, rather it is with respect to explaining international politics as the outcome of cooperation between individuals, institutions, and/or states. For example, the Apollo-Soyuz mission opened the way for cooperation in the space domain between the United States and Russia. Similarly, UNCOPUOS set in place the cooperative foundations for space behavior found in the OST. While long-term, extensive cooperation remains an elusive goal, even today, the idea that space security will require institution building and some sense of shared responsibility in the fate of the domain suggests a range of policy options that differ than those adhering to a world primarily based on power politics. Added to these concerns is the increasing relevance of commercial space, requiring more coordination and cooperation in the space domain, if only not to ruin the domain for the use of all.

Finally, chapter 2 illustrated the importance of prestige and identity. Acquiring status as a leader in space and a global leader overall remains important to the United States. Constructivist scholars often view themes like these, ones focused on identity, as an important component in explaining international relations. Like the other theoretical perspectives, constructivism is a worthwhile approach to consider as one tries to understand how the United States interacts with others to create shared meanings about the space domain and their roles in it. These shared meanings are often manifest in norm creation, in addition to the recognition of the roles and identities of the various actors in space. For a constructivist, these interactions and norms help determine the types of preferences and policies we see in the space domain, as well as shed insight on the range of outcomes likely faced by the international community in dealing with the myriad space issues that require attention.

What these various perspectives on international politics offer is a window into how we think about the space domain in general. Is space a sanctuary? Is it competitive and militarized? Is space a warfighting domain? One's thoughts on these views on space tend to be shaped by one's thoughts on how relations between states and other actors play out on Earth. Who are the dominant actors in the space domain? How are their preferences shaped? How do the combinations of actors and preferences shape the types of outcomes we see in space?

The next section examines the perspectives on international politics in greater detail, focusing on the actors, preferences, and outcomes most associated with these perspectives. Included in this section is a discussion of how various schools of thought relate to each perspective. The examination moves then to the post–Cold War and current period of international politics to provide a more applied approach to understanding how these perspectives matter for our understanding of US space policy in general. The chapter concludes with a few points on how the future of international politics may impact US space policy in the decades to come.

Space as a Zero-Sum Competitive Environment

If one begins by thinking of the international system as anarchic, that is a world without sovereign authority beyond that of a nation-state, one might conclude that states are left on their own to fight for and defend their own interests. In this type of world, might makes right and states that can build the most strength are the states most likely to ensure their survival and well-being. Historically, states have sought to guarantee their survival and achieve their aspirations through the building of strong military and economic capabilities. Through the present day, it is the measure of these capabilities in various ways that lead to the description of states as superpowers, great powers, regional powers, and minor powers, or some variant of these categories.

The United States and USSR, as two superpowers, fought for these capabilities in equal measure. In the space domain, both sides recognized the potential of space as the ultimate high ground. Space could be used to spy unfettered, facilitate the delivery of ICBMs, and perhaps even be used for launching nuclear weapons. By the end of the Cold War, the two sides managed to limit militarized competition in space but not so much that states failed to see the value in using space for military purposes. As noted at the end of the Cold War, the United States was able to use space-based assets to support its efforts in compelling Iraq to reverse its territorial gains achieved through the latter's invasion of Kuwait. In this case, the use of force became the final arbiter between states with respect to policy preference and the subsequent outcome (i.e., a liberated Kuwait).

Given the significant implications surrounding the effects of anarchy and power, realist approaches tend to focus on what is viewed as an inherent outcome of these system-level characteristics: the security dilemma. Since there is no overarching authority to assist states in their defense, states must take it upon themselves to provide their own security if they expect to survive in an international system of powerful states. Historically, this meant using military and economic capabilities to do so. The problem with the buildup of one's military and economic capabilities, however, is that such behavior is likely to threaten other states, particularly one's neighbors. Consequently, states also are pushed into efforts to maintain a balance of capabilities against others. This sort of balancing takes place among all states but tends to be most notable among the major powers, those states with the most military and economic capabilities. These powers include states that have enough capabilities and resources to threaten the security of regions and other parts of the globe in the modern era. The United States and the USSR during the Cold War, and the European powers prior to the two world wars, stand out in this regard. Today, this description would certainly include China.

The focus here is on the strength of states because for realists, states are the dominant actors in the international system. While international institutions and other actors may play a role from time to time, they are only able to do so because states allow this to happen. So even in large institutions like the United Nations, state preferences are what drive decision-making and subsequent action. Given this focus, combined with anarchy, one should expect that stronger states are likely to shape international interactions and strongly influence the behaviors of weaker states. One can see that such an environment leads to individual states seeking to get ahead at the expense of others. States are either seeking to acquire power or are engaging in behaviors that keep them from losing power relative to others. It is relative power and changes to relative power that provide the context for understanding international politics from the realist perspective as it is these changes that impact the problem of the security dilemma.

The realist literature is rich in history. Building from a long tradition that goes back to Thucydides and his study of the Peloponnesian War and working through more modern authors such as E. H. Carr in the early 20th century, and Hans Morgenthau and Kenneth Waltz more recently, realist theorists have spent a considerable amount of time focused on the intractable nature of the security dilemma and humankind's insatiable quest for power. As states have become the primary actor at the system level, the difficulties of managing the security dilemma and balancing states' quests for power usually has left humanity in a constant state of warfare, or at least in a constant state of preparation for it. Thus, there is considerable attention paid to how states can translate resources into offensive and defensive capabilities in an environment that typically is viewed as competitive in the best of times and conflictual in the worst of them.

If interactions between states are competitive and/or conflict laden, then states are unlikely to have much in the way of mutual trust, creating a lot of uncertainty when it comes to the others' intentions. In international politics broadly defined, this lack of uncertainty makes it challenging for states to be sure that something done that appears benign today will stay that way tomorrow. So, what does this all mean for space? As we saw during the Cold War, it tends to imply that space, not unlike other interactions, is likely to be contested, with a focus on self-help, limited cooperation, and an inability to know the intentions of the other. This sort of context underpinned the Cold War history discussion, where the United States and the Soviet Union viewed space in terms of competition and self-help, with the idea that space could be utilized in ways that would provide an advantage over the adversary.

In some sense, this approach to international politics is straightforward. As seen with the development of ICBMs in the utilization of rocket capabilities, the United States perceived itself at a disadvantage over the USSR, helping to create the perception of a missile gap. This perceived missile gap subsequently established an acute security dilemma for the United States. On the other side, the Soviet Union also found itself in a precarious situation when the missile gap was exposed, leaving the USSR in a position where its own security dilemma became acute. When considering the efforts of both countries to employ nuclear weapons in space, while also using space for other military aspects, including basing on the Moon and potentially other types of weapons, it became challenging to understand each other's short- and long-term intentions in the space domain. With respect to the potential for space to be used in ways that can harm an adversary, in the realist world, successful efforts all equate to a zero-sum game. If one country is better in space, it must come at the expense of another. Again, the Cold War exemplifies this perfectly in terms of how both the Soviet Union and the United States thought about space and the space race.

As relevant as security was during the Cold War, prestige was also important. From a realist perspective, prestige matters because it comes from a place of power. In other words, one gains prestige by having more power and ability to coerce and have others do what you want them to do, while preserving your state's own freedom of action. It is worth noting here that realist theory diverges somewhat with some scholars focusing on the influence of power politics at the individual and state levels, and while others focus solely at the international or system level. The distinctions between what's often labeled as a classical versus systemic approach play out in understanding the primary drivers of behavior for states. At the individual and state levels, the quest for power is viewed as a characteristic of self-interested people; at the system level, the quest for power is driven by the state of anarchy that forces states to gain power. For the purposes of this chapter, it is not terribly important in parsing the differences further beyond noting that the distinctions provide opportunities to consider the influence of power politics across three different levels of analysis (i.e., the individual, state, and international system), which can be important from the standpoint of policymaking.

Many policymakers subscribe to a realist approach in international politics, contending that a lot of the interactions in space make the domain particularly sensitive to the ideas of zero-sum politics, misperception, and the difficulties of understanding intent. One school of thought that derives from realism is that of space nationalism. Space nationalists propose a realist-based policy with America flexing its muscles in space to dominate potential adversaries and control the space domain. The prominence of space nationalism has increased over the last several years, due in part to advocacy for more strident US actions in space but also due to a perceived increase in threatening actions taken by Russia and China in space. These aggressive moves have allowed space nationalists to claim the high ground in the debate over the status of the space domain to argue that space is now a warfighting domain rather than a sanctuary.

As James Clay Moltz notes in his discussion of space nationalism, its roots are firmly grounded in the realist tradition of international relations. "As applied to space, realism makes the assumption that human behavior is essentially static and unchanging, meaning that the prevalence of Machiavellian notions of duplicity, power seeking, and brutality are likely."[1] Under this assumption, space nationalists have argued that the US must seek to expand its own power in space to undercut the growing power of others and thus the threat that they potentially pose. Cooperation is unlikely to occur unless it is with like-minded allies and in a way that gives strategic advantage to the United States. Finally, like traditional realists, space nationalists discount the potential for international organizations and regimes to regulate the domain.

But how much power is enough? Just as defensive and offensive realists differ in terms of the amount of military might that is necessary for a state to pursue, Moltz identifies a more strident version of space nationalism that he calls "space hegemony." From this perspective, it is not enough that the US has space power to confront potential adversaries in space; rather, the US must seek to dominate and control space completely. This view is best encapsulated in Everett Dolman's *Astropolitik* where he introduces the "astropolitical dictum" of "Who controls low-Earth orbit controls near-Earth space. Who controls near-Earth space dominates Terra. Who dominates Terra determines the destiny of humankind."[2] Further, because of the benevolent nature of democratic states, the United States has a unique moral duty to dominate to prevent otherwise malcontent states from doing so. While such an extreme approach is both unlikely and not advocated by many, the example of space hegemony provides a sense of the types of space policies realist approaches can lead to or be perceived to encompass by others, reinforcing zero-sum politics and potentially creating the very conditions space nationalists are trying to avoid.

Space as a Cooperative Environment

Beginning again with the idea that the international system is anarchic, one can ask, Are the conclusions drawn by the realist approach the only options in terms of understanding the influence of anarchy? Liberal theorists contend that there are a range of other choices available besides ones that lead to zero-sum politics, misperceptions, and continuous preparations for conflict. Despite all these conditions being present during the Cold War, for example, Apollo-Soyuz still managed to happen, the US and USSR did not actually go to war, and space remains devoid of weapons of mass destruction. Thus, even in the most challenging times, states seemingly manage to cooperate. A liberal approach to international relations focuses on understanding the factors that help facilitate cooperation despite the presence of anarchy.

Like the realist approach, liberal theory also diverges among those who examine individual- and state-level explanations and those that look at systemic approaches. At the individual and state levels, liberal theory pushes back against the idea that self-interest typically leads to conflict and power seeking. Among individuals there can be mutual benefit realized through absolute gains. These are gains that everyone gets through cooperation, often more than they might get by pursuing self-help. The development of the nation-state exemplifies this way of thinking. Individuals and groups have sacrificed a significant degree of self-help to cooperate with others to realize absolute gains provided by the development of the state as the dominant political unit. Liberal theorists then can explain state-level preferences not as those forced on it by system-level characteristics (i.e., anarchy), but rather by the aggregation of the actors' interests at the domestic level. In other words, a state's foreign policy and international politics are a function of domestic actors' negotiated positions that lead to some sort of state-level exhibited preference(s). As it relates to US space policy, preferences are the result of negotiations between the actors examined in the previous chapters.

As one might expect, the system of government will have some influence on what these preferences look like. Democracies will differ from autocracies, for example. Thus, liberal theorists contend that the distinctions between the preferences of the United States, China, and Russia in space are driven by the different preferences likely found across the types of governments they represent. Even democracies can differ from each other. The countries that make up the European Union (EU), for example, prefer multinational approaches to space policy, therefore, making the European Space Agency a stronger actor than their individual countries as a result.

Neoliberal explanations counter those of systemic realism in seeking to explain behavior at the system level. Despite conceding the reality of anarchy, neoliberal scholars contend that most states do not spend all their energies on acquiring military capabilities to defend themselves against others. States, as a function of the preferences of their domestic actors, must also expend resources on improving the lives of their citizens. This means states must balance between "guns and butter" in their policymaking, developing both military and economic capabilities.[3]

What about the security dilemma and the need for self-help? Neoliberal scholars point to the development of institutions and regimes as ways to mitigate the most harmful effects of anarchy. Institutions like NATO and the UN provide avenues for collective security, limiting acts of aggression. International regimes like those of nuclear nonproliferation are helpful in establishing rules and norms that help states reliably predict the behavior of others in the international system. Through the provision of information, transparency, norm and rule creation, and sanctioning, states can utilize institutions to increase absolute gains for members versus focusing so much on relative gains in an unregulated, anarchic environment. In the space domain, the ITU provides for a regulated approach to the distribution of orbital slots and the radio frequency spectrum. Organizations such as Arabsat and Eutelsat help with the coordination of satellite telecommunications among their members. These examples, along with other conventions such as the Liability and Registration Conventions and of course the OST, which offers a normative framework for state behavior as agreed on by the international community, all point to the ideas of systemic-level cooperation and the mitigation of anarchy through organizations and regimes.

While neoliberalism can offer insight on how the security dilemma is addressed beyond balancing and power politics, it can also expand to consider cooperation more broadly, such as that which occurs in the arena of the international political economy. This aspect becomes important when we consider international politics and the rise of commercial space. While competition inevitably exists between commercial actors, states have a vested interest in agreeing on rules and regulations that will help commercial industry achieve gains that can benefit all. As discussed next, however, one of the problems today is that the negotiations among states are proving difficult given the lack of historical foundations to build upon.

In sum, liberal approaches to international politics focus on cooperation as the key to interstate relations. In ameliorating the security dilemma, liberal scholars point out that unilateral efforts to build capabilities and acquire power are limited in their effects. Essentially, security cannot be guaranteed by any one state alone despite an anarchical system. Thus, institutions and regimes provide a more effective way to achieve these aims, providing for absolute gains over relative ones. This systemic-level approach is consistent with the preferences of states and their domestic-level inputs that tend to prioritize a focus on economic well-being and stability over power seeking, intense security dilemmas, and constant conflict. One other thing worth pointing out is the acknowledgment by some of the factors that may help or hinder international cooperation. Like-minded states, often identified in terms of shared types of governance and shared policy preferences, tend to make cooperation easier. As mentioned previously, this means that democratic states may find it easier to cooperate with each other than cooperating with autocratic states. Such an idea hearkens back to Immanuel Kant who suggested that liberal states would eventually work together to create a federation of republican states that could enjoy a "perpetual peace" among them.[4] In today's system, such an ideal is often categorized in terms of a "democratic peace," the notion that democracies seek to cooperate and are unlikely to engage in violent conflict versus each other. In the context of space policy, this idea remains meaningful when considering which states are viable partners for cooperation and which are not.

This liberal tradition is exemplified in a space internationalist approach. Unlike space nationalism, space internationalists contend that cooperation is the only way forward to both escaping the security dilemma and to increasing the possibility of the absolute gains space can provide in security and economics. Space internationalists point to past cooperation and current institutional development in the space domain as indicators that such an approach is both pragmatic and flexible enough to deal with the range of uncertainty present in international politics. Like liberal theory more broadly, there is also the idea that like-minded states can work together for shared gain. As mentioned previously, the Artemis Accords perhaps are symbolic of this. With respect to collective security, intelligence sharing (including space intelligence) agreements like the Five Eyes arrangement between Australia, Canada, New Zealand, the UK, and the United States, and space domain awareness cooperation among states all fit within this context. Even if states pursue militarization, scholars like Hitchens and Johnson-Freese offer that diplomacy and strategic restraint, positions consistent with liberal institutionalism, can help mitigate the tendency to move toward more aggressive types of space policies and actions that will appear as seeking dominance in space.[5]

Space as a Shared Meanings Environment

Anarchy is what states make of it, suggests constructivist scholar Alex Wendt.[6] In other words, even if one takes anarchy at face value, there is no predetermined outcome for states in terms of their derived preferences or behaviors. Constructivists contend that preferences and behaviors are a part of an interplay between actors, the result of which shape what actors think about others as well as what they think about themselves. The Cold War, for example, need not have played out as it did; the policy outcomes resulted from the processes of learning that shaped the dynamics between the United States and USSR, for better or worse. It was for the better in that the two sides learned how to compete without engaging in nuclear war. As we know, the space race provided one mechanism for this competition. Along these lines, the shared experiences of the threat of nuclear annihilation made it easier to agree that weapons of mass destruction in space were not in their shared interest. In terms of being for the worse, the inability to trust each other, part of a longer historical relationship between the United States and the Soviet Union, made it difficult to effectively resolve the security dilemma between them. This inability to trust led to the mass production of tens of thousands of nuclear warheads and sizable conventional military buildups throughout most of the Cold War.

International politics for constructivists is about intersubjective and shared meanings. These meanings come to shape behavior through the process of norm creation and identity building. To understand international politics, one must understand relationships between actors. Structure, and in this case the condition of anarchy associated with it, does little to shed light on behavior by itself. This perspective on international politics stands out from the other two in that the types of outcomes predicted by realism and liberalism may also be explained through this approach. The international system can be conflict prone as realism suggests if the relationships in the system are viewed by the actors as zero-sum. Stated differently, if states are socialized to adhere to the informal or formal rules of a self-help system (e.g., power politics), then these are the types of behaviors one can expect. Similarly, if states are socialized to adhere to the informal or formal rules of a cooperative system (e.g., peaceful conflict management), meaning that states value absolute gains more, then states should behave accordingly.

One can see this most clearly in the identification of realist or liberal-oriented interactions across the international system. In Western Europe, for instance, state interactions take place on the foundation of shared meanings regarding cooperation and multilateralism. Consider, for example, the previously discussed establishment of the ESA, which allows states in Europe to pool resources with the goal of gaining collective benefits. In doing so, they can share advanced technology without concerns that states will use this knowledge for the purposes of creating military advantages directed toward other members. This did not magically happen; the patterns of behavior we observe in the EU are the result of decades of learning and socialization (including through warfare), where states in the EU have come to share similar ideas regarding governance and security. Learning and socialization have resulted in mutual trust and the elimination of the security dilemma among its members. Such patterns of interaction stand in contrast to what we observe in the Middle East today, where countries like Israel and Iran have a long history of distrust and rivalry, leading to an inability of states in the region to develop deep patterns of cooperation that could move the region away from conflict and toward collective security. In this latter case, the development of space-based capabilities is viewed with distrust rather than shared understandings that these capabilities will be used in ways not threatening to the other state. In August 2022, for example, Iran placed an advanced remote sensing satellite on orbit with launch assistance from Russia. While Iran claims the satellite will be used for nonmilitary purposes, its potential military uses will undoubtedly cause concern in Israel since it will allow Iran to spy with much greater precision than it has been able to previously. Such concerns would not be likely if the same scenario took place between the United Kingdom and France.

Although the focus here is on state interactions, constructivist approaches are not limited to states as the only significant actors in the international system. As seen in the space domain, there are different actors that impact space policy; constructivist explanations for outcomes can consider individual-, state-, and system-level explanations. This should not be terribly surprising as the concepts of shared intersubjective meanings and norms are important across all levels.

For constructivists, international politics is also a story about identities, particularly how states view themselves and each other. The importance of identity is why the United States so often references the concept of leadership. The United States views itself as a leader in the space domain, in part because it also has viewed itself as the leader in the current international system. Identity is relational, however. In this case, how do others view the United States with respect to leadership? What does it mean to be a leader in the space domain, specifically?

Once established, identities can be a challenge for international interactions as patterns of behavior between actors can take time to change, particularly if they are formed from a troubled history. Prestige is also a part of the identity calculus for states. It is perceived by many states that having space-based capabilities provides a level of prestige that makes it enticing to develop a domestic space enterprise. This has led to even small states seeking to become space powers, symbolizing to the international community that they are a technologically advanced and capable country with a "seat at the table" when it comes to the international politics of space.

To review, constructivism explains international politics through an understanding of relationships and identity. Informal and formal rules and norms guide behavior in the international system and can shape environments that range from conflict to cooperation. The structure of the system and the presence of anarchy does not guarantee any particular outcome. System actors create outcomes through a process of learning and the creation of shared meanings. For space policy, this has very important implications given the limited number of norms and rules agreed upon in space and the potential for interactions in space to move toward a power politics model, one based on cooperation, or something in between.

Schools of thought underpinned by the constructivist approach to international politics tend to gravitate toward seeking the development of norms as agreed on by most of the states in the international system. Moltz offers the social interactionist school of thought as one example, arguing that behaviors and outcomes are conditioned by the interactions of states over time.[7] In short, states can learn from their relationships with each other and develop norms that lead to predictable patterns of behavior. This is what one can observe in chapter 2 in the discussion regarding the United States and USSR, where the two sides learned over time what actions were and were not acceptable or what would possibly lead to war.

Over the last few years, several actors have tried to push for the development of space norms. In the United States, efforts by the DOS and DOD have sought to increase the rhetoric in support of peaceful norms as well as actions that reinforce these norms. Cassandra Steer identifies efforts by the EU to socialize an International Code of Conduct for Outer Space Activities, as well as a draft Treaty on the Prevention of the Placement of Weapons in Outers Space and of the Threat or Use of Force against Outer Space Objects (PPWT), in ad-

Table 11.1. Theoretical perspectives: actors, preferences, outcomes, and schools of thought

Theory	Primary Actors	Preferences	Outcomes	Schools of Thought
Realism	States	Shaped by perceived self-help system Power politics Strong military capabilities	Competitive and conflictual Limited agreements Sovereignty of weak states remain under threat Security dilemmas are heightened	Space nationalism
Liberalism	States Institutions	Cooperation Information sharing Importance of absolute gains	Interdependence and shared values Influence of international rules and law Bad behavior sanctioned Security dilemmas mitigated	Space institutionalism
Constructivism	States Institutions Other Non-state actors	Relationship driven Reliance on norms Shaped by identity	Varies based on relationships and strength of norms Difficult to change patterns of behavior quickly because of entrenched identities	Social interactionism

dition to private efforts to develop a Manual on International Law Applicable to Military Activities in Outer Space (MILAMOS).[8] All these activities point toward efforts to socialize behavior and expectations in space. While largely unsuccessful to this point, it is worth noting that socialization and norms will develop like they have in other areas of international politics. The question that remains unanswered is what types of socialization and norms will they be?

Table 11.1 provides a reference summarizing the differences across theoretical perspectives in terms of the focus of explanation, the central actors, the shaping of preferences, the types of outcomes expected, and the associated schools of thought. Again, these are ideal types, useful for conceptualizing about how theory helps shape thoughts and expectations in international politics. To this end, the schools of thought are also examples; the space literature is not lacking different categorizations but typically they are defined along similar dimensions. Policymakers tend not to think about these categorizations directly but do have a sense for them in their own feelings about how the world works. This is important for space policy as it drives their own decision-making and policy formulation based on what they believe is in the realm of possibility.

Now we turn to how theory helps shape ideas about the reality of politics in the space domain, beginning with a look at the post–Cold War period, followed by an examination of the current and short-term future.

Theoretical Perspectives in the Post–Cold War Period

As we pointed out in chapter 3, the beginning of the post–Cold War period is not clearly defined with respect to dates, as the end of the Cold War was more of a process than an event. For our purposes in this chapter, we pick up with the post–Cold War period beginning on August 2, 1990, and make an argument for this period ending at least by January 11, 2007. The first date represents the beginning of the Persian Gulf War, and the latter date the first Chinese ASAT test. Both are important in that they set the boundaries by which international politics in the space domain departed from the status quo. We have discussed already the utilization of space-based assets in the attack on Iraqi forces to force the withdrawal from Kuwait. Whether or not the moniker of the first space war is fitting, the extensive use of the space domain as an enabler is without question, ushering in a new era in the ability of the United States to project power globally. This power projection capability, coupled with the fall of the Soviet Union, placed the United States in the unique position of being the sole superpower, with no immediate rivals. Often described in systemic terms as unipolarity, the United States was unchallenged in its global influence and unmatched in its military and economic capabilities.

What did this unipolar period mean for international politics and space? This is where theoretical approaches influence one's thinking on such a question. Realists viewed unipolarity as transient; states would seek to balance against the United States, lest they suffer domination in the international system. Liberals argued that, given the US penchant for multilateralism and the desire to establish a liberal international system, there would be a new global order that would put pressure on states to democratize, embrace capitalism, and strengthen human rights norms.[9] In a sense, the new system would represent the "end of history," a condition where humanity would find the perpetual peace Kant hypothesized.[10] All of this, of course, rested on a foundation of US power, which begged the question of whether realism or liberalism was better at explaining the post–Cold War environment. As the dominant actor in the international system, the United States alone had the power to shape and influence the international system according to its preferences. Without these overwhelming material capabilities, would this new global order be possible?

Part of this global order included the continuation of space as a sanctuary. By the mid-1990s, the United States looked as if it would be the only significant actor in space for some time to come. Embracing space as a sanctuary would help ensure US leadership in space while it continued to focus on the maintenance of a liberal international order on Earth. US space policy options throughout the post–Cold War period were numerous. Yes, there was a path-dependent aspect to policy, making it easier to continue with some of the trends we observed at the end of the Cold War. Yet, there were also new opportunities in the international arena in the military, civilian, and commercial sectors that could be explored. These new opportunities begin in the 1990s with increased cooperation with Russia and China. NASA purchased contracts with the Russian space agency Roscosmos to supply the ISS and keep scientists from pursuing other endeavors, specifically those that might increase the likelihood of proliferation in other countries. US agreements with China provided opportunities for the latter to launch American commercial satellites throughout the 1990s. In short, it appeared that the United States engaged in policies that offered a more cooperative approach to space and one that would continue the sanctuary ideal. Perhaps the epitome of this cooperation was the Space Station Intergovernmental Agreement in 1998 that established the ISS as a cooperative project between the United States, Russia, Canada, Japan, and the members of the ESA, a total of 15 countries. Together, the ISS members would cooperate in maintaining a permanent human presence in space for peaceful purposes.

As previously discussed, this sanctuary status allowed the United States to build its space infrastructure without the security concerns that might come from rival states. From a realist viewpoint, however, this favorable environment would inevitably change. As such, realist-minded policymakers and scholars suggested that the United States take advantage of the relatively peaceful environment to prevent challengers from gaining influence in the space domain. Scholars like Everett Dolman, for example, argued that a continuation of US dominance in space could only be maintained by controlling access to space, in essence using military capabilities to prevent the rise of threats to US space superiority, as well as preventing the use of space assets for projecting power.[11] This latter set of capabilities would be invaluable to any potential challenger of the United States with respect to the global system. Other scholars like John Klein offered that the United States use its space superiority to serve as a guardian of the space domain, acting in similar ways to the US Navy by keeping space open for freedom of navigation for states and commercial entities alike.[12]

The debate on cooperation or dominance in space was interrupted by the terrorist attacks on the United States in 2001, effectively ending a collective focus for the United States on the space domain and arguably loosening the grip on space dominance for the country. The War on Terror utilized space-based capabilities for ISR, but resources poured into terrestrial requirements, leaving US space to develop at a much slower pace than previous periods. In terms of international politics, the War on Terror also created cleavages that were difficult to undo. After the invasion of Iraq, many states in the international community, including America's closest European allies, began to question whether US global dominance was a force for stability and the best guarantor of the liberal order, the same order that it had such a significant role in creating. This questioning of role and identity by others highlights a constructivist approach to explaining attitudes during this period. The United States was viewed as willing to go it alone, eschewing international norms and abrogating treaties like the ABM Treaty to do so. If the United States would not meet its own international commitments, some wondered, what was the likelihood that others dissatisfied with the current system would do so? In short, the War on Terror changed the relationship between the United States and the international community in ways that were difficult to predict with respect to future cooperation and the maintenance of the liberal order.

Russia and China, for example, made it clear that their international political preferences were not consistent with those of the United States. Militarily, both countries followed suit with their own anti-terrorism campaigns, using counterterrorism arguments to enhance their own regional military capabilities. Liberal norms regarding human rights were de-emphasized in the name of security. On the civilian space side, we noted in chapter 3 that the *Columbia* disaster brought about the end of shuttle operations, further diminishing US space leadership and leaving the country dependent on Russia for crewed spaceflight. In 1999, a US congressional report implicated China in stealing sensitive nuclear and space technology, allowing the country to accelerate its own nuclear and space programs. Going into the 2000s, there appeared to be a transition away from the path of cooperation to one that looked more like traditional power politics.

This transition began to solidify with Russian efforts to rebuild its space program under President Vladimir Putin and China's efforts to implement a range of policies across all space sectors to catch up to the United States. All these actions tended to be viewed by the three space powers in zero-sum terms; either they were competitive in diminishing the distance between US capabilities, prestigious in terms of their achievements or both. China's 2007 successful ASAT launch, removing a defunct weather satellite, illustrates all three attributes. The launch made clear that China could threaten US assets and demonstrated advanced Chinese technology. In terms of power politics, it is likely China ushered in a new era of international space relations with this test.

The test also demonstrated the extent to which international norms have influence in the space domain. The kinetic kill ended up creating large amounts of debris, violating what were thought by many in the international community to be established norms with respect to maintaining the space environment. While China was not punished per se, its international reputation did seem to be affected, at least in terms of other actors pointing out the shortsightedness of its actions. This loss of reputation was limited, however, as China continued to build partnerships in its establishment of the Asia Pacific Space Cooperation Organization (APSCO) with Bangladesh, Mongolia, Pakistan, Peru, Thailand, Iran, Turkey, and Mongolia in 2008. What was missing in this case, something identified by liberal scholars as essential for cooperation, was an earlier institutional presence capable of influencing China's behavior in its consideration of employing the ASAT weapon. The OST and other UN agreements did not prove strong enough in this regard.

Theoretical Perspectives in the Current Period and Beyond

China's march toward space parity with the United States seemed to validate earlier realist assessments that there would be a challenge to long-term US dominance. China's actions in space have been consistent with other interactions across different domains and issue areas. Its economic and military capabilities throughout this current period have continued to grow to such an extent that it is difficult to consider today as like the environment that existed immediately after the Cold War. Beyond China's rise, there has been a change in the international politics of space that also suggests the international system is no longer in the same post–Cold War period. The number of space actors has increased significantly in the current era. States may still be the most dominant players, but they are far from dominant in the space enterprise given the importance of institutions, the rise of non-state actors, and the overall importance of space for both security and commercial interests.

Another change one can observe in the current period is the US response to the shifting international political landscape. Rather than accepting the existence of power politics completely, US policy has attempted to hedge, leaning forward on cooperation despite the rise in competition and the change in the international system. With the election of President Obama in 2008, the United States began to move away from its unilateral approach to international politics to reclaim its global leadership. This was an intentional effort on the part of the Obama administration, which believed the United States was in danger of losing its traditional role. While acknowledging the challenges brought on by a dynamic international system, the administration chose to pursue international engagement as its global strategy, reconnecting with allies and extending partnerships to those supporting the liberal order. This approach extended into space policy as well, reflected in the administration's 2010 National Space Policy and its 2011 National Security Space Strategy. Both documents sought to expand international cooperation, develop commercial industry through international agreements, and further commit the United States to the peaceful use of space. Acknowledging that the space domain has become "congested, contested, and competitive," both documents also reasserted US leadership in the space domain, calling on others to act responsibly and in ways consistent with established international agreements and norms. International cooperation with US allies in the areas of space situational awareness, satellite communications, and ISR became a centerpiece of the US approach to space security.

Despite rising competition with China, the United States has remained committed to international engagement in the space domain. With the Trump administration, however, the United States simultaneously increased its competitive behavior with states like Russia and China, thus seeking stronger cooperation with allies while intensifying the competition with its rivals. This pattern is something constructivism seems well capable of explaining, as there are opportunities to build on gains with those that share similar perspectives while hedging against those that threaten the status quo. This pattern has continued in the current Biden administration as well, with competition taking place both in terms of capabilities and rhetoric. With respect to the latter, the difference between the major spacefaring nations today seems to be in their interpretation of what it means to commit to broader norms and agreements originating from the OST. the United States and its competitors claim to support established principles, while also claiming the other to be in violation of the same principles. These differences have led to stalemates on further international agreements in space between China, Russia, and the United States (i.e., the proposed PPWT), while there has been increased cooperation among other allies and partners.

To this end, the United States has pursued a broad range of international space policies, both bilateral and multilateral in nature. Since 2018, it has agreed to arrangements with Japan, between the Japan Aerospace Exploration Agency (JAXA) and NASA on lunar space exploration; with the EU on security and civil space activities, including on the Galileo global navigation system for better interoperability and on the Copernicus Earth observation program. In the Asia-Pacific region, the United States has pursued opportunities to work with South Korea, Vietnam, and Thailand on space policy collaboration, including dialogue to open commercial opportunities for US businesses and in the case of Vietnam, the sharing of space-based data to assist with maritime safety, security, and environmental protection. As mentioned in chapter 8, the United States remains active in the United Nations' engagements on rules and norms for the long-term sustainment of space activities. Of all these efforts, the crowning achievement to date with respect to international cooperation in the current period is the Artemis Accords.

With the goal of collaborative deep-space exploration, initially signed in 2020, the Artemis Accords demonstrate a commitment by over 20 states at present to build upon the principles of the OST for the peaceful use of space. This includes support for policies that provide for open international standards to increase interoperability, the sharing of scientific data, protecting historic sites and artifacts, deconfliction, sustainable resource utilization, and debris mitigation. In all, this multinational recognition of the need for strengthening the framework of the OST suggests strong support for an international space regime that emphasizes absolute gains over relative ones. Notably, however, Russia and China, two of the largest space actors, remain outside of this collaborative effort.

The international system in 2023 is quite different than it was in 1992. The United States, once the sole superpower with space superiority, now finds itself still as the strongest state in the space domain but far from alone. From the near-peer competition represented by China to the need to support the access to space for the smallest commercial actor, the United States is being challenged on many levels with respect to maintaining leadership and helping to keep the space domain stable for use by all. Rather than accept the power politics model of international relations that other states seem to be emulating, the United States today continues to foster international cooperation and support the establishment of norms that will enable the free and fair use of the space domain for those embracing responsible behavior in their pursuit of security and stability. That said, the United States still possesses military capabilities that inherently create security dilemmas for other actors, regardless of intent. As a result, other countries have taken actions to reduce this security dilemma, only they have done so not through institutions, which as liberal theorists contend, could produce absolute gains in security by all. Rather, they have chosen to try and match US capabilities, leading to continued misperceptions and distrust by all sides.

Constructivists would likely point out that unless the relationships between these actors change, their views on what is or is not possible in terms of policy will remain fixed and change will prove difficult. As pointed out previously, the United States is limited in its legal ability to cooperate with China, therefore, China must look elsewhere for partners or go it alone. For China's part, it has taken a mixed approach, seeking out some partnerships—for example, with Russia—while also committing the resources to embark on unilateral efforts, like its Chang'e 5 lunar exploration mission and the development of plans for its own space station. Given these limitations on cooperation, the current and short-term future looks to remain an environment that will be competitive for some states, cooperative for others, and overall dynamic to the point that policy will remain difficult for all actors in the space domain. So too then will be the challenge to keep space as a sanctuary with the widespread cooperation needed to maintain its sustainability.

Conclusion

With so many moving parts, the international politics of space are difficult to discern. International relations theories provide some insight into what is possible and why. Realism points to the intractable nature of the security dilemma caused by the condition of anarchy in the international system, the result of which leads to power politics and a zero-sum worldview. Certainly, some of the relationships observed mirror these expectations. Liberalism, however, offers a different take on the amelioration of the security dilemma through institutions and cooperation. Again, in an assessment of the current period, this manifested in a wide range of partnerships between the United States and others. These partnerships tend to take place among those sharing similar values and goals, such that interests like commercialization become important policy inputs alongside security issues.

Finally, constructivists offer that the differences seen in the international system are based on the intersubjective relationships between these actors in the space domain. Anarchy need not lead to power politics. States that share values and goals create norms and over time these norms reinforce behaviors such that they become predictable and valued in and of themselves. Shared meanings of the concept of sanctuary and the peaceful use of space, for example, make it easier to cooperate on a range of space endeavors. Shared identities do as well. Without these shared meanings, states are prone to mistrust and suspicion, reinforcing the security dilemma predicted by realism. Unfortunately, such is the nature of international politics and the theories used to explain it.

Summary Points

- Theories of international politics provide explanations for state behavior in the international system, given the state of anarchy that exists.
- Realism focuses on power politics, zero-sum outcomes, and relative gains as the basis for state interactions.
- Liberalism offers possibilities of mitigating the security dilemma and facilitating cooperation through institutions.
- Constructivism emphasizes the importance of relationships and norms of behavior for understanding international politics.
- International relations in the space domain tend to reflect international politics more broadly.
- The current system exhibits patterns of cooperation and competition in the space domain; theory is helpful in uncovering why these patterns exist.

Discussion Questions

1. Which theoretical perspective do you think best explains international politics in the space domain currently? Why?
2. The Artemis Accords represent a very important cooperative agreement, yet there is not a strong institutional foundation to go along with it. What insights might theory provide with respect to their possibility of success in the long term?
3. Beyond the legal constraints, what types of changes would need to take place between the United States and China such that cooperation would be more likely? How could these changes be implemented?
4. What are some of the implications of change in the international system between the beginning of the post–Cold War period and today as it relates to space policy?
5. How can the development of military capabilities in the space domain not be viewed as threatening to another actor? What are the implications of developing military capabilities with respect to competing schools of thought in the space domain?

Further Reading

Hickman, John, "International Relations and the Second Space Race Between the United States and China," *Astropolitics*, vol. 17, no. 3 (2019): 178–90.

Knorr, Klaus, "On the International Implications of Outer Space." *World Politics,* vol. 12 no. 4 (1960): 564–84.

Moltz, James Clay, "The Changing Dynamics of Twenty-First Century Space Power." *Journal of Strategic Security,* vol. 12, no. 1 (2019): 15–43.

Sheehan, Michael, *The International Politics of Space* (London: Routledge, 2007).

12

Commercial Space Policy

While the US government is still carrying out any number of activities in space today, a strong argument can be made that the most exciting developments are coming about because of and by the growing commercial space industry. For evidence, look no farther than SpaceX. The rebellious upstart has pushed for more than 20 years now to shake up the space industry and its list of achievements is indeed remarkable. In 2008, it launched the first privately developed orbital rocket, Falcon 1. This success paved the way for contracts from NASA first to send cargo to the ISS and later to send astronauts. SpaceX soon fought the US government for the right to compete for valuable national security space launches—and won. It has pioneered reusability with its Falcon 9 rocket whose booster stage returns to Earth, landing either on land or on a drone ship at sea. As a result, the costs of space launch have dropped precipitously. And all this pales in comparison to its achievement of being the first private company to launch humans to orbit, first on behalf of NASA and later on commercial tourist flights.

The pace at which SpaceX has both developed new technology and succeeded comes despite great skepticism early on whether such things would be possible at all for private companies. Though the US government had anticipated a commercial space market as early as the 1980s, commercialization and even privatization of government space assets did not seem to be plausible considering the costs of the technology, the cost of launch, and the lack of a market base and customers willing to bear those costs. A burst of interest in satellite internet in the late 1990s also failed to materialize into anything substantial. Thus, when SpaceX and others established themselves in the early 2000s, there was little confidence that they could succeed where others had failed. It was often remarked that the quickest way to become a millionaire in the space industry was to start out as a billionaire.

This chapter will explore the rise of the commercial space industry as well as the policy regimes that have supported it and the policy challenges that it now poses. However, prior to beginning, it is important to define what we mean by "commercial space." There is no doubt that companies have built and profited from space technology going all the way back to the 1950s. While the difference between the typical government model of contracting out and commercialization will be fleshed out later, here, we take the definition of commercial space to be private actors, primarily businesses and corporations, selling space services that they themselves have substantially invested in and developed. While companies like SpaceX and Blue Origin tend to get most of the attention in discussions of commercial space, a number of other companies have also developed in the past two decades including Planet and Maxar, both of which provide space-based imagery services; Virgin Galactic, which focuses on space tourism; OneWeb, which is building out a satellite constellation to provide communication and internet services; and a whole host of other companies that, while they may not be building rockets and satellites, are providing support services and other technologies to enable these activities. The sum of this activity is a growing economic sector that today is worth $424 billion with some projections seeing it as high as $1 trillion by 2040.[1]

The Path to Commercial Space

1950s–1980

Given the still rather speculative nature of space technology along with its incredibly high cost, the focus of space development in the early period of space history was squarely on governments and in particular the US government. This is not to say that private companies were not involved in the development and operating of space technology—companies like North American, Boeing, and Lockheed played key roles in the development of everything from rockets to the capsules that took astronauts to the Moon to the space shuttle. The method through which this happened was government contracting. The US government (and other governments) have a long history of providing money to private actors and organizations to carry out certain tasks. The military provides funding to companies to build airplanes and missiles, for example. Space was no different in this early period.

Private companies played a large role in the major space programs of the 1960s—Mercury, Gemini, and Apollo. Recall from chapter 6 that NASA's early organizational culture was heavily influenced by the engineering and research backgrounds of its early employees. As such, NASA engineers were deeply involved in the development of the rockets and the vehicles they sent to space, though it was private companies that built the actual hardware. Astronauts spent a significant amount of time not just visiting these companies but consulting with them as they built the hardware.

One exception to this pattern of commercial activity was with communications satellites. In the early 1960s, the Kennedy administration proposed passing legislation to set up the Communications Satellite Corporation (Comsat), which would work with both private corporations and foreign governments to set up a communications satellite system. While the US government founded this organization and appointed early leadership, it operated as a business would need to—raising money, selling stocks, and selling services. Comsat, in turn, worked with 17 other countries to form the International Telecommunications Satellite Consortium (Intelsat) in 1964. The first satellite built and operated under this framework, Early Bird, was launched in 1965.

The role of private companies as primarily contractors shifted as the development of the space shuttle began in the early 1970s. With NASA facing a significant amount of retrenchment with a loss of funding and employees, private companies became further involved in early developmental efforts. Through a series of requests and contracting stages, several companies submitted proposals for a space transportation system based on standards laid out by NASA and other executive branch officials, most importantly the OMB. In 1972, NASA awarded the space shuttle contract to North American.

For both NASA and other government agencies involved with space activities, contracting out became the norm for developing, building, and—in some cases—operating space technology. The government would provide most of the funding while private companies would develop and build the required system. Companies would typically receive a certain percentage over the total cost of the program as their profit (something called a cost-plus contract). It is important to note that this is quite different than the commercial model of the company investing its own resources into the space systems. Although private companies were profiting from space ventures, they were doing it on behalf of the government rather than through their own actions.

1980–2000

As space technology began to mature in the early 1980s, the US government actively began contemplating how it might support a growing commercial space sector. With Comsat already operating the Intelsat system, Landsat newly privatized (previously discussed in chapter 10), and greater involvement on the part of private companies in launching rockets and the new space shuttle, it soon became apparent that the government needed to be more proactive in setting commercial policy. The urgency for such policy was further demonstrated in 1982 when Space Services Incorporated requested permission to launch its suborbital Conestoga I rocket in what would be the first private rocket launch in the United States. However, no single agency appeared to have the ability to approve of the launch and many departments and agencies seemed to have a role to play. While approval was granted and the launch successful, the incident demonstrated that the government needed to be more proactive in setting standards for launch, identifying responsible agencies, and providing the private sector clarity.[2] The result of both presidential and congressional interest was the Commercial Space Launch Act of 1984, which not only recognized that the private industry could carry out activities such as launch and operation of satellites but centralized authority over commercial space launch in the DOT.

While the 1984 act was important in recognizing that private space actors could exist in the United States, a market for such activities still didn't exist. As part of a deal to support the space shuttle in the 1970s, the DOD agreed to launch all its payloads on the shuttle. Outside of the DOD, there still existed little demand for launches that would carry private satellites to orbit. Thus, in the early 1980s, there was little to no private rocket business to be had. This situation changed with the 1986 *Challenger* disaster. With limited shuttle operations, the DOD was forced to look elsewhere for launch services. Additionally, then-president Ronald Reagan also determined that the shuttle would not be used for commercial launches in the future. The result was the beginning of a private launch market but one which still faced significant headwinds. Not only were the costs of building a rocket quite high, but insurance costs were as well. The Commercial Space Launch Amendments Act of 1988 helped to reduce these costs—the law limited the potential losses of a company with the government agreeing to pay claims in excess of $500 million should a launch fail.

The result of these policy initiatives was growing responsibility for space launch on the part of private companies. Lockheed Martin built on the Atlas family of rockets, first developed in the 1950s as an ICBM and later used in early space launches; while Boeing evolved the Delta family, which also found its origins in 1960s ballistic missiles. These companies soon found themselves in a growing international launch market, competing with the ESA's Ariane and Soviet and Chinese commercial launches. As a result, Congress remained concerned about commercial launch policy throughout the 1990s, particularly as NASA shifted its efforts toward the ISS.[3] In addition to requiring that the US government use American launch services in most cases, it also directed NASA to make economic development a goal of the ISS program and to study ways in which it could utilize commercial services and providers to a greater degree.

In addition to these laws pertaining to launch, Congress also passed the Land Remote Sensing Policy Act of 1992. Recall from chapter 10 that this law was a reaction to the failed privatization of Landsat. In addition to making changes to that program, this law also sought to build on remote sensing language in the 1984 legislation that had proved to be a challenge to the potential of a private remote sensing market. Under the 1984 bill, imagery had to be sold at the same price to all potential customers; this was amended in 1992 allowing companies to charge differential rates thereby allowing American remote sensing companies to compete with foreign companies that could charge lower rates.

The sum of these efforts was to set the stage for a growing commercial space market. Even with these initiatives in place, the biggest hurdle remained: cost. Satellites were still largely unique pieces of equipment built in one-off fashion and rockets because they were thrown away with each launch, kept costs high. Interest from dot-com billionaires in the late 1990s in launching satellite constellations to provide internet from space briefly suggested that a larger launch market would soon appear, but most of these efforts also evaporated along with the billionaires' stock prices.

2000–Today

Though a substantial commercial space industry did not arise as quickly as policymakers expected, the early 2000s marked a turning point where several trends collided to create the conditions for success. First, for a company to decide to provide a product or service, there must be a need. With the space shuttle largely filling launching needs for NASA and NASA largely uninterested in a commercial space industry that might threaten its own role, there was little need for launch services outside of those that were being provided by Boeing, Lockheed Martin, and other international options. Because rockets were expendable, launch costs remained high, keeping other potential companies from entering the market to begin with. Those that did attempt to provide innovative launch solutions often failed. This situation changed in 2003 with the destruction of the shuttle *Columbia*.

As we have detailed elsewhere in this book, *Columbia*'s breakup as it reentered Earth's atmosphere on February 1, 2003, signaled the end of an era for the aging space shuttle system. The Bush administration's VSE sought to retire the shuttle in 2010 to make way for a new rocket system. While the plan was for this system to replace the shuttle and allow the United States to return to the Moon, then-NASA administrator Michael Griffin decided to provide some funds for the development of commercial alternatives, particularly for taking cargo to the ISS, through the COTS program. This decision represented a small opening for new space companies as well as the possibility of breaking NASA's stranglehold on launch services for civilian space programs, though NASA was still determined to control the most significant element: human spaceflight.

While the dot-com bubble burst reduced some people's portfolios, other winners in the new economic market remained—some of whom had a longstanding interest in space exploration. Recall from chapter 9 that internet billionaires Jeff Bezos and Elon Musk founded two companies that proved to be pivotal: Bezos's Blue Origin in 2000 and Musk's SpaceX in 2002. With both companies focused on developing reusable rockets, not only had a market opened, but the ability to reduce costs as well. Their efforts also coincided with another one to encourage the development of reusable rockets: the Ansari XPRIZE.

Established in 1996, the XPRIZE aimed to incentivize private space travel by offering a $10 million prize to the first group that could launch humans into space on a reusable spacecraft twice in two weeks. In 2004, Scaled Composites, a group funded by Microsoft cofounder Paul Allen, launched SpaceShipOne to the edge of space twice to win the prize. With that success, British billionaire Richard Branson bought the SpaceShipOne design and declared his intentions to establish Virgin Galactic as the first space tourism company. Branson's purchase made him the third billionaire to enter the space arena in the early 2000s.

SpaceShipOne's success not only signaled that private spaceflight was possible, but that it was coming with little in the way of policy to restrain it. While earlier laws regulated rockets more generally, the laws did not envision humans launching on them. Members of Congress turned to considering what new policies might be necessary to ensure safety and oversight while also trying to avoid overregulating a new industry in such a way that would prevent its ultimate success and survival. The result was the Commercial Space Launch Amendments Act of 2004, which extended the commercial launch indemnification regime first introduced in the 1980s, limiting liability for launch providers, located regulatory authority for private spaceflight in the FAA's Office of Commercial Space Transportation, and established minimal safety requirements that essentially required "spaceflight participants" to provide informed consent for their trips. The bill also established a "learning period" for commercial spaceflight that specifically limited any further safety regulations to allow the commercial industry to grow and further understand what type of government involvement would be necessary.

SpaceX's success in 2008 led to additional COTS awards from NASA and a financial lifeline for the company. While the company quickly moved on from its first-generation Falcon 1 to the Falcon 9, it also developed the Dragon module that would first take cargo to the ISS and later astronauts. In 2010, with delays for Constellation becoming greater, the new Obama administration proposed ending Constellation altogether and moving toward fully commercial alternatives for LEO. This approach would have allowed NASA to focus on long duration spaceflight and leave LEO to commercial companies. Members of Congress with strong space interests as well as senior NASA leaders were hesitant to trust these new companies as well as critical of policy moves that would reduce money flowing to their districts. As a result, some elements of Constellation were retained and repurposed for a new SLS along with expanded plans for commercial crew programs. The opening of a market that began with the failure of the space shuttle had come to fruition, opening a path for SpaceX and other emerging space companies.

Falcon 9's success also allowed SpaceX to compete for other national security launches. Despite this capability, the US Air Force initially balked at allowing SpaceX to bid against ULA (the joint effort of Lockheed Martin and Boeing) for launches. Prior to being able to bid for a national security launch, the Air Force typically certifies that a rocket meets its requirements; despite submitting documentation, the Air Force delayed issuing such a certification. In 2014, SpaceX sued the US government to force this certification. At the time, SpaceX also cited that its rockets would be significantly cheaper than ULA's at a cost of $60 million per launch compared to $400 million.[4] In 2015, the lawsuit was settled, the Falcon 9 was certified, and SpaceX was allowed to compete for the valuable national security launches.

Though SpaceShipOne's first flights in 2004 seemed to indicate a quick development of commercial spaceflight, much like the 1980s, the market failed to flourish right away. With launch indemnification and the learning period set to expire in 2016, Congress once again considered legislation to extend both policies. In addition to addressing both areas, the US Commercial Space Launch Competitiveness Act of 2015 also took up the utilization of space resources. Recall from our discussion of the OST in chapters 1 and 2 that the OST prohibits states from claiming any celestial body as their own. However, some argue that space resources located on the Moon, asteroids, or other planetary bodies might be used without a state claiming sovereignty. The bill allowed American companies to do just that despite concerns it may violate the OST. Proponents argued that establishing such rights, while perhaps not relevant now, would be an important part of encouraging not only space commercialization, but giving the United States a favorable commercial advantage when the time came.

Though most of the focus thus far has been on the area of launch, technological shifts in satellite technology also enabled the increasing commercial uses of space. Though smaller satellites had long been thought about and designed, the cost of space launch made it imperative that satellites work when they got to space. As such, they were often overdesigned with a focus on redundancy and resiliency that served to increase the cost to build them. With launch costs reduced, satellite companies could take a chance on cheaper satellites that used off-the-shelf technology since, on the off chance they failed, the companies could afford to send other ones to replace them. One company that pioneered this approach was Planet. Established in 2010, it developed a series of cubesats called Doves that could be launched in batches and placed in LEO to provide routine imagery of Earth. The ability to launch large batches of small, cheaper satellites into orbit not only helped reduce the overall cost of operating in space, but also led to an increasing demand for launch services, further enabling the reduction of costs.

While skepticism of commercial space ran high at the beginning of the 21st century, today, just more than two decades later, there is little doubt that commercial space companies can succeed and are important to US ambitions in space. While NASA continues to build its own large rocket, the SLS, for a return to the Moon, not only is it wholly dependent on commercial companies for launch services to LEO, but it is increasingly reliant on those same companies for the Artemis Moon program. NASA has already contracted with SpaceX to adapt its developing Starship system to serve as a lunar lander with the prospect of further commercial contracts to come. In the area of military and national security space, Congress has increasingly pushed the USSF to purchase services from commercial companies rather than build and operate its own satellite systems. Space tourist launches are even becoming a regular occurrence—not only has Blue Origin launched several groups on short suborbital jaunts, SpaceX has sent one commercial crew into Earth orbit (the Inspiration 4 mission) and two others (Axiom-1 and Axiom-2) to the ISS, the first private missions to do so. In the near future, Boeing will test its Starliner system with NASA astronauts on board and SpaceX will demonstrate further commercial firsts, planning the first private spacewalk. Though it took many decades longer than initially envisioned for a commercial space industry to develop, it has now become vitally important for future US space policy.

State of Commercial Space Today

While some of the achievements of the commercial space industry have already been noted, this section briefly summarizes the major capabilities as they stand in the early 2020s and what is expected to develop in the next three to five years. It is important to note, however, that though several of these capabilities are planned to come online soon, predicting the pace of space development is notoriously difficult. Further, commercial companies can change and adjust their direction far quicker than government agencies such as NASA or the military as they are not subject to legislation and appropriations that can take several years to work out. Thus, what space companies may be planning today for tomorrow, might never develop, might take longer to develop, or evolve into something totally new and unforeseen.

Human Spaceflight

Human spaceflight is by far the most difficult task associated with space exploration. Not only does the rocket need to work perfectly (or nearly so), but even more complicated systems are needed to keep a crew alive and well. Further, its difficulty and the need to ensure that it will work only increases the cost of space exploration, meaning that historically, only three states have been able to carry it out. When SpaceShipOne flew in 2004, it was the first privately developed system to carry humans to space though the journey was a short suborbital one. Recall from chapter 1 that suborbital flights are those where the rocket does not travel high enough or fast enough to reach Earth's orbit. These flights typically reach the lowest parts of space for somewhere around five minutes before returning to land.

While Virgin Galactic worked to turn the SpaceShipOne system into a commercial one, SpaceX built on its early success with its Dragon cargo capsule to develop a crewed version that could carry astronauts into LEO. Working under NASA's Commercial Crew Program, SpaceX first launched two NASA astronauts to orbit in May 2020 with its Demo-2 mission. The Dragon capsule was docked with the ISS for more than 60 days before returning its two astronauts home. As of mid-2022, SpaceX has carried out seven crewed missions, five for NASA and two private flights. In 2022, SpaceX announced an expansion of their human spaceflight program. Working with Jared Isaacman, a billionaire who previously helped fund and organize the Inspiration 4 mission, the Polaris program is designed to expand commercial spaceflight capabilities and develop new technologies. Among the goals of the Polaris program is to fly the Dragon capsule to its highest orbit yet, carry out the first private spacewalk, and conduct the first crewed mission of SpaceX's new Starship.

With Falcon 9 and its Dragon capsules carrying out work in LEO, SpaceX is also developing its next generation launch system, Starship. Designed to carry more than 100 metric tons to Earth orbit, SpaceX intends it to be a fully reusable vehicle capable of carrying both humans and cargo to LEO, the Moon, and Mars. To do so, SpaceX is also developing orbiting fuel depots, which will be able to refuel Starship for journeys beyond Earth orbit. The company has been developing and building early prototypes out of its South Texas facilities while awaiting FAA approval for orbital flight tests from there.

In addition to SpaceX, several other companies are also investing heavily in human spaceflight. In the summer of 2021, Virgin Galactic and Blue Origin battled to be the first suborbital space tourism company with successful launches. On July 11, Virgin Galactic flew its first official mission with passengers as it carried its owner, Richard Branson, and five others on a suborbital trip with an advanced system based on SpaceShipOne. Nine days later, Jeff Bezos participated in Blue Origin's first suborbital tourist mission with its New Shepard rocket. While both reaching suborbital heights, the two systems are significantly different. While New Shepard follows the model of a traditional rocket launch and flight, Virgin Galactic follows the SpaceShipOne model where the rocket, Unity, is attached to the belly of a mothership, WhiteKnightTwo, which carries it aloft before it is dropped and activates its own rocket to take it into space. The rocket then returns to Earth and lands as a modern airliner does. Virgin Galactic is planning even larger versions of its system. Blue Origin is also developing a heavy-lift rocket, New Glenn, which will also be able to carry passengers into orbit in the coming years.

Two other companies are also involved in human spaceflight. Boeing has participated in NASA's Commercial Crew Program along with SpaceX developing its Starliner, which is designed to be launched on a variety of rockets. Like most major space projects, Starliner has suffered its share of delays with its first orbital mission in 2019 unable to dock with the ISS due to several problems upon launch. A second uncrewed test occurred in 2022, allowing Starliner to proceed with a crewed test later. Though it also initially participated in the Commercial Crew Program, Sierra Nevada's Dream Chaser ultimately lost out on later funding but that has not stopped its development. Unlike the module-type systems that SpaceX, Boeing, and Blue Origin have developed, the Dream Chaser looks much like a mini–space shuttle. Launched on an Atlas V rocket, it could carry between three and seven passengers to space and return to land as a glider like the space shuttle. It currently intends to fly by 2025.

While much of the focus is on getting humans to and from orbit, there has also been a question of what they will do once they're there. NASA is currently planning to operate the ISS with its international partners until 2030 but does not intend to develop a replacement, instead relying on the commercial industry to fill the gap. Much as retiring the space shuttle opened a market for commercial launch, retiring the ISS is also opening a market for commercial space stations. In 2021, a group of companies including Blue Origin, Sierra Nevada, and Boeing announced its intention to develop a station called Orbital Reef, which is described as a "mixed-use business park."[5] In addition to providing early funding for Orbital Reef, NASA has also given money to support the development of two other commercial options, one developed by Nanoracks and another by Northrop Grumman.[6] Another company, Axiom Space, is also developing its own space station concept and hopes to launch in 2025. While it is unlikely that all these projects will ultimately launch, commercial space stations are following a similar developmental path as commercial launches.

Launch Services

If you can launch humans into space, the same rockets can provide general launch services. In addition to SpaceX's Falcon 9 and Starship when it comes online, ULA also provides a significant number of services with its expendable Delta and Atlas launch vehicles. They are also in the final stages of developing a new reusable rocket, the Vulcan, a two-stage heavy-lift rocket that is designed for national security launches as well as human spaceflight with the Starliner and Dream Chaser. Vulcan also takes advantage of engines developed by Blue Origin who itself is developing a heavy-lift competitor, New Glenn. New Glenn is also a two-stage heavy-lift rocket designed for reusability. Though initially projected to be in operation by 2020, the rocket has experienced delays with test launches coming no earlier than 2024.

In addition to these large launch systems, several other companies are working on smaller launch vehicles that primarily launch small satellites including Rocket Lab, Astra, Firefly Aerospace, and Relativity Space. These companies are all relatively newer startups that are focusing on technological innovation in the launch industry with improvements including 3D printing, responsive launch, new materials and fuels, and further improvements to reusability. Given that most of these rockets are less powerful than SpaceX's or ULA's, their main focus is on the launch of smallsats and cubesats, sometimes carrying several of them to orbit per launch.

Communication

As noted previously, lower launch costs are enabling greater use of space-based systems and services. One of these areas is in communications. While communications were one of the first identified uses for satellites, in the late 20th century, communications satellites tended to be large and expensive undertakings and therefore carried out by a select few companies and countries.

In the late 1990s with the dot-com boom, some companies proposed newer satellite constellations to provide global internet services leading launch service companies to believe that a large influx of demand was coming. When that petered out, the market for such services did as well. The most recent cycle of innovation in the space market has instead been led by cheaper launch services, which have incentivized more companies to consider communications systems once again.

Communications satellite systems can be built in several ways. As science fiction author Arthur C. Clarke noted in the 1940s, a minimum of three satellites placed in geostationary orbit would be needed to provide a global communications system. Because of the difficulty of reaching that orbit and other technological challenges in the early space age, the first communications satellite systems, such as those pursued by Comsat, were a higher number of satellites placed in MEO. Because satellites placed in MEO would not appear stationary over a single point on Earth, more satellites, typically in the 20–30 range, are necessary. Like satellites in MEO, satellite systems in LEO would require even more satellites to provide global coverage since they would be orbiting Earth at a faster pace than satellites in GEO or MEO. Though more would be necessary, however, communications satellites in MEO and LEO provide lower latency meaning that signals would not have to travel as far from Earth to a satellite and back to Earth.

While Comsat started to build a communications system in the 1960s, other communications systems developed along the way including Inmarsat to provide satellite communications to those at sea, Iridium, and specialized military systems. Additionally, satellites were also developed to distribute television (DirecTV and Dish notably) and radio (Sirius and XM, which later merged to form SiriusXM). Though satellites were initially being developed to provide portable telephone services in the 1980s and 1990s, their expense and the development of cheaper ground-based cell services reduced consumer demand for such pricey phones.

Today, communications systems are being developed to provide internet service globally to allow users to tap into an even wider array of services in the Internet of Things. Partially as a means of funding the development of Starship, SpaceX introduced its Starlink system in 2015. Starlink intends to launch a constellation of over 40,000 small satellites into LEO to provide internet and communications services globally. Launched in batches of approximately 50 from its Falcon 9 rocket, as of mid-2022, there are more than 2,300 Starlink satellites in orbit, which provide service to more than 250,000 customers. In addition to Starlink, OneWeb is also in the midst of building its own constellation. Founded in 2012, OneWeb's predecessor, WorldVu, initially was in talks with SpaceX to develop the constellation. After a falling out and a bankruptcy, OneWeb reemerged under British ownership to start deployment of more than 650 LEO satellites. With more than 200 satellites in orbit in 2022, its launch cadence was significantly affected when Russia's invasion of Ukraine forced it to cancel a planned launch on a Russian Soyuz rocket. Finally, Bezos's Amazon is also planning a constellation called Kuiper designed to include more than 3,000 LEO satellites. Though launches have yet to begin for Kuiper, in 2022, Amazon purchased launch services not just from Blue Origin but from ULA and Europe's Ariane rocket for the satellites.

While the development of multiple communications systems is a positive step toward economic competition and the development of a market, the sudden influx of so many satellites in LEO has been a particular concern. Not only does it increase congestion in an already crowded area of space, but it also raises concerns of space traffic management. The astronomy community has also showed significant concern as so many satellites moving at such a quick pace often interferes with the viewing of the night sky. Starlink, OneWeb, and Kuiper have all stated that they are making changes to minimize the effect that their satellites will have on astronomy, for instance, by utilizing darker paint to minimize the light being reflected off them. The operators have also had to develop ways to remove the satellites from orbit at the end of their lives to ensure that congestion and debris is kept to a minimum.

Remote Imaging

One final area of commercial development in space is remote imaging. Though Landsat pioneered this area of space utilization in the 1970s, viewing Earth from space has proven to be vital in a number of areas. The first most obvious area of interest is in terms of national security. Recall from early space history that the Eisenhower administration was primarily concerned about developing satellites to spy on the Soviet Union without fear of being shot down. As such, these types of surveillance systems were among the first kinds of satellites to be developed. As we will see later, national security concerns continue to color policy for remote imaging and often hamper the fuller development of a commercial market.

In addition to national security concerns, looking at Earth from space now provides a whole host of information that has proven vital. Remote imaging assists farmers in planning and managing crops, provides valuable geologic data on things like volcanoes, provides monitoring in case of natural disasters like hurricanes, floods, and earthquakes, helps identify locations and sources of rare Earth minerals, helps state and local governments plan and manage development, and monitors major changes on Earth connected with global climate change. All these functions provide significant economic impact.

While Landsat historically provided most of this data for civilian and commercial use in the United States, the US reluctance to improve the resolution of its imagery often forced other countries to develop their own remote sensing systems like France's SPOT. Remote imaging companies also sprung up elsewhere given strict regulations in the United States regarding how sensitive commercial remote sensing imagery could be and who it could be sold to. With changes to those policies in the 1990s and the technological improvements already noted in this chapter, the United States experienced a resurgence in this commercial area with companies like Planet and Maxar developing constellations of satellites to provide commercial services that added to already established companies like EOSAT. As remote imaging services have become more widely available and cheaper, this type of data is being put to even greater use monitoring commerce, traffic patterns, and even violations of fishing laws. Individual consumers now have access to both updated and frequent imaging services via apps like SpyMeSat that allow consumers to purchase archived satellite imagery as well as task satellites for new imagery.

As these systems have improved in terms of sensitivity and availability, even the United States government is beginning to purchase more remote imagery data than ever before for national security purposes. In 2022, the NRO, the premier agency involved in collecting national security remote imaging data, awarded three commercial companies (Maxar, Planet, and BlackSky) major five-year contracts.[7] This represents a significant shift in the national security community away from developing and operating satellites to purchasing commercial services. Despite this, there is still likely to be a heavy government presence in operating these types of satellites given the highly classified nature of surveilling other states and ensuring national security. Further, the use of commercial satellite imagery in the war in Ukraine only reinforces the importance of having a strong commercial satellite industry in the United States.

Policy Challenges

Though the United States government has tried to anticipate commercial space developments since the 1980s, the increasing commercialization of space continues to pose new and unique challenges to legal and regulatory frameworks. Here, we will consider four of these: regulating commercial space, further privatization of American space activities, the increasing power of private space companies in setting and influencing international norms, and the increasing tension between security and commerce.

Regulations

In the earliest commercial space legislation in the 1980s, Congress placed responsibility for regulating commercial launch with the DOT, specifically, the FAA. Given this early action, regulations on commercial launch are perhaps the most well-outlined regulations regarding commercial space. Though the FAA and its Office of Commercial Space Transportation have had a set of launch regulations for some time, they very much assumed both a lower launch cadence and more traditional expendable launch vehicles. As such, they have proved to be rather outdated given the nature of new commercial space companies, many of which would like to move faster in obtaining permission for launch with new types of technology.

As a result of these pressures, the FAA updated and streamlined its commercial launch regulations in late 2020. Taking four different sets of regulations and combining them into one, these updates were intended to make it easier for commercial launch companies to complete the licensing process. Among the revisions were rules allowing multiple launches on the same type of rocket under a single license rather than requiring companies to obtain a license for each launch. Additional changes regarding safety also offer companies multiple ways of ensuring safety goals rather than having a specific method prescribed to them. Even though the space industry has been supportive of the new regulations, both they and others recognize there are further challenges.

One such challenge has been the review of new launch and landing sites. To date, most commercial companies have launched out of previously established sites like Kennedy Space Center, Vandenberg Space Force Base in California, and NASA's Wallops facility in Virginia. However, SpaceX has been seeking permission from the FAA to launch orbital test flights of Starship from its South Texas facility in Boca Chica. To grant permission, the FAA must, among other things, assess the suitability of the site to ensure that launches will not harm the surrounding area or the people around it. This is particularly important given that rocket launches are not guaranteed to be successful each time. While the FAA gave permission to carry out suborbital test flights of Starship test articles, it also began an environmental review in November 2020 to determine whether an orbital launch could take place. The report was expected to be issued within a year but by the time it was finally released in June 2022, it had already been delayed half a dozen times. While the report found no significant environmental impact, it did specify more than 75 actions that SpaceX must take to ensure full environmental protection. More importantly for SpaceX though, delays in the report also amounted to delays in its Starship program as it could not conduct further tests.

For a company seeking to move fast in its rocket development, the delays have been difficult, and Musk has not been quiet in stating his frustration with both the FAA and continued regulatory hurdles. Musk has also demonstrated a willingness to violate federal regulations on occasion. In December 2021, a high-altitude test flight of a Starship prototype violated its FAA license and launched anyway leading to an FAA investigation and increased oversight on future launches. These difficulties only go to demonstrate the difficulty that government regulations have in keeping up with fast-moving technological developments despite the 2020 revisions.

Launch is only one aspect of commercial spaceflight requiring regulations—with the growth of space tourism, questions are also being raised about future regulatory regimes. While laws like the Commercial Space Launch Competitiveness Act of 2015 extended the space tourism learning period until October 2023, the industry has worked to develop voluntary standards in the meantime. Even in the absence of government regulations, commercial companies have placed a premium on the safety of their passengers given that one launch failure could be very detrimental to their overall business model. While it is difficult to forecast what specific regulations may result if the learning period is not extended, a report to Congress by the Congressional Research Service highlights several other policy issues facing lawmakers in this area including whether any fees or taxes should be assessed on tourist trips and larger environmental impacts.[8]

As highlighted in chapter 8, the FAA is not the only agency involved in regulations. The FCC oversees and licenses the operation of satellites in space. The FCC has jurisdiction in this area because satellites must use specific radio band waves to send and receive data, and the FCC oversees these types of communications. As part of its licensing requirements, it also considers debris remediation plans to ensure that satellites can be removed from their assigned orbits at the end of their lifetimes and that satellite operators have processes in place to ensure a minimum of debris creation. The DOS is also involved in enforcing ITAR. The purpose of these regulations is to keep potentially sensitive data and equipment from being exported to bad actors. Over time, ITAR regulations have proven to be rather troublesome for space operators depending on whether satellites and rockets are classified as munitions. When they are, much more scrutiny is placed on American companies that are launching satellites on foreign rockets or American companies that are launching foreign satellites.[9] Finally, under policy shifts during the Trump administration, the Department of Commerce is also projected to be the lead agency not only in terms of tracking objects in space (SSA) but developing regulations for space traffic management.

The number of actors and issues involved makes an already complicated policy area even more difficult to maneuver through. As such, some have proposed centralizing authority for regulating space activities in one office or organization to prevent duplication of effort and to further streamline regulatory activities. Key questions to be considered with such a move would be whether responsibility should be centralized in an organization that already exists or whether an entirely new entity would be required, the type of funding that may be required, where the expertise for such regulatory authority is located, and who that organization should report to. Historically, when offices are located further "down" in agency hierarchies, they do not have as much authority or receive as much attention. While having such divided regulatory authorities may have worked in an era where space activities were not as frequent or numerous, such a situation may become increasingly untenable soon.

The challenge facing commercial space policy moving forward, then, is whether regulations can keep pace with the companies themselves. Government bureaucracies tend to move slowly and cautiously—something that is heightened when multiple government agencies are involved. While the regulatory environment has been key to promoting commercial space to this point, policy will need to quickly adapt to further encourage innovation and development.

Privatization of Space

While we defined commercial space previously in this chapter, *commercialization* of space is often confused with *privatization* of space. Privatization is a distinct process that involves the transitioning of services from being provided by the government to being provided by the private sector. One easy example is picking up trash. Some cities pick up trash from their residents using trucks bought by the city and operated by city employees. However, other cities purchase those services from private companies who provide their own trucks and employees. Proponents of privatization argue that providing services in this way helps reduce costs for governments while improving the level of service to citizens. While this might be true in some areas, there is also some contentious debate over privatization. If there is only one company that can provide a particular service, cost may go up in the absence of competition. Additionally, there might be a question of equity, whether a business will serve all citizens equally—this question arises particularly when adoption or other social services are provided by religiously affiliated groups. Finally, it may not be appropriate to privatize some services such as the military and law enforcement.

Undoubtedly, the line between commercialization and privatization can be blurry. In the 1980s, there were some proposals to privatize the space shuttle by selling it to a contractor to operate on behalf of the US government. Though the *Challenger* disaster scuttled that proposal for a time, much of the servicing of the shuttle was eventually outsourced to private contractors. With respect to the state of human spaceflight today, some may argue that it has already been privatized as SpaceX, and soon Boeing, provide flights to the ISS on behalf of NASA. While this is true, it does not meet the full definition of privatization as SpaceX and Boeing are flying *NASA astronauts* to the ISS and not their own employees acting on behalf of the government. For American human spaceflight to be fully privatized, these launch service companies would also fly their own astronauts who would in a sense be acting as representatives of the United States. Similarly, to take other government space activities private would require the government to stop contracting for, building, and operating satellites and instead rely on private companies to completely provide those services. Existing government-owned and -operated satellites would also likely be sold off to private companies to fulfill the privatization paradigm, much as Landsat was in the 1980s.

While there is little doubt that military- and national security–related space activities are unlikely to be privatized, there has been some suggestion that the success of the private industry may at some point mean that civilian space policy including human spaceflight might be fully privatized. In this ideal-type hypothetical, the money budgeted for NASA would instead be given to a private company or companies that would then carry out American policy. While this might save money given the cost savings that private space companies are introducing, it also introduces some deep questions about the nature of democratic governance. In a democratic republic such as the United States, a key defining feature is that elected representatives of all citizens get to determine what the United States does and how it does it. If money is simply given to a private company that receives a vague direction of what to do with it, is that democratic? Even aside from questions about democracy, would a SpaceX astronaut, even if they had an American flag on their spacesuit, stepping foot on Mars have the same effect that a NASA astronaut working for the American government directly would? Much of the international pride and prestige that comes from operating a high-technology and expensive activity like spaceflight comes from the fact that the *country* is directly carrying it out. While some would respond that it takes the country to support and grow a company like SpaceX, is it really the same thing?

While privatization of space is unlikely to be fully carried out in the near or medium term, the increasing success of private space companies will likely lead to further consideration of how cost might be reduced to the federal government. Though cost is a major motivating factor for privatization efforts, it is not the only one nor the primary one particularly in a democratic country.

Rising Power of Commercial Companies

As we have discussed elsewhere in this book, one of the major issues facing the space domain today is the status of international law, rules, and norms. When the OST was signed in the 1960s, it only contemplated a role for states, much as other international law historically has. Over the ensuing decades, non-state actors have been an increasingly powerful force around the world. Globalization has not only made the world "smaller," but it has increased the economic power of multinational corporations whose activities span the globe. These increasingly powerful companies can influence terrestrial politics in any number of ways including moving their businesses from one country to another depending on favorable government policy, funding campaigns, influencing elected officials, or conducting public relations campaigns. Technological changes have aided this growing power as things like the internet have further enabled non-state actors to have increasing power in the global arena.

Though private companies and other non-state actors have amassed much power over the past decades, international law has generally been defined and advanced enough so that they have had minimal impact on it. The ways in which companies, countries, and international law interact, in other words, have been firmly established. In space, this situation is somewhat different. Not only is international law not as well defined, but the technological power and capabilities that private companies have on hand is substantively different. While we generally treat launching something into space as a relatively common occurrence, those somethings can be incredibly powerful with many of their capabilities hidden. The fear that the Soviet Union had launched a weapon into space in the guise of Sputnik helped contribute to the public panic that followed it. The seriousness of such capabilities has continued to pressure governments to find ways not only to defend against attacks from space but to carry them out in the form of kinetic and non-kinetic weaponry. That private actors have the power and capability to not only launch a rocket but launch something potentially dangerous gives them significant power. To be sure, we are not saying or arguing that companies like SpaceX or Blue Origin intend to do such a thing on their own behalf. Both companies have expressed nothing but peaceful intentions for the future exploration and settlement of space. But the fact that they *could* is still important.

Even aside from recognizing the sheer power of the private companies' capabilities, the fact that they are going to space and operating in space more frequently gives them more of an opportunity than many states to influence the ways in which actors interact in space. For instance, in late 2021, China expressed its displeasure to the United Nations when a Starlink satellite flew too close, it believed, to its space station. In the absence of internationally recognized rules regarding space traffic management, SpaceX will play a vital role in establishing the types of behaviors that will be considered routine and acceptable by how it chooses to respond in such a situation. In areas of the space domain that are entirely unexplored or new, the first actors there and what they do will exert a large influence on what happens next, something often called the first-mover advantage. As these companies make more expansive plans and carry them out, the chances that a private company is the one that makes the first move increase giving it further power in setting norms of behavior.

Even with this growing power, some argue that the OST is enough to regulate the behavior of private actors in space since states must oversee their behavior. To date, that has been the case. However, there is nothing that keeps any company in a country—if another country offers a company a more favorable operating environment, they could easily move. Though each scenario would come with a significant cost, it would afford private companies even more freedom of action to assert their power. This would also be the case as private companies look to creating settlements on the Moon or Mars that would limit the states' authority.

As the importance of space continues to increase, the international community will have to face these questions much as the United States must consider the proper role for commercial companies in its own space policy. While states may still be the primary actors, novel ways of granting them standing in the international arena may be necessary along with ways of enforcing rules of behavior outside state boundaries.

Commercial Services and National Security Concerns

In February 2022 as Russia invaded its neighbor Ukraine, the evolving power of space, even for countries that do not have a significant presence there, became increasingly clear. In the weeks and months leading up to the invasion, remote imaging showed a major military buildup by Russia on the Ukrainian border. This imagery, often provided by commercial companies, was used not just to confront Russia on its military intentions but to "pre-but" misinformation anticipated to come from Russia. As the invasion got underway, this imagery was key in alerting Ukraine as to where the forces were coming from, where they were going, and what capabilities they were bringing with them. The commercial imagery proved so pivotal that Ukraine's vice prime minister pleaded with these companies to provide them more.[10]

But that was not the end of Ukraine's use of space. Just as Russia invaded, they conducted a cyberattack on one of the more heavily used communications satellites system in the region, Viasat, knocking out key communications for Ukraine. On February 26, a Ukrainian official tweeted at Elon Musk requesting Starlink terminals and access, and within hours, Musk responded that "Starlink service is now active in Ukraine. More terminals en route." With hundreds of Starlink terminals in Ukraine, not only were communications and internet access restored, but the Ukrainian military has been using the system to carry out drone attacks. In response, Russia threatened SpaceX and its Starlink satellites arguing that they represented legitimate military targets given their use for military purposes.[11] In early 2023, SpaceX moved to restrict how the Ukrainians use Starlink to potentially limit threats to the system as a whole.[12]

This situation has demonstrated how countries that are not space powers can still have space power via the commercial sector. While commercial services have no doubt strengthened US national security and military capabilities, the use of commercial assets in such a way does introduce serious complications. In response to the introduction of Starlink in Ukraine, Russia threatened SpaceX with attacks, which Musk also predicted was a likelihood in a later tweet. Despite sustained cyberattacks against the Starlink system, SpaceX has been able to resist Russia's efforts to jam and hack it. Thus, one of the dangers of using such assets for military purposes is that, despite international law restricting attacks on civilian systems, they now become targets for attack. Russia, in doing so, would argue that such attacks are legitimate because of the way in which Starlink is being used, but others would also argue against the attack since military use is not the only, or even the most significant, use of Starlink.

From the perspective of the military, US or otherwise, using commercial assets seems like a no-brainer—it adds to their capabilities, increases resiliency and redundancy of their capabilities, is often cheaper, and can cause the adversary to think twice about attacking space assets that are commercially owned and operated. It complicates the calculations of an adversary that, if it wishes to degrade military capabilities, must attack a much larger number of satellites. At the same time, though, it increases the danger of operating in space for the companies themselves and may make them less likely to work with governments if they know that their systems may be targeted or attacked. This may be reinforced if attacks are likely to incur a significant economic cost for the company or even impose economic loss on the industry or country.

While this scenario poses a particularly thorny international problem, there are additional issues when it comes to national security and domestic space policy. This has played out particularly in the area of remote sensing. With the increasing capabilities of commercial remote sensing satellites, some companies may choose to sell sensitive imagery of military or otherwise noteworthy sites or activities to potential hostile actors. Given this, US regulations have often considered how to both protect US national security and encourage a commercial market. Unfortunately, these regulations have not always been successful at doing so. In the late 1980s and 1990s, regulations took the form of limiting the resolution of imagery as well as who could purchase such products. Difficulty in receiving operating licenses along with these rather strict rules encouraged emerging companies to move their business elsewhere to avoid such onerous restrictions. While restrictions have been relaxed over time, the damage was already done—by the start of the 21st century, the US remote imaging market was falling behind.

As foreign remote sensing companies not only advanced their technology but made it available worldwide, such strict regulations no longer made sense—if good images were available elsewhere, prohibiting American companies from advanced imaging simply served to hurt the market. As such, remote sensing restrictions have been loosened and streamlined over the past two decades with the latest updates coming in 2020. The latest regulations allow US companies to operate under the "bare minimum of conditions" if their capabilities are no better than foreign competitors.[13] While this doesn't necessarily encourage American companies to pursue advanced technologies, it also doesn't unfairly restrict them either.

In both situations, quickly advancing commercial systems are challenging long held ideas about national security and military use. Given the deep relationship between space and national security, the entrance of independent, commercial actors adds another degree of complexity to space policy. While the US (and other countries) must look to protect its own security, this must also be balanced with encouraging a viable economic market, which can further contribute to both economic and national security.

Conclusion

Commercial space companies have proven to be perhaps the most important factor in space policy moving forward. Regulations and policy regarding them have not kept pace with the rapid development and even faster expansion of capabilities. The US government is likely to struggle soon as it tries to not just keep up with the space industry but get ahead of it. There is also a danger that the growing importance of the commercial space industry prevents the adoption of policy that is necessary to also ensure good behavior. Their growing economic power and the increasing reliance of the US government on them, puts members of the commercial space industry in an increasingly powerful and influential position.

While this chapter has outlined the state of commercial space and the policy challenges moving forward, perhaps the larger theme is how to integrate commercial space into governmental space activities and how to balance competing concerns. Both NASA and the US military have historically been reticent in allowing commercial providers to take over areas that they had previously controlled. This is due to several different factors including the protection of activities that they traditionally dominated (along with the appropriations that came with them), the inability of the commercial sector to deliver on earlier promises, lack of trust in new companies with new ways of doing business, and uncertainty about the status of commercial actors in space. As a result, it has only been in the past ten years or so that private companies have had a real opportunity to demonstrate what they can do. With an increasing pace of activities in both military and civilian space policy, both NASA and the US military are now trying to transition to using more commercial services, but the process has not been easy. Established processes and procedures are often major roadblocks, and fast-moving commercial companies do not like the long timelines government agencies often operate on. There will likely continue to be a period of transition as both sides understand what the other brings to the table.

Even with the rise of commercial providers, the US government must still balance national security concerns with economic concerns. While bolstering an emerging market is an important policy goal, it may often conflict with national security needs. The commercial market may not be able to provide all the capabilities the military needs nor should they—things like missile warning and nuclear command and control may prove too important to be left to private companies. At the same time, relying more on commercial capabilities leaves questions about their status in times of conflict open. While the US can work with its international partners to try to shape the future of international law in space, it must also recognize that commercial companies are increasingly forces of their own, which will complicate the policy landscape. Even with all these open questions, there is no question that commercial space is opening up the domain to all sorts of new players and increasing public excitement about what will come next.

Summary Points

- Although legislation and regulations in the 1980s began to foster a commercial space industry, the high cost of launch and the lack of customers beyond the government hampered its development.
- Commercial space development was spurred in the early 2000s by the establishment of new space companies like SpaceX and Blue Origin that began pursuing reusable rockets. They were helped by plans to retire the space shuttle and the forced creation of ULA.
- Though NASA and the US military were initially hesitant to use the new space companies, the reduced costs that they offered along with an increasing number of demonstrated capabilities has enabled the rapid growth of the commercial space sector.
- Commercial space companies today now offer a range of services from human spaceflight (orbital and suborbital), space-based internet and communications, and remote sensing.
- Challenges in commercial space policy today include the regulatory environment, the extent to which space is or should be privatized, how to integrate increasingly powerful commercial companies into the space domain, and implications to national security of commercial space companies and capabilities.

Discussion Questions

1. What factors have enabled commercial space development? What did the government do to create the appropriate policy environment?
2. Are there parallels between the space history of the Soviet Union and the United States and today's emerging commercial competition between SpaceX and Blue Origin? What are the benefits of competition?
3. Should space exploration be privatized? Why or why not?
4. How can companies influence international laws and norms of behavior?
5. Are commercial satellites appropriate targets in times of conflict if they are being used for military purposes? Why or why not?

For Further Reading

Clarke, Arthur C., "Extra-Terrestrial Relays: Can Rocket Stations Give World-Wide Radio Coverage?" in *Exploring the Unknown Volume III: Using Space,* John Logsdon ed. (Washington, DC: NASA). https://history.nasa.gov/SP-4407/vol3/cover.pdf.

Davenport, Christian, *The Space Barons: Elon Musk, Jeff Bezos, and the Quest to Colonize the Cosmos* (New York: Public Affairs, 2019).

Garver, Lori, *Escaping Gravity: My Quest to Transform NASA and Launch a New Space Age* (New York: Diversion Books, 2022).

Slotten, Hugh R., *Beyond Sputnik and the Space Race: The Origins of Global Satellite Communications* (Baltimore, MD: Johns Hopkins University Press, 2022).

13

Major Issues in US Space Policy

In this book we have covered a range of aspects concerning US space policy. The complexity of the space enterprise continues to grow over time as the number of actors and the importance of space for states grow in the security and economic arenas. While one book cannot cover every aspect of space, we have endeavored to provide insight on policymaking across the entire policy process. Additionally, we have presented an introduction to the actors involved in space policy, providing a sense of their roles and responsibilities. Still, there is a need to discuss further some of the practical problems associated with major issues in space policy. This final chapter revisits many of the problems previously identified, but with an eye to understanding the basic dilemmas of the problem, the policy options available, and the implications of those options. There are currently five issues we think merit more discussion given their importance to space policy. First, is the problem of how to address space debris. Second, is the problem associated with the slow development of the space regime in terms of stabilizing norms. Third, is the problem regarding the limitations currently present with respect to commercialization and resource exploitation in outer space. Fourth is the problem of the policy issues associated with the US-China rivalry. Fifth is the last issue examined in terms of actual policy, which is US leadership.

There are common themes associated with these major issues. First, none of the issues can be addressed solely through domestic space policy. As space is part of the global commons, space policy is inherently linked to the behavior and policies of other actors beyond one's borders. Second, there is a sense that these problems are very acute and time-sensitive. In other words, there is often a perception among policymakers that if not addressed soon, US policy will be behind the curve and have to be responsive instead of proactive. Third, these issues are complicated domestically in terms of how they are perceived. US policymakers tend to ascribe all these issues to concerns regarding US national security and strategic interests. This perspective biases the types of policy decisions made on these issues. In the remainder of this chapter, we explore each issue along these themes.

Space Debris

It did not take long after the launch of Sputnik to realize that the space domain might soon suffer from the effects of space debris. While there was no immediate concern about the routine creation of debris since space access remained limited to a small number of actors, the fouling up of LEO through weaponization was something that both the United States and USSR did not want to happen. As we saw previously, this mutual interest led to a host of agreements that created policies for both states to shape their behavior in ways that limited the likelihood of such an outcome. Today, the relationship between spacefaring nations is such that the developed norms of the Cold War still mostly remain in effect.

The space debris issue has come to the fore over the past decade as the number of actors responsible for littering space has increased tenfold. Traveling at speeds of over 17,000 miles per hour, even the smallest piece of debris can cause considerable damage to spacecraft. Spacewalks on the ISS have been curtailed or canceled due to the danger of debris in orbit. Additionally, the ability to track debris is increasingly complicated. While there is a robust space domain awareness network, with the United States and its partners able to track over 30,000 pieces, this is still a fraction of what is on orbit that could potentially be lethal to people or satellites.

In terms of response, the United States remains proactive in developing policy that limits the amount of space debris created. Owners of satellites in the United States are required to de-orbit their platforms responsibly by shifting them to what are called graveyard orbits or having them burn up upon atmospheric reentry. These regulations can facilitate debris mitigation in the United States and are driven by a robust, regulatory structure led by actors like the FCC. Most recently, the United States also decided to unilaterally impose a moratorium on debris-creating behaviors in space though this action is mostly an effort to prevent or reduce the number of anti-satellite weapons tested in space. It is the hope of the United States that its self-imposed moratorium will put pressure on other states to conform, potentially developing a norm that reinforces this behavior over time. As discussed later, US leadership presents its own challenges in terms of getting others to cooperate on space problems. Some states like Germany and Japan, have agreed to the moratorium, leading to at least some success on this US effort. That stated, it is easy to agree to a moratorium on something for which a state has no plans of carrying out. Additionally, as we have seen, Russia, India, and China have conducted relatively recent ASAT tests, all creating additional debris in LEO and all at least for now unwilling to give up their right to do so.

With space debris, the challenge will always be one in which domestic policy alone does not effectively solve the problem. While the US not creating debris is meaningful given its role and importance in space, it is much less consequential and arguably irrelevant if the other major spacefaring powers are not invested in mitigating the space debris problem. This is certainly the case regarding the problem of debris creation in the first place, but it is also the case in terms of cleaning up what already is there. Legally, debris remains the property of the state that put the asset into orbit, thus countries are not free to simply remove the debris of another country without permission. This legal obstacle likely will remain important even as technology evolves that may help mitigate the problem. JAXA, for example, has worked on what is known as a space vacuum cleaner, a space vehicle that could essentially "sweep up" debris that has accumulated in LEO. If this system worked, would other states accept the system being employed? While the US might be accepting of this capability given its close relationship with Japan, what if Russia developed such a system? What policy might the US enact to deal with this issue? Seemingly, the removal of debris would be good for all, yet any space asset that can remove debris likely would be capable of doing harmful things as well. This is the conundrum that awaits even a technological solution to the problem of space debris. Who gets to do the cleaning up? In short, while the space debris problem requires both sound domestic policy and international cooperation, there will be difficulties in achieving success on both debris creation and removal.

Time appears to be limited in addressing the debris problem as well. The amount of debris has reached levels that are making tracking almost impossible. There is uncertainty on the number of debris needed for LEO collisions to cascade into a Kessler effect, creating a sense of urgency by many in the United States and others in the international community to deal with the problem before it's too late. Yet, there is not uniformity when it comes to the space debris debate. There are many that subscribe to the "big space" approach that suggests it is difficult to appreciate the size and amount of literal space that exists in these orbits. Pictures that show how congested the domain is, for example, do not portray an accurate relationship between the amount of area that exists between objects, creating false impressions on policymakers about the urgency needed to address the debris problem. For those taking this approach, the Kessler effect is something that may happen, but not a guarantee and not seemingly something that will take place soon.

For policymakers, why does this matter? Access to space is first and foremost among space policy priorities since nothing else is possible in the domain without it. If the debris problem is immediate, then it must be addressed as an urgent global problem. If the debris problem is less definite in terms of its actual effects over the short and medium term, then other policy priorities can move ahead of it. The fact that there is a great deal of uncertainty on the debris problem, added to a lack of concern by many of the major space powers in creating debris, and the challenge of a mutually agreeable plan to deal with existing debris, the sense of urgency from the US perspective is unlikely to create significant changes in others' behaviors. Does this mean the United States needs a more aggressive effort to effect change in global debris mitigation? How would the United States do this? Is there a unilateral approach that could be taken, akin to a space nationalist perspective? For example, should the US employ its own debris removal system despite the international legal challenges that come with it? After all, if the United States only targets debris that is potentially harmful to all space actors or debris it "owned," could this type of policy set the stage for a new norm of space debris maintenance led by the United States?

In the space debris case, there is a clear connection between US security and strategic concerns, and as such it is not surprising that US policymakers and military leaders seem to view debris from this lens. The access to space is paramount for the US projection of power. Any debris problem that would limit this capability reduces the US ability to operate its military most effectively. The larger issue in terms of perceiving the debris problem from this perspective is whether this limits the range of options open to policymakers. Going back to the idea of a "space vacuum cleaner," would it be better, for example, to consider a commercial approach instead of a state-driven one? Would a national policy that promoted private development work in ways that a US government project could not? Or, tying this to the theme of independence, could the US and the international community reach an accord that would allow an international consortium the opportunity to do this for the benefit of everyone? While the US interest at present centers on security, it is also the case that there has been strong policy promotion of private enterprise as well; the mitigation of debris would be a necessary requirement for a fully developed commercialized space sector that operated alongside military and civil space missions for the United States.

Commercialization of Space: Space Traffic and Resource Exploitation

US space policy is in place to deal with commercialization of space as it relates to the use of LEO. Much of commercial space to this point focuses on satellite communications and the general provision of information through GPS via space-to-ground data links. Space launch is also fairly regulated at this point, with good coordination between the government and private industry on a range of safety and procedural issues that have effectively supported a broader US approach to maintain access to space for commercial, civilian, and military use. Commercialization in LEO, however, is just the beginning for many private space visionaries. As we learned in chapters 9 and 12, SpaceX, Blue Origin, and other commercial actors have plans for space that go well beyond this. For US space policy, the speed of commercialization is challenging the government's ability to keep up with it.

Even within LEO, the development of smaller microsatellites has begun to put thousands of satellites in orbit over small periods of time, creating challenges for STM. The policy decision to move STM away from the USSF to the Department of Commerce just as this revolution in satellite technology is taking off leads to questions regarding the ability of the Department of Commerce to effectively manage a growing area with less resources available than the DOD. One policy approach to consider will be whether the Department of Commerce can work with interagency partners like the DOS to facilitate international cooperation in STM. Potentially, it could do so in a more benign way than a military organization could, but this is only accurate to the extent that policy allows for coordination and cooperation with other states and actors that have the potential to place so many satellites in orbit over very small periods of time. SpaceX, for example, launched over 3,000 small satellites into LEO in just over three years to start its satellite internet constellation Starlink beginning in 2019. While a domestic actor governed by current US regulatory policies, we should expect non-US actors to try and accomplish similar goals, placing additional burdens on the US to maintain its domain awareness with an effective STM mechanism in place. The China Academy of Space and the China Aerospace Science and Industry Corporation (CASIC), for example, are reported close to completing two new commercial facilities that would allow for the introduction of more than 400 small satellites into LEO each year.[1] These efforts are in addition to the government's state-owned enterprise, the China Satellite Network Group, that plans on rivaling SpaceX's Starlink with a 13,000 internet satellite constellation of its own in LEO over the next few years.[2] Given the lack of cooperation with China on space issues, US policy will face difficult choices on the path ahead if space domain awareness and traffic management remain paramount to broader US interests.

Beyond LEO, what is the role of US regulatory policy on behaviors concerning the Moon, Mars, and other celestial bodies? What will the US government allow domestic actors to pursue for private, commercial gains, for example? Can a private company mine and sell minerals from celestial bodies? How consistent will international and domestic law be on these sorts of issues? What if other states do not regulate their commercial actors similarly? Will US policy create limitations that hurt US commercial interests relative to other states? As recently as 2015, US policy, namely the Space Resource Exploration and Utilization Act, provides for citizens' rights to collect resources from asteroids, for example, but such a policy may not be consistent with the OST, depending on whether one determines resource collection to be an expansion of sovereignty into space.[3] Other policy implications arise from this in terms of US governmental responsibilities for the actions of commercial entities in outer space. The situation becomes even more complicated when human spaceflight is involved. Do private actors have the right to launch human spaceflight missions beyond LEO? What policy criteria must be in place to allow for this? US regulations and broader policy is not where it needs to be to effectively deal with these issues.

Examining this issue from the perspective of our three main themes provides a bit more insight into the nature of the problem. In terms of the limitations of domestic policy to solve the problem, we have already noted that the US is not in an ideal position with respect to space policy to coordinate as effectively as it needs to handle the enormous increase in commercial space traffic likely to occur in the next several years. There is no widespread standard for STM at present to include the involvement of commercial actors as well as those of state authorities. There have been conversations and discussions of best practices, some of which have been adopted; yet there remains a fairly decentralized approach to the problem, providing a host of gaps in regulation with no actors coming to the fore to provide effective and sustainable policy solutions.[4]

With respect to commercial mining and resource exploitation beyond the STM problem in LEO, one can observe similar policy limitations. Here the issue is even more challenging given the possible inconsistency between domestic and international law. If states take a unilateral approach to regulating their own actors' resource extraction behaviors, it will become increasingly more difficult to reach agreement on commercial actor behavior beyond LEO. If the US is too restrictive in its approach, it potentially weakens US commercial activity relative to international competitors. Given this dilemma, US policy is best served by identifying possible points of coordination with other international actors, both state and non-state to provide a robust framework from which all actors operate. As explored later, this likely requires aspects of leadership that may currently be lacking among the international community.

In terms of the acute nature of timing for commercial space policy, is it possible that the US is too late to shape the landscape, forcing policy to be reactive rather than proactive? There are two ways to examine this theme in the context of commercialization. First, as it relates to LEO, policy will have to settle for being responsive moving forward. As the most established part of commercial space, non-state actors are in the best position to help US policymakers identify and support best practices so that new policies facilitate and stabilize what's already in place in terms of good behavior. For commercial space, the goal would seem to involve policymaking that allows the industry to be self-sufficient.[5] Efforts to hinder this goal likely limit overall commercial growth for the United States, both in terms of nonhuman and human spaceflight endeavors. In essence, this de facto condition reduces the likelihood of big changes to US commercial policy in LEO as we look into the future. The only exception to this would be something like STM as we have discussed, given the implications of poor management for everyone. Second, there is still time to shape the future of commercial space beyond LEO; time is of the essence, however. Technology makes it feasible to develop an acute need for property and resource rights, as well as a wide range of international law covering problems beyond the range of Earth. It will be much more difficult to negotiate such policies if there is a realized interest versus one that only is imagined. For US policymakers, being proactive as quickly as possible is the only way to ensure that the international legal regime that develops in the context of these new endeavors will be driven by US actions versus ad hoc efforts by others that will not be easily reversed. Returning to the example of property rights on the Moon for example, there may be a policy case to be made that some rights may be afforded to states (e.g., the presence of Moon bases) and by extension commercial actors without violating the non-appropriation principle as put forth in the OST and the Moon Treaty—the latter has not been accepted as binding international law.[6] As explored in the US-China rivalry section that follows, such a concern is likely to happen sooner than later. For US policymakers wanting to shape that agenda, time is of the essence.

Finally, what is one to make of the connection between commercialization and national security? Here the evidence is clear that US policymakers view a very close relationship between the two. We know, for example, that the US military uses commercial communications satellites for a great deal of its communications. This relationship alone creates an interdependence between the commercial sector and national security. Similarly, the US has relied on SpaceX to launch government payloads, again creating additional ties between the two sectors. If one thinks about national space policy more broadly, calls for a whole of government and whole of nation approach to maintaining US leadership and reputation as a preeminent space power will likely keep the connections between commercial space and national security space a priority for policymakers over the long term. This does not come without its own challenges, however. The importance of national security for space policy may create disadvantages for US commercial areas similarly to how it did for the remote sensing industry over many decades. As the US seeks to develop norms consistent with the international liberal order, policymakers may limit the freedom of action of its commercial sector to shape the behavior of other actors. In part, these sorts of policy decisions will be driven by the schools of thought on space policy with which future leaders associate themselves.

The Creation of Space Norms: Not All Norms Are Preferred Norms

In revisiting these schools of thought, one set of major policy challenges involves efforts to push for particular types of norms. We have spent time in previous chapters on liberal norms, those that US policy has embraced for the prescription of behavior in the space domain. Unfortunately, these norms need not be the ones that the US prefers. Referring to the discussion of ASATs, what is the current norm regarding their use? US policy and preferences are for them not to be used. The US is pushing for a norm proscribing debris-producing tests. Despite this push, one can state with confidence that there is either no norm regarding their use, or that ASATs are normatively okay to use. Even the US has only focused on debris-producing tests, suggesting that tests capable of not producing debris would be acceptable. Would this include debris guaranteed to burn up in the atmosphere? Would this include other possible ASAT tests that do not produce debris?

Sticking with our themes, the creation of space norms requires more than domestic space policy. It is obvious that those who seek cooperation through institutionalist approaches must gain that cooperation through rule-sharing, transparency, and a commitment to focusing on mutual benefits. Domestic policy must be consistent with international arrangements. Even those policymakers ascribing to a space nationalist position, however, can only push international norms through domestic policy via example. We note later that such an example is unlikely to be successful without some degree of leadership. Space nationalists would likely have an easier time of norm creation as their approaches to policy would lead to a much more self-interest-oriented environment. Current norms like those associated with the OST would soon be replaced with norms reflecting more balance of power–type relationships. Space nationalist policies increase the presence of competition and conflict-driven norms like arms races and territorial appropriation of celestial objects. This is an important point for those interested in the influence of space policy beyond US borders. While there may be gains relative to others by the promotion of policies that value sovereignty and space supremacy, there is also a potential cost as it relates to the long-term patterns of behavior that likely erode opportunities to cooperate even when it is in states' best interest to do so. One further point to make is that the issue of norm creation works in both directions. While the focus here is on US policymaking and its role in both domestic and international space, when it comes to norm creation, the behaviors of other states also impact US policymakers' willingness to adopt policies reflected in any of the schools of thought.

Given the limited ability of the United States to create international norms alone, many efforts associated with norm creation stem from the generation of principles and rules of space behavior in multilateral forums such as UN-COPUOS. In 2019, for example, the US agreed to support a limited working group to focus on the development of guidelines for promoting long-term sustainability in space, the sharing of best practices and the building of space capacity, and awareness for emerging space powers. This working group includes Russia and China and reflects a US commitment to synchronize domestic and international priorities on space norms. There is still a long way to go with multilateral efforts, as much of these arrangements take a great deal of time to work through. Additionally, while states can come to an agreement on principles, these do not and will not translate into rules or norms without a clear translation into policy. For the US and others, it is the policy creation based on these agreements that remain a challenge.

While norm building takes time, space actors continue to proceed with the pursuit of their interests. In doing so, actors organically create their own norms and standards of practice. This organic process can be in direct competition with the more deliberate attempts to create rules and norms, leading to a sense of urgency for policymakers to do something in order to stand a greater chance of influencing the direction of norm development. In some ways, this can help explain unilateral activities by the US to shape norms despite the limitations of such efforts. In July 2021, for example, Secretary of Defense Lloyd Austin provided guidelines for the armed forces on norms for space military activities. The memo identified five tenets of responsible behavior to include: operating in, from, to and through space in a professional manner with due regard for others; limiting the generation of long-term debris; avoiding harmful interference; maintaining safe separation and trajectory; and communicating to enhance safety and stability of the domain.[7] This policy directive represents the type of effort to deliberately shape the norm environment now versus having it shaped externally without much input.

It is also noticeable that the DOD is the organization driving the development of norms although it is not terribly surprising, demonstrating once again the tendency to associate space policy with national security. When it comes to norm development, there undoubtedly is a holistic, whole of government approach. We observed the DOS in this role in chapter 8. Yet the rationale for much of the efforts to develop norms is based on the need to protect US space infrastructure, particularly that associated with the US ability to project power. What is interesting in this case is to think about why the DOD would pursue this type of policy without any credible commitments by other military actors in the domain to do the same. Perhaps one reason is that it represents a symbolic commitment to a set of norms consistent with the OST and a broader liberal international order. The other reason is a bit more cynical in that the DOD memo does not restrict much in the way of accomplishing any objectives it may have in space, including the use of weapons if those weapons do not produce long-term debris. In any event, such a policy does allow the US an opportunity to call out other actors' behaviors and signals a willingness to adhere to standards that appear responsible.

The US-China Rivalry: Space as a Microcosm

One actor that the US likes to call out in terms of irresponsible behavior in the space domain is China. Its space operations often are discussed as part of a larger challenge to the international order; it is this larger challenge that serves as the basis of the US-China rivalry. Having slowly evolved over the decades since the end of the Cold War, the rivalry has now reached a critical stage for both countries and from a US perspective, dominates much of US foreign policy, including those associated with space. Given what we observed in the Cold War context, this should not come as a surprise. However, there are many problems in the space domain for which even less competition between the US and China would not solve. With both Russia and India using ASATs and seemingly unyielding in their efforts to continue testing them, other states have become more interested in using space-based capabilities to hold other states' assets at risk, project power, or augment other aspects of their national security enterprise. That stated, it is China's rapid development of space-based capabilities, tied to its unique relationship with the United

States, that puts policymakers in the dilemma of how best to engage with this competitor.

In terms of interdependence, the US and China are tied in ways that make it difficult to separate space relations from other aspects. As rivalries often illustrate, what one state does in the relationship directly influences the policy actions of the other. While this undoubtedly is the case for security, it is also the case in terms of economic and technological development in space as well as space exploration. China's nascent commercial space market is an outgrowth of the economic competition between the two countries. The inherent tension between the two approaches to the market is exemplified in the commercial sector, with China tying its commercial activities to state-owned enterprises (SOEs) while the US approach remains more in the hands of private companies. It is this difference that has driven a strong reluctance by the United States to open any avenues for cooperation with anything associated with China in space. As a policy initiative, this is most pointedly exhibited in the Wolf Amendment, discussed in chapter 5. It is also exhibited in the ITAR and EAR restrictions, discussed in chapter 8. The fear is that any cooperation will lead to the theft of intellectual property, with technological know-how being put toward unfair advantages in all space sectors. It is precisely this US reliance and its advantages in space that China seeks to emulate because of America's success. Thus, for as long as the broader US-China competition exists, so too will it exist in the space domain. In some sense, this is helpful; as China becomes more dependent on space for the well-being of its domestic population, its leaders should have a more vested interest in making sure it is accessible and stable. In this way, the ability of both the US and China to operate freely in space are tied to each other's willingness to enact policies that help preserve the domain for future use.

This perspective should be consistent with civilian space exploration as well. While cooperation between NASA and CNSA is forbidden, the two agencies and thus the two countries remain interdependent to some extent given their inherent rivalry. Therefore, as it relates to policy, one observes a certain "tit-for-tat" approach to the agendas and implementation of space policy by both countries. China's long-range plans for Moon exploration and exploitation (namely in the possibility of Helium-3 mining) has put pressure on the United States to speed up its own return to the Moon as well. Both sides have made policy statements that offer their Moon agendas as the first stage of implementing a more ambitious policy to marshal resources for the human exploration of Mars. The US efforts to support the ISS and a follow-on commercial space station with trusted partners has led to China's policy initiative to build its own permanent Tiangong space station. Both countries likely will compete for the possibility of building further partnerships independent of each other as a result. One direct policy question for the US then will be how to address its relationships with those states that might be interested in cooperating with both countries in the space domain, like the United Arab Emirates (UAE). The UAE recently signed both the Artemis Accords and an agreement with China to develop a lunar rover.

Beyond making policy decisions interdependent, the rivalry also creates a sense of immediacy for US policymakers, lest they are forced to create policy that is in reaction to Chinese advances versus setting the agenda. One can view the American self-imposed moratorium on debris creation in this way. While US statements offer this initiative as an effort to set the agenda, the earlier actions by Russia, India, China, and arguably even the United States have made this a response to a problem versus addressing a potential problem that has yet to arise. Across commercial, civilian, and military space, US officials have struggled to get ahead of the curve in ways that are reminiscent of the Cold War space race. Like the Soviets, China seems to be making the space headlines, only interrupted by occasional successes of US commercial space. While there is an argument to be made that this is not necessarily bad for US space policy, it will be unlikely that US commercial space can directly compete with China and the CNSA. This inability to compete leaves the need for more initiative from NASA, such as we have seen with its partnership with SpaceX, but as we observed in earlier chapters, domestic constraints make this more challenging than what the CNSA faces in China.

Nonetheless, the fears that the space race has not only begun between China and the US but may be passing American policymakers by is very real. This is evident in some of the data published across various US governmental organizations. The DIA, for example, publishes a report titled *Challenges to Security in Space* that identifies and discusses the main threats to US space capabilities from rivals such as China, Russia, Iran, and North Korea. The most recent publication makes evident the speed at which China's space programs have gained ground on the US in terms of being viewed as a peer competitor. The country has utilized its resources to improve and develop across the entire range of spacefaring activities, with its goal of becoming a space power as equally capable as the United States.[8] One interpretation of the DIA report suggests much of its space-based capabilities place it close to achieving this goal.

The interdependence and immediacy of the US-China rivalry also influences US policymakers' perception. Most of the time there is a tendency to associate space problems with those involving national security or strategic interests, even if the association is tenuous or several degrees removed from such interests. Consider, for example, China's efforts to conduct lunar missions with its Chang'e 4 probe and Yutu 2 rover on the far side of the Moon. Rather than viewing these actions in terms of competition in space exploration, many US policymakers and space leaders express concerns of threat and strategic advantages. "We must be very concerned that China is landing on the Moon and saying: 'It's ours now and you stay out,'" claimed Bill Nelson, NASA's administrator in an interview for a German newspaper in July 2022.[9] This statement was made despite any claims by China to this effect. Others have echoed similar claims, with many believing that since China does not officially separate out its military and civil exploration programs, anything that it does must have a military purpose. Thus, if humans are going to be sent to the Moon, they may create a base there. As a result of this association, even NASA budgets have become tied to the bigger security and strategic concerns of the US in lieu of focusing on exploration for its own sake. In 2019, for example, the House version of the NASA fiscal year 2020 budget called for NSC and interagency cooperation on assessing China's exploration capabilities and their potential threats to US assets in space,[10] illustrating just how concerned US policymakers are on all aspects of China's space enterprise when it comes to US security and strategic interests.

China's space gains are viewed by the United States as zero-sum in significant part because of the developed global competition between them. Like the Cold War, space has become a microcosm of the broader competition between the US and China. This raises several policy questions for the US. First, are there any lessons to be learned from the earlier competition between the US and USSR? How did US policymakers deal with space policy in the broader Cold War competition? Does space provide an opportunity for competition that is not as intense as what may occur on Earth like it did in the 1950s, or is space too important today such that competition in space may become more intense than what happens terrestrially? One way to get at these answers is to look at both current policy and actual behaviors. Given the interdependence of policymaking between the two countries, combined with a sense of urgency and association with national security, it is difficult to imagine policymaking tending toward initiatives that push the two sides back toward sanctuary anytime soon.

US Space Leadership: Erosion or Resurgence?

Although initially lagging the USSR in the early days of the Cold War, the US soon became the clear leader in space after the Apollo missions. This leadership has continued through to the present day. Currently, however, the US has found itself in a position where its leadership is no longer a certainty. While the US-China dynamic is one part of this challenge, there is also a combination of factors and other actors that present a challenge. The push by the EU for an international space code of conduct, is one example. The rise of non-state actors is also another. In short, the erosion of US leadership is likely to come from multiple and perhaps subtle changes to the space enterprise versus a significant one.

What is leadership? Before understanding what it is that the US is challenged with, it is helpful to conceptualize leadership and perhaps even provide a more tangible sense of how such an attribute is recognized. Stewart Ingersoll and Frazier, for example, offer a useful discussion.[11] While pointing out that leadership tends to be intuitive, they note a lack of consensus involving the term despite many overlapping similarities in definitions. Building upon an understanding of power common to most definitions, Stewart Ingersoll and Frazier settle on leadership as a particular type of influence and define the concept as "the act of eliciting cooperation toward or acceptance of shared objectives and a means through which to achieve them amongst members of a group.[12]" This sort of definition naturally connotes that leaders must have followers, others who are willing and able to act on behalf of the objectives and methods specified by the leader.

One additional set of criteria specified by Stewart Ingersoll and Frazier and useful for this discussion is that associated with what they label indicators of leadership. These indicators represent observable facets across the conceptual range of leadership activities. They offer five indicators: process-initiation, issue framing, interest consideration, institution development and deployment of power.[13] While Stewart Ingersoll and Frazier are specifically interested in broader patterns of global and regional order, these indicators remain useful to discuss in the context of space domain leadership as they tend to reflect extensions of similar concerns with respect to patterns of global and regional interactions. The United States, for example, often references the international liberal order in its expectations of behaviors for space actors. There is an implicit assumption involved here that the US is both the leader of the international liberal order and as such has a similar role to play in extending those expectations to space. The EU's proposed international conduct for space also follows a similar set of expectations; while perhaps not the leader of the international liberal order, its standing as a leader offers a similar vantage point from which to engage international space behaviors.

The five indicators are interesting to observe from the position of the United States. The first two indicators should be familiar given their importance in the policymaking process and are actions the United States has taken historically in shaping the activities going on in the space domain. One can think of the beginning of the Apollo missions in terms of initiation and, more recently, DOD narratives of space as a warfighting domain as a framing behavior. Interest consideration is more difficult to think about in the context of US leadership. On one side, US policy lends strong support for the protection of freedom of access to space and the peaceful use of space for everyone. On the other side, the US seems to show little interest consideration for its competitors. For example, how much does US policy reflect legitimate security concerns by China with respect to the US projection of power? While this question may seem irrelevant for policymakers, it is the security dilemma between China and the US (i.e., shared security interdependence) that makes this problem an important one. Also, it is not just about foreign policy. US policy on domestic commercial remote sensing capabilities historically limited an entire commercial space sector, weakening the US by comparison to others in the international community.

The last two indicators, institution development and deployment of power should also resonate with those evaluating US leadership. Institution development has a mixed track record for the United States. It has supported UNCOPUOS and the OST regime but has historically not pushed strongly for formal institutions, rather engaging more in partnerships and less formal arrangements. The newly initiated Artemis Accords, however, are a strong indicator of recent preferences for institutionalization. Deployment of power needs little elaboration, given how much we know the US relies on space for power projection, but also how active the US has been in using its capabilities to provide space domain awareness and enhancing its overall space infrastructure resiliency. It is also worth pointing out here as well, however, that for quite some time, the US ability to deploy power was limited and dependent on Russian rocket engine capabilities, undoubtedly tarnishing perceptions of US leadership.

Having identified what international leadership looks like, can one categorize the US as having it? President Kennedy's Moon speech would seem to qualify as a good start to US space leadership. The speech shaped the agenda for US space to be sure, but it also influenced the Soviet Union's plans, forcing the USSR to also consider attempting a Moon mission. With the USSR's failure and the Americans' success, the international community soon looked to the US as the technological leader and most advanced space nation. While this alone might not have ensured the US of a leadership role, subsequent programs like the shuttle and missions like Voyager 1 and 2 provided a level of transparency and exposure that no other state could emulate. On the military side, the dominance of US space-based capabilities brought to the fore the unique attributes of US space and its impact on international security. The US leadership in developing a "coalition of the willing" in the Gulf War was only enhanced by its ability to project power from space. Thus, there remained a connection between a broader, US global leadership and its position in the actual domain.

It is important to distinguish, however, between leadership and power. The latter does not infer the former, even if power is required for leadership.[14] US space policy initiatives to date have reached to achieve both power and leadership. Power has been viewed as necessary to deter and dissuade potentially disruptive actors in the space domain, while also providing a foundation for leadership in terms of shaping the international agenda and convincing others that US preferences for the space domain are consistent with what is best for stability in the broader global order. The most recent US space policy and subsequent organizational policy documents all reflect this sort of outlook.

Despite indicators to the contrary, there is some sentiment that the US is losing ground in its quest for long-term leadership in space. We have mentioned China's efforts to develop its own partnerships and multilateral space arrangements as it competes in a much wider rivalry with the United States. It has brokered agreements with states like Brazil and Nigeria for ground-based tracking stations and satellite sales, for example. Other actors like the ESA, however, also represent challenges to leadership. In this latter case, those states that tend to want a more sanctuary approach to space may find EU and ESA policies much more consistent with such a goal, given the US association with space is often dominated by security interests. This includes policies that bring together cooperation through agreements like the sharing of Galileo satellites as an alternative or supplement to American-provided GPS.

Another possibility for the space domain is that there will be a lack of leadership moving forward. With an increasing diversity of actors in the domain and the rise of non-state actors, it may very well be an impossible task for any actor, including the US to maintain leadership in the domain. US leadership has not only been challenged in various areas of the space enterprise already, it also is lacking in some areas. For example, the development of legal regimes to deal with the expansion of space-based activities has not kept apace, creating uncertainty. To its credit, the United States has tried to shape international policy through setting an example, with efforts to strengthen its own domestic space laws. Even in these cases, as we have seen throughout this book, policymakers have struggled to keep up in the domestic arena, sometimes overregulating and many times under-regulating behavior. This likely creates mixed signals for the international community.

Returning to our thematic discussion, US leadership only works with the recognition that there is a strong interdependence between its behaviors as a leader and others' willingness to follow. As far back as 2000, US leadership in the space domain showed considerable limitations with respect to states willing to follow US preferences. For example, the United Nations General Assembly's adoption of the resolution, "Prevention of an Arms Race in Outer Space." that year found the US on the outside of a strong consensus, with only the US, Israel, and Micronesia abstaining from the vote. Subsequent resolutions on preventing space arms races have been strongly supported by the international community, with the United States and a few other smaller countries remaining as outliers. While there may be domestic political reasons to remain outside of such consensus, this sort of position calls into question the type of leadership, if any at all, that the US has in particular areas. Such positions also point to the difficulty with which other actors believe they can rely on the United States for stable policy initiatives that are not strongly subject to changes in US presidential administrations.

Like the other major issues, leadership in the space domain is definitely a time-sensitive proposition, at least if it is the case that leadership is eroding. One need to look no further than the changes to space policy in the Trump and Biden administrations as evidence of this concern. President Trump's reestablishment of the National Space Council after 24 years of dormancy reflected a pressing need to restore US space leadership in the face of historic change in the domain.[15] President Biden's US Space Priorities Framework also recognizes the rapid acceleration of the space domain and the challenges it presents to US leadership, calling for a need to establish clear goals and objectives to pursue if leadership is to be maintained. In sum, the US ability to maintain its leadership in space will come down to its willingness and determination to act in ways that defend this role but do so in recognition of the changes coming quickly in the domain. Presidential policy is not the only place this urgency is recognized. House Resolution 6391, the US Leadership in Space Act of 2021, is one example of congressional efforts to jump-start domestic initiatives that will seek to ensure US leadership across the space enterprise for years to come.

For the United States, leadership is critical for long-term stability and benefits gained from space, while leadership relies on US power. Given this dependence, the association between US leadership and national security, whether it is global or in the space domain cannot be understated. It is US power that underpins its global position and allows it to pursue its preferences for a liberal order that it might not otherwise be able to do. That stated, to the extent that its national security preferences in the space domain remain consistent with liberal norms, we should expect many states to follow US leadership. In the security domain, this seems to be playing out, with US partners and allies developing more integrated space operations and policies, joint civil sector exploration initiatives, and increased international commercial ventures. The opposite, however, is just as likely to be true. If national security preferences are inconsistent, if militarization becomes too dominant of a force in US space policy, for instance, it is possible to envision an erosion of US leadership.

Conclusion: US Space Policy and the Way Ahead

As humanity enters its eighth decade of space-based activities, the path ahead is one fraught with challenges and complexity. The ubiquity of space for US space policy means that there will always be a multitude of factors involved in the entire policy process. In this book we have tried to identify the nature of this process along with the challenges that US actors face, as well as the roles that they play in solving these challenges. While much of US history has seen space policy as a top-down driven process from the president to the public, the direction of influence in the 21st century is likely to move back and forth as civilian exploration and commercialization become as important to space policy as the security component has been.

Policymaking for space is not just an endeavor handled domestically. It requires both domestic and international considerations to maintain access and freedom in the space domain. US policy alone cannot guarantee these conditions, just like it cannot guarantee its security in the domain. As the long-recognized leader of the international system, the United States will likely continue to be the most prominent actor with respect to establishing norms of behavior in space, as well as influencing the type of dynamics we expect to see in space between states and even among non-state actors. Additionally, given US leadership in space technology and space exploration, US policy is likely to set standards in those areas as well. However, it will not be the only powerful actor, nor the only one capable of leadership. As we look across the scope of this book, it is this evolving dynamic that has created a sense of urgency for US policymakers to act; actors engaged in the space enterprise are unlikely to wait patiently for the United States to figure out what it wants as a country before they carry on with their own agendas. Some have no intention of letting the US lead. This sense of urgency makes it imperative that policymakers spend time and energy thinking through the myriad challenges posed by this quickly evolving domain. These challenges include those we discussed, those that remain uncovered, and those unknown and yet to come.

To deal with these challenges, part of the policy response by the United States has been to increase the bureaucratic elements that deal with space policy, both in terms of regulation and agendas, and implementation. On one hand, this is likely to be very helpful as the number of actors involved in implementing policy potentially increases the opportunities for success. Alternatively, the growth of bureaucratic actors makes it difficult to coordinate and cooperate effectively. This difficulty means that the integration of policy must be a priority if it is to have a positive impact on policy outcomes. The establishment of this priority is a difficult task for any actor and certainly for one that is so dependent on the domain.

While we have not dealt with all the possible challenges and actors that affect US space policy, we have offered a discussion of the most influential in the context of the development of the US space domain enterprise since the Cold War through to the present day. Not unlike the current space environment, both US policy and organization for space remain incredibly dynamic. What is enduring are a range of themes that help inform policy and how the US best organizes to carry out the policy process. To this end, we have noted the tension between security and prestige, the interdependence of space actors and policy, both domestically and in international politics. We have also highlighted the context in which policymakers seem bound to make decisions, one in which choices are time-sensitive and intimately tied to national security. These are the themes we believe help inform the nature of US space policy today. They are not the only themes to be sure, but an awareness of them should help provide better situational awareness surrounding current US space policy.

Summary Points

- The rapid increase in space debris is complicating efforts to better track it, and the technology being developed to remediate it can be seen as a weapon. While individual countries like the US can take steps unilaterally, the nature of space as a global common will require larger cooperative solutions to an issue that threatens all those wishing to exploit space.
- Commercialization of space continues at a rapid pace. While it has provided the US many significant advantages in the space domain, challenges remain including regulatory schemes, space traffic management, and how to keep American space companies competitive internationally.
- In the absence of treaties and other stronger forms of international law, norms of behavior will continue to develop in the space domain. The

US will need to consider which norms to encourage and how to go about doing so with international partners and potential adversaries.
- China's rapid accumulation of space capabilities presents a challenge to the US ability to project power, military and otherwise, through the space domain. At the same time, China's increased presence there means that there is increased interdependence between the US and China in space.
- Though the US has largely been the acknowledged global leader in space since the end of the 1960s, this is by no means assured going forward. Whether the US remains in this position going forward will significantly influence many of these other space policy issues in the future.

Discussion Questions

1. Of the issues discussed in this chapter, which do you believe is the most important one for the US to address in the next five years? Why?
2. When negotiating norms of behavior internationally, what do you believe US priorities should be? In other words, what should the US goal be in negotiating with other countries?
3. How might domestic issues and politics influence how the US treats these space issues?
4. Should the United States focus so heavily on the space activities of China? Why or why not?
5. Should the US prioritize maintaining leadership in space? How important is this to US success in space?

For Further Reading

Defense Intelligence Agency, *2022 Challenges to Security in Space: Space Reliance in an Era of Competition and Expansion,* March 2022, https://www.dia.mil/Portals/110/Documents/News/Military_Power_Publications/Challenges_Security_Space_2022.pdf.

Daniels, Matthew, "The History and Future of US-China Competition and Cooperation in Space," Johns Hopkins Applied Physics Lab, 2021, https://www.jhuapl.edu/Content/documents/Daniels-Space.pdf

McClintock, Bruce, Katie Feistel, Douglas C. Ligor, and Kathryn O'Connor, "Responsible Space Behavior for the New Space Era," *Rand,* 2021, https://www.rand.org/pubs/perspectives/PEA887-2.html.

NOTES

Chapter 1. Introduction

1. Office of the Director of National Intelligence, "National Security Space Strategy."
2. Kingdon, *Agendas, Alternatives, and Public Policies.*
3. Mintrom, "Policy Entrepreneurs and the Diffusion of Innovation."
4. Lindblom, "The Science of Muddling Through"; Wildavsky, *The Politics of the Budgetary Process.*
5. Baumgartner and Jones, *Agendas and Instability in American Politics.*
6. Smith, "Typologies, Taxonomies, and the Benefits of Policy Classification."
7. Kertzer and Zeitzoff, "A Bottom-Up Theory of Public Opinion about Foreign Policy"; Jeong and Quirk, "Division at the Water's Edge: The Polarization of Foreign Policy."
8. The Space Foundation, "Global Space Economy Rose to $447B in 2020, Continuing Five-Year Growth."
9. Trevithick, "China's Historic Mission to the Dark Side of the Moon Is About More Than Science."
10. Union of Concerned Scientists, "UCS Satellite Database."
11. International Telecommunications Union, "Regulation of satellite systems."
12. UNOOSA, "Treaty on Principles Governing the Activities of States in the Exploration and Use of Outer Space, including the Moon and Other Celestial Bodies."

Chapter 2. The Space Race during the Cold War

1. McDougall, . . . *The Heavens and the Earth*, 26.
2. Richelson, *America's Space Sentinels*, 8.
3. Here we use the term "manned missions" intentionally. While Valentina Tershkova's spaceflight in 1953 generated some interest in American women also going to space, NASA did not begin training women until 1978. In 1983, Sally Ride became the first American female in space, securing a future for human spaceflight and greater gender equality in the US space program.
4. Kennedy, "Excerpt from the 'Special Message to the Congress on Urgent National Needs.'"
5. Moltz, *The Politics of Space Security*, 108.
6. McDougall, . . . *The Heavens and the Earth*, 380.
7. Ackmann, *The Mercury 13.*
8. Sylvester, "John Glenn and the Sexism of the Early Space Program."
9. Shetterly, *Hidden Figures.*
10. Jackson, "Ed Dwight Was Going to Be the First African American in Space. Until He Wasn't."

11 Odom and Waring, *NASA and the Long Civil Rights Movement.*
12 Foster, *Integrating Women into the Astronaut Corps.*
13 Erickson, *Into the Unknown Together,* 9.
14 Erickson, 179.
15 Moltz, *The Politics of Space Security,* 116.
16 Moltz, 11.

Chapter 3. Post–Cold War Space History

1 Logsdon, *Ronald Reagan and the Space Frontier.*
2 Easton and Frazier, *GPS Declassified,* 125.
3 Paxton, "'Faster, Better, and Cheaper' at NASA."
4 ISS National Laboratory, "In-Orbit Activities: The ISS as a Research Platform."
5 Davenport, *The Space Barons.*
6 Krebs, "Orbital Launches of 2018."
7 Gould, "Space for a Space Corps? Congress Lays Groundwork for Controversial Plan."

Chapter 4. Presidents and Space Policy

1 Albrecht, *Falling Back to Earth;* Hogan, *Mars Wars.*
2 Neustadt, *Presidential Power and the Modern Presidents.*
3 Canes-Wrone, Howell, and Lewis, "Toward a Broader Understanding of Presidential Power."
4 Rogowski, "Presidential Influence in an Era of Congressional Dominance."
5 Handberg, "Human Spaceflight and Presidential Agendas."
6 Edwards, Barrett, and Peake, "The Legislative Impact of Divided Government."
7 Cohen, "Presidents, Polarization, and Divided Government."
8 Lewis, *Presidents and the Politics of Agency Design.*
9 Eshbaugh-Soha, "The Politics of Presidential Agendas."
10 Logsdon, *After Apollo?*
11 Reagan, "Remarks at the National Forum on Excellence in Education in Indianapolis, Indiana."
12 Edwards, *On Deaf Ears.*
13 Lazarus and Thornton, "Bully Pulpit? Twitter Users' Engagement with President Trump's Tweets"; Miles and Haider-Markel, "Bully Pulpit or Ham Radio? Receptive Audiences and Presidential Messages."
14 Bolton and Thrower, "Legislative Capacity and Executive Unilateralism."
15 Ostrander and Sievert, "The Logic of Presidential Signing Statements."
16 Office of the Director of National Intelligence, "2011 National Security Space Strategy."
17 Department of Defense, "Defense Space Strategy Summary."
18 Conley and Whitman Cobb, "Presidential Vision or Congressional Derision? Explaining Budgeting Outcomes for NASA, 1958–2008."
19 Wood and Waterman, "The Dynamics of Political Control of the Bureaucracy"; Moe, "Control and Feedback in Economic Regulation"; Golden, *What Motivates Bureaucrats?*
20 Garver, *Escaping Gravity.*
21 Garver, *Escaping Gravity.*

22 Morgan, "The National Space Council."
23 Logsdon, *Ronald Reagan and the Space Frontier.*
24 Morgan, "The National Space Council."
25 Morgan.
26 Morgan, "The National Space Council."

Chapter 5. Congress and Space Policy

1   McCubbins and Schwartz, "Congressional Oversight Overlooked."
2   Whitman Cobb, *Unbroken Government*; Dodd, "Making Sense out of Our Exceptional Senate"; Sinclair "The '60-Vote Senate'"; Lee and Oppenheimer, *Sizing Up the Senate*; Shepsle, Van Houweling, Abrams, and Hanson, "The Senate Electoral Cycle and Bicameral Appropriations Politics."
3   Fenno, *Congressmen in Committees*; Sinclair, "The '60-Vote Senate.'"
4   Dodd, "Making Sense out of Our Exceptional Senate."
5   Shepsle, Van Houweling, Abrams, and Hanson, "The Senate Electoral Cycle and Bicameral Appropriations Politics."
6   Walker, "Setting the Agenda in the US Senate."
7   Deering and Smith, *Committees in Congress.*
8   Mayhew, *Congress: The Electoral Connection*, 2nd ed., 5, 16.
9   Steinberg, "Space Policy Responsiveness."
10 Machay and Steinberg, "Influence of Industry on Legislative Behavior toward NASA."
11 Nail, "NASA Report Says It Has Big Economic Impact on Florida."
12 Dodd and Schott, *Congress and the Administrative State.*
13 Huber, Shipan, and Pfhaler, "Legislatures and Statutory Control of the Bureaucracy."
14 Moltz, *The Politics of Space Security.*
15 Weingast and Moran, "Bureaucratic Discretion or Congressional Control?"; Wood and Waterman, "The Dynamics of Political Control of the Bureaucracy"; Carpenter, "Adaptive Signal Processing, Hierarchy, and Budgetary Control in Federal Regulation."
16 MacDonald, "Limitation Riders and Congressional Influence over Bureaucratic Policy Decisions."
17 *Department of Defense and Full-Year Continuing Appropriations Act, 2011,* Public Law 112–10.
18 Conley and Whitman Cobb, "Presidential Vision or Congressional Derision?"
19 Whitman Cobb, *Unbroken Government.*
20 Whitman Cobb, "Who's Supporting Space Activities?"
21 Burbach, "Partisan Rationales for Space."
22 Machay and Steinberg, "NASA Funding in Congress."

Chapter 6. NASA

1   Downs, *Inside Bureaucracy*: 20.
2   Bendor and Moe "An Adaptive Model of Bureaucratic Politics": Bendor, Taylor, and Van Gaalen, "Politicians, Bureaucrats, and Asymmetric Information"; Niskanen, "Bureaucrats and Politicians."
3   Burrows, *This New Ocean.*
4   McCurdy, *Inside NASA,* 4.

5 McCurdy.
6 The Planetary Society, "Your Guide to NASA's Budget."
7 McCurdy, *Inside NASA*.
8 McCurdy.
9 Garver, *Escaping Gravity*.
10 Banks and Weingast, "The Political Control of Bureaucracies under Asymmetric Information."
11 Albrect, *Falling Back to Earth*; Hogan, *Mars Wars*
12 Roulette, "At Pivotal Points in Human Spaceflight, NASA Undergoes Major Reorganization."
13 Vaughn, *The Challenger Launch Decision*
14 Foust, "Blue Origin Sues NASA over Human Landing System Contract."
15 Roulette, "The Senate Just Advanced the Beef between SpaceX and Blue Origin."
16 Gohd, "Senate Directs NASA to Choose Another Company to Build a Lunar Lander: Report."
17 Carpenter, *Reputation and Power*.
18 Motel, "NASA Popularity Still Sky High"; Etkind and McGuinness, "Decade of Excellence: NASA Named Best Place to Work 10th Year in a Row."
19 Lambright, "James E. Webb: A Dominant Force in 20th Century Public Administration."
20 McCurdy, *Faster, Better, Cheaper*, 1.
21 Ward, "Faster, Better, Cheaper Revisited."
22 Lambright, "Relay Leadership and Long-Term Policy," 6.

Chapter 7. Military Space and the Intelligence Community

1 White House, "United States Space Priorities Framework."
2 White House, "National Cislunar Science and Technology Strategy."
3 Department of Defense, "Fact Sheet: 2022 National Defense Strategy."
4 Department of Defense, *Space Policy*, 10.
5 Department of Defense.
6 Department of Defense, *Space Executive Agency*.
7 US Army, "About the Command."
8 US Fleet Cyber Command/US Tenth Fleet, "Strategic Plan 2020–2025."
9 Hays, "Space and the Military," 152.
10 Hays.
11 Spires, *Beyond Horizons*, 277–278.
12 Department of Defense, *Functions of the Department of Defense and Its Major Components*.
13 Not all of the IC is subject to joint oversight with the DOD, but the secretary of defense does have authority over those IC members that fall within the DOD hierarchy.
14 Broad, "Spy Satellites' Early Role as 'Floodlight' Coming Clear"
15 Coats, "Worldwide Threat Assessment of the US Intelligence Committee."

Chapter 8. The "Other" Actors

1. Office of Space Affairs, "Key Topics."
2. ITAR is governed by US Federal Code, 22 U.S.C. 2778 of the Arms Export Control Act of 1976 and Executive Order 13637 that delegates statutory authority to the secretary of state.
3. Highfill, Jouard, and Franks, "Updated and Revised Estimates of the U.S. Space Economy, 2012–2019."
4. U.S. Code Title 51, Chapter 50702.
5. *Consolidated Appropriations Act, 2021.*
6. Raimondo, "The Department of Commerce Budget in Brief: Fiscal Year 2023."
7. US Department of Energy, "Energy for Space: Department of Energy's Strategy to Advance American Space Leadership (FY 2021–FY 2031)."
8. U.S. Department of Transportation, "About DOT."
9. Federal Communications Commission, "FCC Opens Proceeding on Servicing, Assembly, & Manufacturing in Space."
10. US Government Accountability Office, "NASA: Assessments of Major Projects."

Chapter 9. Nongovernmental Forces

1. Harwood, "Soyuz Flight to Space Station Sets Stage for NASA Transition to US Crew Ships."
2. Berger, "A Politician Who Said Politicians Shouldn't Run NASA Wants to Run NASA."
3. Launius, "Public Opinion Polls and Perceptions of US Human Spaceflight."
4. Manchester, "Hill.TV poll: Majority of Americans approve of 'Space Force'"; Rasmussen Reports, "33% Support Creating a National 'Space Force.'"
5. Erwin, "Space Force Idea Lacks Public Support, Survey Reveals."
6. Sparks, "Space Force: To 37% and Beyond."
7. Whitman Cobb, "'It's a Trap!' The Pros and Mostly 'Khans' of Science Fiction's Influence on the United States Space Force."
8. Nadeau, "Examining Public Support for Space Exploration Funding in America."
9. Nadeau; Whitman Cobb, "Who's Supporting Space Activities?"
10. Whitman Cobb, "Stubborn Stereotypes."
11. Burbach, "Partisan Rationales for Space."
12. Whitman Cobb, "Who's Supporting Space Activities?"
13. Steinberg, "Space Policy Responsiveness."
14. Engineering and Technology History Wiki, "Milestones: TIROS-1 Television Infrared Observation Satellite, 1960"; Wikipedia, "Project Echo."
15. Brooks, Grimwood, and Swenson, *Chariots for Apollo.*
16. Akens, *Saturn Illustrated Chronology.*
17. Clary, "50th anniversary of Satellite Telstar Celebrated."
18. While there would still be a commercial market for American launch providers to serve in private and international companies, the absence of government satellites meant that the potential market would be far smaller than expected.
19. Mellow, "The Rise and Fall of Iridium."
20. The Space Foundation, "Global Space Economy Rose to $447B in 2020, Continuing Five-Year Growth."

21 Daniels, "The History and Future of US-China Competition and Cooperation in Space."
22 Open Secrets, "Industry Profile: Defense Aerospace."
23 Open Secrets, "SpaceX"; Open Secrets, "Blue Origin."
24 Davenport, "Blue Origin's Loss to SpaceX on the Lunar Lander Contract May Get Congress to Do Something It Hadn't Done Before: Give NASA Extra Money."
25 Open Secrets, "Boeing Co."; Open Secrets, "Lockheed Martin"; Open Secrets, "Northrop Grumman."
26 Open Secrets, "SpaceX"; Open Secrets, "Blue Origin."
27 Mack, "SpaceX Starship Launch Proposal Draws Vocal Support, Some Criticism in FAA Hearing."
28 Schwartz, *The Value of Science in Space Exploration*.
29 NASA, *NASA Spinoff*.
30 Howell, "NASA efforts Had a $65 Billion Economic Impact Last Year, Agency Report Shows."
31 Cabana, "Kennedy Space Center Director: Space Coast Future Is Bright."
32 Machay and Steinberg, "Influence of Industry on Legislative Behavior toward NASA."
33 Conley and Whitman Cobb, "Presidential Vision or Congressional Derision? Explaining Budgeting Outcomes for NASA, 1958–2008."
34 Redmond, "Bernie Sanders Is Dragging Elon Musk and Jeff Bezos over Their Billionaire Space Race"; Wall, "Oregon Congressman Proposes New Space Tourism Tax."
35 Carroll, "A Comprehensive Definition of Technology from an Ethological Perspective," 126.
36 Jones, "The Recent Large Reduction in Space Launch Cost."
37 Wyatt, "Technological Determinism Is Dead, Long Live Technological Determinism."
38 Moltz, *The Politics of Space Security*.
39 McDougall, . . . *the Heavens and the Earth*.
40 McDougall, "Technocracy and Statecraft in the Space Age—Toward the History of a Saltation."
41 Cameron, "China's Silos: New Intelligence, Old Problems"; Kurekin et al., "Operational Monitoring of Illegal Fishing in Ghana through Exploitation of Satellite Earth Observation and AIS Data," 293.

Chapter 10. Civilian Space Policy

1 NASA, "Near-Earth Object Observations Program."
2 Dreier, "How NASA's Planetary Defense Budget Grew by More Than 4000% in 10 Years."
3 NASA, "History."
4 NASA.
5 Folger, "Landsat: Overview and Issues for Congress."
6 Folger.
7 Folger.
8 Folger.
9 Normand, "Landsat 9 and the Future of the Sustainable Land Imaging Program."
10 Normand, 9.

11 Normand.
12 Davis, "History of the NOAA Satellite Program."
13 Cirac-Claveras, "Weather Satellites."
14 Cirac-Claveras.
15 Cirac-Claveras.
16 Cirac-Claveras.
17 Wang and Degol, "Gender Gap in Science, Technology, Engineering, and Mathematics (STEM)"
18 Baram-Tsabari and Yarden, "Quantifying the Gender Gap in Science Interests"; Nosek, Smyth, Siriam et al., "National Differences in Gender-Science Stereotypes Predict National Sex Differences in Science and Math Achievement."
19 Weitekamp, *Right Stuff, Wrong Sex*; Ackmann, *The Mercury 13*.
20 Garver, *Escaping Gravity*.
21 Chang, "Jeff Bezos' Rocket Company Accused of Toxic Culture and Safety Issues"; Bartels, "SpaceX Faces Sexual Harassment Allegations from Five Former Employees."
22 Maher, *Apollo in the Age of Aquarius*.
23 Maher.
24 Grush, "The Mission to Break Barriers to Space Travel for People with Disabilities."
25 Whitman Cobb, "For All (Wo)mankind."
26 Older, "Satellite Surveillance Can Trace Atrocities but Not Stop Them."
27 The Planetary Society, "Your Guide to NASA's Budget."
28 Sambaluk, *The Other Space Race*.
29 Van Allen, "Is Human Spaceflight Obsolete?"
30 Spice, "Grief Tweets About Mars Rover Sound Very 'Human.'"
31 Van Allen, "Is Human Spaceflight Obsolete?"
32 White House, "Space Policy Directive-3, National Space Traffic Management Policy."
33 Foust, "Report Endorses Giving Commerce Department Responsibility for Space Traffic Management."
34 Foust, "Senators Push for Action on Space Traffic Management."

Chapter 11. International Politics and National Security Space Policy

1 Moltz, *The Politics of Space Security*, 24.
2 Dolman, *Astropolitik*.
3 Powell, "Guns, Butter, and Anarchy."
4 Kant, *Perpetual Peace and Other Essays*.
5 Hitchens and Johnson-Freese, *Toward a New National Security Space Strategy*.
6 Wendt, *The Social Construction of International Politics*.
7 Moltz, *The Politics of Space Security*.
8 Steer, "Global Commons, Cosmic Commons."
9 Ikenberry, *Liberal Leviathan*.
10 Fukuyama, *The End of History and the Last Man*.
11 Dolman, *Astropolitik*.
12 Klein, *Space Warfare*.

Chapter 12. Commercial Space Policy

1. Sheetz, "The Space Industry Is on Its Way to Reach $1 Trillion in Revenue by 2040, Citi Says."
2. Steptoe, "United States Government Licensing of Commercial Space Activities by Private Enterprise."
3. Knipfer, "Congress and Commerce in the Final Frontier (Part 1)."
4. NBC News, "Elon Musk's SpaceX Sues Government to Protest Military Launch Monopoly."
5. Chappell, "Blue Origin Says It Will Build an Orbiting Mixed-Use Business Park in Space."
6. Howell, "NASA Awards $415 Million for Private Space Stations amid ISS Transition Questions."
7. Hitchens, "NRO Keeps 3 Vendors for Commercial Imagery with New 10-Year Contracts."
8. Morgan, "Commercial Human Spaceflight."
9. Foust, "One Nation, Over Regulated: Is ITAR Stalling the New Space Race?"
10. Lin-Greenberg and Milonopoulos, "Boots on the Ground, Eyes in the Sky."
11. Reuters, "Russia Warns West: We Can Target Your Satellites."
12. Roulette, "SpaceX Curbed Ukraine's Use of Starlink Internet for Drones—Company President."
13. Foust, "Commerce Department Releases Streamlined Commercial Remote Sensing Regulations."

Chapter 13. Major Issues in US Space Policy

1. Jones, "New Chinese Small Sat Manufacturing Capacity Could Have International Ramifications."
2. Jones, "China Is Developing Plans for a 13,000-Satellite Megaconstellation."
3. Hampson, "The Future of Space Commercialization."
4. Sadat and Siegel, "Space Traffic Management: Time for Action."
5. Weinzierl. "Space, the Final Economic Frontier."
6. See Leib, "State Sovereignty in Space: Current Models and Possible Futures."
7. Austin, "Tenets of Responsible Behavior in Space."
8. Defense Intelligence Agency, *Challenges to Security in Space*.
9. Both, "Bad News About the Moon."
10. Howell, "Are China's Moon Missions a Threat to the US? Space Experts Don't Think So."
11. Stewart Ingersoll and Frazier, *Regional Powers and Security Orders*.
12. Stewart, p. 74.
13. Stewart, p. 78.
14. Nye, *The Powers to Lead*.
15. National Space Council, "Renewing America's Proud Legacy of Leadership in Space."

BIBLIOGRAPHY

Ackmann, Martha, *The Mercury 13: The Untold Story of Thirteen American Women and the Dream of Space Flight* (New York: Random House, 2003).

Akens, David S., *Saturn Illustrated Chronology* (Washington, DC: NASA, n.d.), https://history.nasa.gov/MHR-5/contents.htm.

Albrecht, Mark, *Falling Back to Earth: A First Hand Account of the Great Space Race and the End of the Cold War* (New Media Books, 2011).

Austin, Lloyd, "Tenets of Responsible Behavior in Space," *Department of Defense*, July 7, 2021, https://media.defense.gov/2021/Jul/23/2002809598/-1/-1/0/TENETS-OF-RESPONSIBLE-BEHAVIOR-IN-SPACE.PDF.

Banks, Jeffrey S., and Barry Weingast, "The Political Control of the Bureaucracy," *American Political Science Review*, vol. 92, no. 3 (1992): 509–24.

Baram-Tsabari, Ayelet, and Anat Yarden, "Quantifying the Gender Gap in Science Interests," *International Journal of Science and Mathematic Education*, vol. 9 (2011): 523–50.

Bartels, Meghan, "SpaceX Faces Sexual Harassment Allegations from Five Former Employees," *Space.com*, December 16, 2021, https://www.space.com/spacex-sexual-harassment-allegations-lawsuits.

Baumgartner, Frank R., and Bryan D. Jones, *Agendas and Instability in American Politics* (Chicago: University of Chicago Press, 1993).

Bendor, Jonathan, and Terry M. Moe, "An Adaptive Model of Bureaucratic Politics," *The American Political Science Review*, vol. 79, no. 3 (1985): 755–74.

Bendor, Jonathan, Serge Taylor, and Roland Van Gaalen, "Bureaucratic Expertise versus Legislative Authority: A Model of Deception and Monitoring in Budgeting," *The American Political Science Review*, vol. 79, no. 4 (1987): 1041–60.

Bendor, Jonathan, Serge Taylor, and Roland Van Gaalen, "Politicians, Bureaucrats, and Asymmetric Information," *American Journal of Political Science*, vol. 31, no. 4 (1987): 796–828.

Berger, Eric, "A Politician Who Said Politicians Shouldn't Run NASA Wants to Run NASA," *Ars Technica*, February 23, 2021, https://arstechnica.com/science/2021/02/a-politician-who-said-politicians-shouldnt-run-nasa-wants-to-run-nasa/.

Bolton, Alexander, and Sharece Thrower, "Legislative Capacity and Executive Unilateralism," *American Journal of Political Science*, vol. 60, no. 3 (2016): 649–63.

Both, Maximilian, "Bad News About the Moon," *Bild*, May 7, 2022, https://www.bild.de/politik/inland/politik-inland/nach-bild-bericht-zoff-um-den-mond-80605038.bild.html?_x_tr_sl=auto&_x_tr_tl=en&_x_tr_hl=en&_x_tr_pto=wapp.

Broad, William J., "Spy Satellites' Early Role as 'Floodlight' Coming Clear," *New York Times,* September 12, 1995, https://www.nytimes.com/1995/09/12/science/spy-satellites-early-role-as-floodlight-coming-clear.html.

Brooks, Courtney G., James M. Grimwood, and Loyd S. Swenson, *Chariots for Apollo: A History of Manned Lunar Spacecraft* (Washington, DC: NASA, 1979), https://www.hq.nasa.gov/office/pao/History/SP-4205/contents.html.

Burbach, David T., "Partisan Rationales for Space: Motivations for Public Support of Space Exploration Funding, 1973–2016," *Space Policy,* vol. 50 (2019).

Burrows, William E. *This New Ocean: The Story of the First Space Age* (New York: Modern Library, 1999).

Cabana, Bob, "Kennedy Space Center Director: Space Coast Future Is Bright," *Orlando Sentinel,* January 25, 2020, https://www.orlandosentinel.com/opinion/guest-commentary/os-op-bob-cabana-kennedy-space-center-rebound-20200125-ukv7zpnrfbgencurzmkaaktxiy-story.html.

Cameron, James, "China's Silos: New Intelligence, Old Problems," *War on the Rocks,* August 12, 2021, https://warontherocks.com/2021/08/beijings-silos-new-intelligence-old-problems/.

Canes-Wrone, Brandice, William G. Howell, and David E. Lewis, "Toward a Broader Understanding of Presidential Power: A Reevaluation of the Two Presidencies Thesis," *The Journal of Politics,* vol. 70, no. 1 (2008): 1–16.

Carpenter, Daniel P., "Adaptive Signal Processing, Hierarchy, and Budgetary Control in Federal Regulation," *American Political Science Review,* vol. 90, no. 2 (1996), 283–302.

Carpenter, Daniel, *Reputation and Power: Organizational Image and Pharmaceutical Regulation at the FDA* (Princeton, NJ: Princeton University Press, 2010).

Carroll, La Shun L., "A Comprehensive Definition of Technology from an Ethological Perspective," *Social Sciences,* vol. 6, no. 4 (2017).

Chang, Kenneth, "Jeff Bezos' Rocket Company Accused of Toxic Culture and Safety Issues," *New York Times,* September 30, 2021, https://www.nytimes.com/2021/09/30/science/jeff-bezos-blue-origin-safety.html.

Chappell, Bill, "Blue Origin Says It Will Build an Orbiting Mixed-Use Business Park in Space," *NPR,* October 25, 2021, https://www.npr.org/2021/10/25/1049077333/blue-origin-space-station-business-park-space-orbital-reef-bezos.

Cirac-Claveras, Gemma, "Weather Satellites: Public, Private and Data Sharing: The Case of Radio Occultation Data," *Space Policy,* vol. 47 (2019): 94–106.

Clary, Gregory, "50th Anniversary of Satellite Telstar Celebrated," *CNN Lightyears Blog,* July 13, 2012, https://lightyears.blogs.cnn.com/2012/07/13/50th-anniversary-of-satellite-celebrated/.

Coats, Daniel, "Worldwide Threat Assessment of the US Intelligence Committee," *Office of the Director of National Intelligence,* May 11, 2017, https://www.dni.gov/files/documents/Newsroom/Testimonies/SSCI%20Unclassified%20SFR%20-%20Final.pdf.

Cohen, Jeffrey E., "Presidents, Polarization, and Divided Government," *Presidential Studies Quarterly,* vol. 41, no. 3 (2011): 504–520.

Conley, Richard S., and Wendy Whitman Cobb, "Presidential Vision or Congressional Derision? Explaining Budgeting Outcomes for NASA, 1958–2008," *Congress and the Presidency,* vol. 39 (2012): 51–73.

Daniels, Matthew, "The History and Future of US-China Competition and Cooperation in Space," *Johns Hopkins Applied Physics Laboratory,* 2020, https://www.jhuapl.edu/Content/documents/Daniels-Space.pdf.

Davenport, Christian, "Blue Origin's Loss to SpaceX on the Lunar Lander Contract May Get Congress to Do Something It Hadn't Done Before: Give NASA Extra Money," *Washington Post,* May 12, 2021, https://www.washingtonpost.com/technology/2021/05/12/cantwell-blue-origin-jeff-bezos-nasa/.

Davenport, Christian, *The Space Barons: Elon Musk, Jeff Bezos, and the Quest to Colonize the Cosmos* (New York: Public Affairs, 2018).

Davis, Gary, "History of the NOAA Satellite Program," *Journal of Applied Remote Sensing,* vol 1, no. 1 (2007): 012504.

Deering, Christopher J., and Steven S. Smith, *Committees in Congress,* 3rd ed. (Washington, DC: CQ Press, 1997).

Defense Intelligence Agency, *2022 Challenges to Security in Space: Space Reliance in an Era of Competition and Expansion,* March 2022, https://www.dia.mil/Portals/110/Documents/News/Military_Power_Publications/Challenges_Security_Space_2022.pdf.

Department of Defense, "Defense Space Strategy Summary," *Department of Defense,* June 2020, https://media.defense.gov/2020/Jun/17/2002317391/-1/-1/1/2020_DEFENSE_SPACE_STRATEGY_SUMMARY.PDF.

Department of Defense, *Directive on Space Executive Agency.* DOD Directive 5160.32. Washington, DC: Department of Defense, 2002. https://www.jstor.org/stable/24783257#metadata_info_tab_contents.

Department of Defense, "Fact Sheet: 2022 National Defense Strategy," March 2022, https://media.defense.gov/2022/Mar/28/2002964702/-1/-1/1/NDS-FACT-SHEET.PDF.

Department of Defense, *Functions of the Department of Defense and Its Major Components.* DOD Directive 5100.01. Washington, DC: Department of Defense, 2020. https://www.esd.whs.mil/Portals/54/Documents/DD/issuances/dodd/510001p.pdf.

Department of Defense, *Space Policy.* DOD Directive 3100.10. Washington, DC: Department of Defense, 2022. https://www.esd.whs.mil/Portals/54/Documents/DD/issuances/dodd/310010p.PDF?ver=s-Go4Bf2eesb8Mog7hkFXw%3D%3D.

Department of Defense and Full Year Continuing Appropriations Act, 2011, Public Law 112–10, 2011, https://www.govinfo.gov/content/pkg/PLAW-112publ10/html/PLAW-112publ10.htm.

Dodd, Lawrence C., "Making Sense out of Our Exceptional Senate: Perspectives and Commentary," *US Senate Exceptionalism,* Bruce I. Oppenheimer, ed. (Columbus: Ohio State University Press, 2002).

Dodd, Lawrence C., and Richard L. Schott, *Congress and the Administrative State* (New York: Macmillan, 1979).

Dolman, Everett, *Astropolitik: Classical Geopolitics in the Space Age* (London: Frank Cass, 2002).

Downs, Anthony *Inside Bureaucracy* (Boston: Little, Brown, 1966).

Dreier, Casey, "How NASA's Planetary Defense Budget Grew by More than 4000% in 10 years," *The Planetary Society,* September 26, 2019, https://www.planetary.org/articles/nasas-planetary-defense-budget-growth.

Easton, Richard D., and Eric F. Frazier, *GPS Declassified: From Smart Bombs to Smartphones* (Omaha, NE: Potomac Books, 2013).

Edwards III, George C., *On Deaf Ears: The Limits of the Bully Pulpit* (New Haven, CT: Yale University Press, 2003).

Edwards III, George C., Andrew Barrett, and Jeffrey Peake, "The Legislative Impact of Divided Government," *American Journal of Political Science*, vol. 41, no. 2 (1997): 545–563.

Engineering and Technology History Wiki, "Milestones: TIROS-1 Television Infrared Observation Satellite, 1960," *ETHW*, June 14, 2022, https://ethw.org/Milestones:TIROS-1_Television_Infrared_Observation_Satellite,_1960.

Erickson, Mark, *Into the Unknown Together: The DOD, NASA, and Early Spaceflight* (Maxwell AFB, AL: Air University Press, 2005).

Erwin, Sandra, "Space Force Idea Lacks Public Support, Survey Reveals," *SpaceNews*, November 30, 2018, https://spacenews.com/space-force-idea-lacks-public-support-survey-reveals/.

Eshbaugh-Soha, Matthew, "The Politics of Presidential Agendas," *Political Research Quarterly*, vol. 58, no. 2 (2005): 257–68.

Etkind, Marc, and Jackie McGuinness, "Decade of Excellence: NASA Named Best Place to Work 10th Year in a Row," *NASA*, July 13, 2022: https://www.nasa.gov/press-release/decade-of-excellence-nasa-named-best-place-to-work-10th-year-in-a-row.

Federal Communications Commission, "FCC Opens Proceeding on Servicing, Assembly, & Manufacturing in Space," *Federal Communications Commission*, August 8, 2022, https://www.fcc.gov/document/fcc-opens-proceeding-servicing-assembly-manufacturing-space-0.

Fenno, Richard, *Congressmen in Committees* (Boston: Little, Brown, and Co., 1973).

Folger, Peter, "Landsat: Overview and Issues for Congress," *Congressional Research Service*, October 27, 2014, https://sgp.fas.org/crs/misc/R40594.pdf.

Foster, Amy E., *Integrating Women into the Astronaut Corps: Politics and Logistics at NASA, 1972–2004* (Baltimore, MD: Johns Hopkins University Press, 2011).

Foust, Jeff, "Blue Origin Sues NASA over Human Landing System Contract," *SpaceNews*, August 16, 2021, https://spacenews.com/blue-origin-sues-nasa-over-human-landing-system-contract/.

Foust, Jeff, "Commerce Department Releases Streamlined Commercial Remote Sensing Regulations," *SpaceNews*, May 19, 2020, https://spacenews.com/commerce-department-releases-streamlined-commercial-remote-sensing-regulations/.

Foust, Jeff, "One Nation, Over Regulated: Is ITAR Stalling the New Space Race?" *Ad Astra*, vol. 17, no. 3 (2005), https://space.nss.org/one-nation-over-regulated-is-itar-stalling-the-new-space-race/.

Foust, Jeff, "Report Endorses Giving Commerce Department Responsibility for Space Traffic Management," *Space News*, August 20, 2020, https://spacenews.com/report-endorses-giving-commerce-department-responsibility-for-space-traffic-management/.

Foust, Jeff, "Senators Push for Action on Space Traffic Management," *Space News*, July 23, 2021, https://spacenews.com/senators-push-for-action-on-space-traffic-management/.

Fukuyama, Francis, *The End of History and the Last Man* (New York: Free Press, 1992).
Garver, Lori, *Escaping Gravity: My Quest to Transform NASA and Launch a New Space Age* (New York: Diversion Books, 2022).
Gohd, Chelsea, "Senate Directs NASA to Choose Another Company to Build a Lunar Lander: Report," *Space.com*, October 19, 2021, https://www.space.com/senate-nasa-second-lunar-lander-contract.
Golden, Marissa Martino, *What Motivates Bureaucrats? Politics and Administration During the Reagan Years* (New York: Columbia University Press, 2000).
Gould, Joe, "Space for a Space Corps? Congress Lays Groundwork for Controversial Plan," *Defense News*, November 21, 2017, https://www.defensenews.com/space/2017/11/21/space-for-a-space-corps-congress-lays-groundwork-for-controversial-plan/.
Grush, Loren, "The Mission to Break Barriers to Space Travel for People with Disabilities," *The Verge*, October 20, 2021, https://www.theverge.com/2021/10/20/22734331/mission-astro-access-disability-zero-g-flight-space-travel.
Hampson, Joshua, "The Future of Space Commercialization," *Niskanen Center*, January 27, 2017, https://www.niskanencenter.org/future-space-commercialization/.
Handberg, Roger, "Human Spaceflight and Presidential Agendas: Niche Policies and NASA, Opportunity and Failure," *Technology in Society*, vol. 29 (2014): 31–43.
Harwood, William, "Soyuz flight to space station sets stage for NASA transition to US crew ships," *CBS News*, October 14, 2020, https://www.cbsnews.com/news/soyuz-nasa-flight-international-space-station-u-s-crew-ships/.
Hays, Peter L. "Space and the Military," *Space and Defense Policy*, Damon Coletta and Frances T. Pilch, eds. (New York: Routledge, 2009).
Henry, Caleb, "FCC Punts Controversial Space Debris Rules for Extra Study," *Space News*, April 23, 2020, https://spacenews.com/fcc-punts-controversial-space-debris-rules-for-extra-study/.
Highfill, Tina, Annabel Jouard, and Connor Franks, "Updated and Revised Estimates of the U.S. Space Economy, 2012–2019," *Bureau of Economic Analysis*, January 2022, http://www.bea.gov/system/files/2022-01/Space-Economy-2012-2019.pdf.
Hitchens, Theresa, "NRO Keeps 3 Vendors for Commercial Imagery with New 10-Year Contracts," *Breaking Defense*, May 25, 2022, https://breakingdefense.com/2022/05/nro-keeps-3-vendors-for-commercial-imagery-with-new-10-year-contracts/.
Hitchens, Theresa, and Joan Johnson-Freese, *Toward a New National Security Space Strategy: Time for a Strategic Rebalancing* (Washington, DC: Atlantic Council Strategy 5, 2016).
Hogan, Thor, *Mars Wars: The Rise and Fall of the Space Exploration Initiative* (Washington, DC: NASA History Division, 2007).
Howell, Elizabeth, "Are China's Moon Missions a Threat to the US? Space Experts Don't Think So," *Space.com*, March 27, 2020, https://www.space.com/china-moon-missions-us-national-security.html.
Howell, Elizabeth, "NASA Awards $415 Million for Private Space Stations amid ISS Transition Questions," *Space.com*, December 3, 2021, https://www.space.com/nasa-private-space-station-design-contracts.

Howell, Elizabeth, "NASA Efforts Had a $65 Billion Economic Impact Last Year, Agency Report Shows," *Space.com*, October 13, 2020, https://www.space.com/nasa-produces-65-billion-dollar-economic-impact-2019.

Huber, John D., Charles L. Shipan, and Madelaine Pfahler, "Legislatures and Statutory Control of the Bureaucracy," *American Journal of Political Science*, vol. 45, no. 2 (2001): 330–45.

Ikenberry, G. John, *Liberal Leviathan: The Origins, Crisis, and Transformation of the American World Order* (Princeton, NJ: Princeton University Press, 2012).

International Telecommunications Union, "Regulation of satellite systems," *ITU*, February 2022, https://www.itu.int/en/mediacentre/backgrounders/Pages/Regulation-of-Satellite-Systems.aspx.

ISS National Laboratory, "In-Orbit Activities: The ISS as a Research Platform," *ISS National Lab*, January 31, 2022, https://www.issnationallab.org/ar2021/in-orbit-activities-iss-research-platform/.

ITU, "Regulation of Satellite Systems," February 2022, https://www.itu.int/en/mediacentre/backgrounders/Pages/Regulation-of-Satellite-Systems.aspx.

Jackson, Shareef, "Ed Dwight Was Going to Be the First African American in Space. Until He Wasn't," *Smithsonian Magazine*, February 18, 2020, https://www.smithsonianmag.com/history/ed-dwight-first-african-american-space-until-wasnt-180974215/.

Jeong, Gyung-Ho, and Paul J. Quirk, "Division at the Water's Edge: The Polarization of Foreign Policy," *American Politics Research*, vol. 47, no. 1 (2019): 58–87.

Jones, Andrew, "China Is Developing Plans for a 13,000-Satellite Megaconstellation," *Space News*, April 21, 2021, https://spacenews.com/china-is-developing-plans-for-a-13000-satellite-communications-megaconstellation/.

Jones, Andrew, "New Chinese Small Sat Manufacturing Capacity Could Have International Ramifications," *Space News*, April 6, 2022, https://spacenews.com/new-chinese-small-sat-manufacturing-capacity-could-have-international-ramifications/.

Jones, Harry W., "The Recent Large Reduction in Space Launch Cost," 48th International Conference on Environmental Systems, July 8–12, 2018, https://ttu-ir.tdl.org/handle/2346/74082.

Kant, Immanuel, *Perpetual Peace and Other Essays*, translated by Ted Humphrey (Indianapolis, IN: Hackett Publishing, 1983).

Kennedy, John F., "Excerpt from the 'Special Message to the Congress on Urgent National Needs,'" *NASA*, May 25, 1961, https://www.nasa.gov/vision/space/features/jfk_speech_text.html.

Kertzer, Joshua D., and Thomas Zeitzoff, "A Bottom-Up Theory of Public Opinion about Foreign Policy," *American Journal of Political Science*, vol. 61, no. 3 (2017): 543–58.

Kingdon, John W., *Agendas, Alternatives, and Public Policies* (Boston: Little, Brown, and Company, 1984).

Klein, John, *Space Warfare: Strategy, Principles and Policy* (New York: Routledge, 2006).

Knipfer, Cody, "Congress and Commerce in the Final Frontier (Part 1)," *The Space Review*, December 10, 2018, https://www.thespacereview.com/article/3619/1.

Krebs, Gunter D., "Orbital Launches of 2018," *Gunter's Space Page*, September 13, 2022, https://space.skyrocket.de/doc_chr/lau2018.htm.

Kurekin, Audrey A. et al., "Operational Monitoring of Illegal Fishing in Ghana through Exploitation of Satellite Earth Observation and AIS Data," *Remote Sensing*, vol. 11, no. 3 (2019).

Lambright, W. Henry, "James E. Webb: A Dominant Force in 20th Century Public Administration," *Public Administration Review*, vol. 53, no. 2 (1993): 95–99.

Lambright, W. Henry, "Relay Leadership and Long-Term Policy: The Case of Mars," *Space Policy*, vol. 42 (2017).

Launius, Roger D., "Public Opinion Polls and Perceptions of US Human Spaceflight," *Space Policy*, vol. 19 (2003): 163–175.

Lazarus, Jeffrey, and Judd R. Thornton, "Bully Pulpit? Twitter Users' Engagement with President Trump's Tweets," *Social Science Computer Review*, vol. 39, no. 5 (2021): 961–980.

Lee, Frances E., and Bruce I. Oppenheimer, *Sizing Up the Senate: The Unequal Consequences of Equal Representation* (Chicago: University of Chicago Press, 1999).

Leib, Karl, "State Sovereignty in Space: Current Models and Possible Futures," *Astropolitics*, vol. 13, no. 1 (2016): 1–24.

Lewis, David E., *Presidents and the Politics of Agency Design* (Stanford, CA: Stanford University Press, 2003).

Lindblom, Charles, "The Science of Muddling Through," *Public Administration Review*, vol. 19 (1959): 79–88.

Lin-Greenberg, Erik, and Theo Milonopoulos, "Boots on the Ground, Eyes in the Sky: How Commercial Satellites Are Upending Conflict," *Foreign Affairs*, May 30, 2022, https://www.foreignaffairs.com/articles/ukraine/2022-05-30/boots-ground-eyes-sky.

Logsdon, John M., *After Apollo? Richard Nixon and the American Space Program* (New York: Palgrave Macmillan, 2015).

Logsdon, John M., *Ronald Reagan and the Space Frontier* (Cham, Switzerland: Palgrave Macmillan, 2019).

MacDonald, Jason A., "Limitation Riders and Congressional Influence over Bureaucratic Policy Decisions," *American Political Science Review*, vol. 104, no. 4 (2010): 766–782.

Machay, Martin, and Alan Steinberg, "Influence of Industry on Legislative Behavior toward NASA," *Astropolitics*, 13 (2015): 205–222.

Machay, Martin, and Alan Steinberg, "NASA Funding in Congress: Money Matters," *European Journal of Business Science and Technology*, vol. 6, no. 1 (2020): 5–20.

Mack, Eric, "SpaceX Starship Launch Proposal Draws Vocal Support, Some Criticism in FAA Hearing," *CNet*, October 18, 2021, https://www.cnet.com/news/spacex-starship-launch-proposal-draws-vocal-support-some-criticism-in-faa-hearing/.

Maher, Neil M., *Apollo in the Age of Aquarius* (Cambridge, MA: Harvard University Press, 2017).

Manchester, Julia, "Hill.TV poll: Majority of Americans approve of 'Space Force,'" *The Hill*, August 8, 2018, https://thehill.com/hilltv/what-americas-thinking/402862-hilltv-poll-majority-of-americans-approve-of-space-force.

Mayhew, David, *Congress: The Electoral Connection*, 2nd ed. (New Haven, CT: Yale University Press, 2004).

McCubbins, Matthew D., and Thomas Schwartz, "Congressional Oversight Overlooked: Police Patrols versus Fire Alarms," *American Journal of Political Science*, 28, no. 1 (1984): 165–179.

McCurdy, Howard E., *Faster, Better, Cheaper: Low-Cost Innovation in the US Space Program* (Baltimore, MD: Johns Hopkins University Press, 2001).

McCurdy, Howard E., *Inside NASA: High Technology and Organizational Change in the US Space Program* (Baltimore, MD: Johns Hopkins University Press, 1993).

McDougall, Walter A., "Technocracy and Statecraft in the Space Age—Toward the History of a Saltation," *The American Historical Review*, vol. 87, no. 4 (1982): 1010–40.

McDougall, Walter A., . . . *The Heavens and the Earth: A Political History of the Space Age* (Baltimore, MD: Johns Hopkins University Press, 1985).

Mellow, Craig, "The Rise and Fall of Iridium," *Air and Space Magazine*, September 2004, https://www.airspacemag.com/space/the-rise-and-fall-and-rise-of-iridium-5615034/.

Miles, Matthew, and Donald Haider-Markel, "Bully Pulpit or Ham Radio? Receptive Audiences and Presidential Messages," *Journal of Political Science*, vol. 41, no. 1 (2013): 7–33.

Mintrom, Michael, "Policy Entrepreneurs and the Diffusion of Innovation," *American Journal of Political Science*, vol. 41, no. 3 (1997): 738–70.

Moe, Terry M., "Control and Feedback in Economic Regulation: The Case of the NLRB," *American Political Science Review*, vol. 79, no. 4 (1985): 1095–116.

Moltz, James Clay, *The Politics of Space Security*, 3rd ed. (Stanford, CA: Stanford University Press, 2019).

Morgan, Daniel, "Commercial Human Spaceflight," *Congressional Research Service*, October 1, 2021, https://crsreports.congress.gov/product/pdf/IF/IF11940.

Morgan, Daniel, "The National Space Council," *Congressional Research Service*, December 12, 2016, https://crsreports.congress.gov/product/pdf/R/R44712/3.

Motel, Seth, "NASA Popularity Still Sky-High," *Pew Research Center*, February 3, 2015: https://www.pewresearch.org/fact-tank/2015/02/03/nasa-popularity-still-sky-high/.

Nadeau, François, "Examining Public Support for Space Exploration Funding in America: A Multivariate Analysis," *Acta Astronautica*, vol. 86 (2013): 158–66.

Nail, Rachel, "NASA Report Says It Has Big Economic Impact on Florida," *Florida Today*, September 25, 2020, https://www.floridatoday.com/story/tech/science/space/2020/09/25/nasa-report-confirms-has-big-economic-impact-florida/3528839001/.

NASA, "Debris Mitigation," *NASA Orbital Debris Program Office*, May 4, 2022, https://orbitaldebris.jsc.nasa.gov/mitigation/.

NASA, "History," *Landsat Science*, May 5, 2022, https://landsat.gsfc.nasa.gov/about/history/.

NASA, *NASA Spinoff*, accessed November 22, 2021, https://spinoff.nasa.gov/.

NASA, "Near-Earth Object Observations Program," *NASA*, May 6, 2022, https://www.nasa.gov/specials/pdco/index.html#neo.

National Space Council, "Renewing America's Proud Legacy of Leadership in Space: Activities of the National Space Council and United States Space Enterprise," *The White House*, January 2021, https://trumpwhitehouse.archives.gov/wp-content/uploads/2021/01/Final-Report-on-the-Activities-of-the-National-Space-Council-01.15.21.pdf.

NBC News, "Elon Musk's SpaceX Sues Government to Protest Military Launch Monopoly," April 25, 2014, https://www.nbcnews.com/science/space/elon-musks-spacex-sues-government-protest-military-launch-monopoly-n89926.

Neustadt, Richard E., *Presidential Power and the Modern Presidents* (New York: Free Press, 1990).

Niskanen, William A., "Bureaucrats and Politicians," *Journal of Law and Econometrics*, vol. 18, no. 3 (1975): 617–43.

Normand, Anna E., "Landsat 9 and the Future of the Sustainable Land Imaging Program," *Congressional Research Service*, November 4, 2021, https://aquadoc.typepad.com/files/crs_report_landsat-9_.future_sustainable_imaging_program_4nov2021pdf.pdf.

Nosek, Brian A., Frederick L. Smyth, N. Sriram, et al., "National Differences in Gender-Science Stereotypes Predict National Sex Differences in Science and Math Achievement," *Proceedings of the National Academies of Science*, vol. 106, no. 26 (2009): 10593–97.

Nye, Joseph S., *The Powers to Lead: Soft, Hard, and Smart* (New York: Oxford University Press, 2008).

Odom, Brian C., and Stephen P. Waring, eds., *NASA and the Long Civil Rights Movement* (Gainesville, FL: University Press of Florida, 2019).

Office of the Director of National Intelligence, *National Security Space Strategy*, January 2011, at https://www.dni.gov/index.php/newsroom/reports-publications/reports-publications-2011/item/620-national-security-space-strategy.

Office of Space Affairs, "Key Topics," *Department of State*, August 26, 2022, https://www.state.gov/key-topics-office-of-space-affairs/.

Older, Maika, "Satellite Surveillance Can Trace Atrocities but Not Stop Them," *Foreign Policy*, January 21, 2020, https://foreignpolicy.com/2020/01/21/sudan-clooney-satellite-surveillance-can-trace-atrocities-but-not-stop-them/.

Open Secrets, "Blue Origin," accessed November 19, 2021, https://www.opensecrets.org/federal-lobbying/clients/summary?cycle=2021&id=D000069501.

Open Secrets, "Boeing Co.," accessed November 20, 2021, https://www.opensecrets.org/orgs/boeing-co/summary?id=d000000100.

Open Secrets, "Industry Profile: Defense Aerospace," accessed November 19, 2021, https://www.opensecrets.org/federal-lobbying/industries/summary?cycle=2020&id=D01.

Open Secrets, "Lockheed Martin," accessed November 20, 2021, https://www.opensecrets.org/orgs/lockheed-martin/summary?id=D000000104.

Open Secrets, "Northrop Grumman," accessed November 20, 2021, https://www.opensecrets.org/orgs/northrop-grumman/summary?id=D000000170.

Open Secrets, "SpaceX," accessed November 19, 2021, https://www.opensecrets.org/orgs/spacex/lobbying?id=D000029147.

Ostrander, Ian, and Joel Sievert, "The Logic of Presidential Signing Statements," *Political Research Quarterly*, 66, no. 1 (2013): 141–53.

Paxton, Larry J. "'Faster, Better, and Cheaper' at NASA: Lessons Learned in Managing and Accepting Risk," *Acta Astronautica* 61 (2007): 954–63.

Powell, Robert, "Guns, Butter, and Anarchy." *American Political Science Review*, vol. 87, no. 1 (1993): 115–132.

Raimondo, Gina M. "The Department of Commerce Budget in Brief: Fiscal Year 2023," *Department of Commerce,* March 2022, https://www.commerce.gov/sites/default/files/2022-03/Commerce-FY2023-BIB-Introduction.pdf.

Rasmussen Reports, "33% Support Creating a National 'Space Force,'" June 22, 2018, https://www.rasmussenreports.com/public_content/politics/general_politics/june_2018/33_support_creating_a_national_space_force.

Reagan, Ronald, "Remarks at the National Forum on Excellence in Education in Indianapolis, Indiana," *The American Presidency Project,* July 21, 2022, https://www.presidency.ucsb.edu/documents/remarks-the-national-forum-excellence-education-indianapolis-indiana.

Redmond, Mike, "Bernie Sanders Is Dragging Elon Musk and Jeff Bezos over Their Billionaire Space Race," *Uproxx,* November 18, 2021, https://uproxx.com/viral/bernie-sanders-elon-musk-jeff-bezos-space-race/.

Reuters, "Russia Warns West: We Can Target Your Satellites," *Reuters,* October 27, 2022, https://www.reuters.com/world/russia-says-wests-commercial-satellites-could-be-targets-2022-10-27/.

Richelson, Jeffrey T., *America's Space Sentinels: The History of the DSP and SBIRS Satellite Systems* (Lawrence, KS: University Press of Kansas, 1999).

Rogowski, Jon C., "Presidential Influence in an Era of Congressional Dominance," *The American Political Science Review,* vol. 110, no. 2 (2016): 325–41.

Roulette, Joey, "At pivotal points in human spaceflight, NASA undergoes major reorganization," *The Verge,* September 21, 2021, https://www.theverge.com/2021/9/21/22685956/nasa-reorganization-spacex-human-spaceflight-heo.

Roulette, Joey, "The Senate Just Advanced the Beef between SpaceX and Blue Origin," *The Verge,* June 9, 2021, https://www.theverge.com/2021/6/9/22457893/jeff-bezos-blue-origin-nasa-spacex-senate-competition-bill-nasa-moon-lander.

Roulette, Joey, "SpaceX Curbed Ukraine's Use of Starlink Internet for Drones—Company President," *Reuters,* February 9, 2023, https://www.reuters.com/business/aerospace-defense/spacex-curbed-ukraines-use-starlink-internet-drones-company-president-2023-02-09/.

Sadat, Mir, and Julia Siegel, "Space Traffic Management: Time for Action," *Atlantic Council,* August 2022, https://atlanticcouncil.org/wp-content/uploads/2022/08/Space-traffic-management_time-for-action.pdf.

Sambaluk, Nicholas Michael, *The Other Space Race: Eisenhower and the Quest for Aerospace Security* (Annapolis, MD: Naval Institute Press, 2015).

Schwartz, James S.J., *The Value of Science in Space Exploration* (New York: Oxford University Press, 2020).

Sheetz, Michael, "The Space Industry Is on Its Way to Reach $1 Trillion in Revenue by 2040, Citi Says," *CNBC,* May 21, 2022, https://www.cnbc.com/2022/05/21/space-industry-is-on-its-way-to-1-trillion-in-revenue-by-2040-citi.html.

Shepsle, Kenneth A., Robert P. Van Houweling, Samuel J. Abrams, and Peter C. Hanson, "The Senate Electoral Cycle and Bicameral Appropriations Politics," *American Political Science Review,* vol. 53, no. 2 (2009): 343–59.

Shetterly, Margot Lee, *Hidden Figures: The American Dream and the Untold Story of the Black Women Mathematicians Who Helped Win the Space Race* (New York: William Morrow, 2016).
Sinclair, Barbara, "The '60-Vote Senate': Strategies, Process, and Outcomes," *US Senate Exceptionalism*, Bruce I. Oppenheimer, ed. (Columbus: Ohio State University Press, 2002).
Smith, Kevin B., "Typologies, Taxonomies, and the Benefits of Policy Classification," *Policy Studies Journal* 30, no. 3 (2002): 379–95.
Sparks, Grace, "Space Force: To 37% and beyond," *CNN*, August 16, 2018, https://www.cnn.com/2018/08/16/politics/space-force-poll/index.html.
Spice, Byron, "Grief Tweets About Mars Rover Sound 'Human,'" *Futurity*, July 30, 2020, https://www.futurity.org/robot-death-tweets-opportunity-rover-2413222-2/.
Spires, David N., *Beyond Horizons. A Half Century of Air Force Space Leadership* (Maxwell AFB, AL: Air University Press, 1998).
Steer, Cassandra, "Global Commons, Cosmic Commons. Implications of Military and Security Uses of Outer Space," *Georgetown Journal of International Affairs* vol. 18, no. 1 (2017): 9–16.
Steinberg, Alan, "Space Policy Responsiveness: The Relationship between Public Opinion and NASA Funding," *Space Policy*, vol. 27, no. 4 (2011): 240–46.
Steptoe, E. Jason, "United States Government Licensing of Commercial Space Activities by Private Enterprise," *Documents on Outer Space Law*, 1985, https://digitalcommons.unl.edu/cgi/viewcontent.cgi?article=1006&context=spacelawdocs.
Stewart Ingersoll, Robert, and Derrick Frazier, *Regional Powers and Security Orders: A Theoretical Framework* (New York: Routledge, 2012).
Sylvester, Roshanna, "John Glenn and the Sexism of the Early Space Program," *Smithsonian Magazine*, December 14, 2016, https://www.smithsonianmag.com/history/even-though-i-am-girl-john-glenns-fan-mail-and-sexism-early-space-program-180961443/.
The Planetary Society, "Your Guide to NASA's Budget," https://www.planetary.org/space-policy/nasa-budget.
The Space Foundation, "Global Space Economy Rose to $447B in 2020, Continuing Five-Year Growth," July 15, 2021, https://www.spacefoundation.org/2021/07/15/global-space-economy-rose-to-447b-in-2020-continuing-five-year-growth/.
Trevithick, Joseph, "China's Historic Mission to the Dark Side of the Moon Is About More Than Science," *The Drive*, January 3, 2019, https://www.thedrive.com/the-war-zone/25781/chinas-historic-mission-to-the-dark-side-of-the-moon-is-about-more-than-science.
Union of Concerned Scientists, "UCS Satellite Database," August 11, 2022, https://www.ucsusa.org/resources/satellite-database.
United Nations Office for Outer Space Affairs, "Treaty on Principles Governing the Activities of States in the Exploration and Use of Outer Space, including the Moon and Other Celestial Bodies," https://www.unoosa.org/oosa/en/ourwork/spacelaw/treaties/introouterspacetreaty.html.
US Army, "About the Command," *US Army Space and Missile Defense Command*, 2022, https://www.smdc.army.mil/ABOUT/.

US Department of Energy, "Energy for Space: Department of Energy's Strategy to Advance American Space Leadership (FY 2021-FY 2031)," January 2021, https://www.energy.gov/sites/prod/files/2021/01/f82/Energy%20for%20Space-DOE%20Space%20Strategy%20Paper%2001-06-2021.pdf.

US Department of Transportation, "About DOT," March 28, 2022, https://www.transportation.gov/about.

US Fleet Cyber Command/US Tenth Fleet, "Strategic Plan, 2020-2025," *US Navy*, 2020, https://www.fcc.navy.mil/Portals/37/FCC_C10F%20Strategic%20Plan%202020-2025.pdf.

US Government Accountability Office, "NASA: Assessment of Major Projects," *GAO*, June 23, 2022, https://www.gao.gov/products/gao-22-105212.

Van Allen, James A., "Is Human Spaceflight Obsolete?" *Issues in Science and Technology*, vol. 20, no. 4 (2004), https://issues.org/p_van_allen/.

Vaughn, Diane, *The Challenger Launch Decision: Risky Technology, Culture, and Deviance at NASA* (Chicago: University of Chicago Press, 2016).

Walker, Jack, "Setting the Agenda in the US Senate: A Theory of Problem Selection," *British Journal of Political Science*, vol. 7, no. 4 (1977): 423-45.

Wall, Mike, "Oregon Congressman Proposes New Space Tourism Tax," *Space.com*, July 22, 2021, https://www.space.com/space-tourism-tax-proposed.

Wang, Ming-Te, and Jessica L. Degol, "Gender Gap in Science, Technology, Engineering, and Mathematics (STEM): Current Knowledge, Implications for Practice, Policy, and Future Directions," *Educational Psychology Review*, vol. 29, no. 1 (2018): 119-40.

Ward, Dan, "Faster, Better, Cheaper Revisited: Program Management Lessons from NASA," *Defense AT&L*, March-April 2010: https://apps.dtic.mil/sti/pdfs/AD1016355.pdf.

Weingast, Barry R., and Mark J. Moran, "Bureaucratic Discretion or Congressional Control? Regulatory Policymaking by the Federal Trade Commission," *The Journal of Political Economy*, vol. 91, no. 5 (1983): 765-800.

Weinzierl, Matthew, "Space, the Final Economic Frontier," *Journal of Economic Perspectives*, vol. 32, no. 2 (2018): 173-92.

Weitekamp, Margaret A., *Right Stuff, Wrong Sex: America's First Women in Space Program* (Baltimore, MD: Johns Hopkins University Press, 2004).

Wendt, Alexander, *The Social Construction of International Politics* (Cambridge: Cambridge University Press, 1992).

White House, "National Cislunar Science and Technology Strategy," November 11, 2022, https://www.whitehouse.gov/wp-content/uploads/2022/11/11-2022-NSTC-National-Cislunar-ST-Strategy.pdf.

White House, "National Security Strategy," November 8, 2022, https://www.whitehouse.gov/wp-content/uploads/2022/11/8-November-Combined-PDF-for-Upload.pdf.

White House, "Space Policy Directive-3, National Space Traffic Management Policy," June 18, 2018, https://trumpwhitehouse.archives.gov/presidential-actions/space-policy-directive-3-national-space-traffic-management-policy/.

White House, "United States Space Priorities Framework," December 2021, https://www.whitehouse.gov/wp-content/uploads/2021/12/united-states-space-priorities-framework-_-december-1-2021.pdf.

Whitman Cobb, Wendy N., "For All (Wo)mankind: Developing a Feminist Space Policy," Paper presented at the annual conference of the Southern Political Science Association, January 2023.

Whitman Cobb, Wendy, "'It's a Trap!' The Pros and Mostly 'Khans' of Science Fiction's Influence on the United States Space Force," *Space Force Journal,* January 31, 2021, https://spaceforcejournal.org/its-a-trap-the-pros-and-mostly-khans-of-science-fictions-influence-on-the-united-states-space-force/.

Whitman Cobb, Wendy N., "Stubborn Stereotypes: Exploring the Gender Gap in Support for Space," *Space Policy,* vol. 54 (2020).

Whitman Cobb, Wendy N., *Unbroken Government: Success and the Illusion of Failure in Policymaking* (New York: Palgrave Macmillan, 2013).

Whitman Cobb, Wendy N., "Who's Supporting Space Activities? An 'Issue Public' for US Space Policy," *Space Policy,* vol. 27 (2011): 234–39.

Wikipedia, "Project Echo," *Wikipedia,* accessed November 17, 2021, https://en.wikipedia.org/wiki/Project_Echo.

Wildavsky, Aaron, *The Politics of the Budgetary Process* (Boston: Little, Brown, 1964).

Wood, B. Dan, and Richard W. Waterman, "The Dynamics of Political Control of the Bureaucracy," *American Political Science Review,* vol. 85, no. 3 (1991): 801–28.

Wyatt, Sally, "Technological Determinism Is Dead, Long Live Technological Determinism," *The Handbook of Science and Technology Studies,* Edward J. Hackett, Olga Amsterdamska, Michael Lynch, and Judy Wajcman eds. (Cambridge, MA: MIT Press, 2008): 165–180.

INDEX

Page numbers in *italics* refer to illustrations.

Advanced Research Projects Agency (ARPA), 39, 119, 145
Agenda setting stage of policymaking, 4–5, 76–77; NASA (National Aeronautics and Space Administration), 126–28
Agnew, Spiro, 88
Allen, Paul, 251
Ansari XPRIZE, 251
Antiballistic missile defenses (ABM), 42; ABM Treaty, 239
Anti-terrorism, 239
Apollo program, 8, 33–35, 51, 77, 123, 181, 204
Apollo-Soyuz mission, 35–37, 225
Arabsat, 231
Ares rocket family, 57
Ariane rocket, 186, 258
Armageddon (film), 203
Armed forces, 145–52. *See also* US Space Force (USSF)
Armstrong, Neil, 32, 34, 213
Army Ballistic Missile Agency (ABMA), 120, 121
Artemis Accords, 162–63, 201, 232, 242, 285
Artemis Moon program, 66, 77, 88, 132–33, 167, 175, 253
ASAT (anti-satellite) capabilities, 43, 61; ASAT tests, 273, 278, 280
Asia Pacific Space Cooperation Organization (APSCO), 240
Assistant secretary of defense for space policy (ASD/SP), 144
Asteroid redirect mission (ARM), 204
Asteroids, 202–4; space resources and mining, 3, 252, 276
AstroAccess, 210

Astronauts: Black Americans, 34, 36; in commercial sphere, 263–64; requirements for, 34; women, 34, 36
Astropolitik (Dolman), 229
Atlantis shuttle, 179
Atlas missiles/launch vehicles, 39, 249, 255, 256
Atomic Energy Act, 168
Atomic Energy Commission, 167
AT&T (telecommunications company), 185
Augustine Commission, 59
Austin, Lloyd, 279
Axiom-1, 67, 179–80, 253
Axiom-2, 253
Axiom Space (commercial space station), 256

Ballistic missiles, 24, 37, 249; antiballistic missile defenses (ABM), 42
Baumgartner, Frank, 7, 135
Bay of Pigs (1961), 32
Berlin Wall, 32, 49
Bezos, Jeff, 58, 60, 188–89, 251, 255, 258
Biden, Joe, 66, 88, 166, 210, 287
Black Americans, 34, 208
BlackSky, 259
Blue Origin: cooperation with NASA, 132–33; lobbying efforts, 189, 190; New Shepard remote landing, 15, 59–60; profit as motivation, 188; reusable rockets, 251; space colonization, 58; space tourism, 169, 255; US responsibility for, 12, 125
Bluford, Guion, 36, 209
Blumenauer, Earl, 193
Boeing (aircraft industry company), 55, 170, 180, 186, 188–89, 190, 249
Bolden, Charles, 86, 210
Branson, Richard, 59–60, 251, 255
Brazil, 286

Burbach, David, 112, 183
Bureaucracy, role of, 116–19, 129, 130–31, 137–38
Bureau of Arms Control, Verification, and Compliance (AVC), 163
Bureau of Economic Analysis (BEA), 165, 166
Bureau of Industry and Security (BIS), 166
Bureau of Oceans and International Environment (OES), 162
Bush, George H. W.: Apollo landing, 20th anniversary, 51, 70; *Challenger* disaster, 88; NASA nominations, 53; National Space Council (NSpC), 87–88; SEI (Space Exploration Initiative), 77, 82, 112, 127; SEI, lack of NASA support for, 135
Bush, George W.: China, condemnation of, 63; NASA nominations, 85; SIG-Space, 87; space shuttle, 109; VSE (Vision for Space Exploration), 57, 70, 82, 84, 112, 187

Canada, 214
Cantwell, Maria, 133
Cape Canaveral, 120
Capitalism, 23, 29, 237. *See also* Commercial space programs; Privatization of space
Carpenter, Daniel, 134
Carr, E. H., 227
Carter, Jimmy, 43, 87
Cassini probe, 214
Central Intelligence Agency (CIA), 31, 153
Challenges to Security in Space (DIA), 282
Chandra X-ray Observatory, 54
Chang'e 4 probe, 243, 283
Chelyabinsk, Russia, 202–3
China: ASAT tests, 63–64, 237, 239–240, 273; Beidou, 207; Chang'e 5 lunar mission, 243; competition/cooperation with US, 8, 55, 67, 75, 101–2, 280–83; intellectual property theft, 63; lunar missions, 283; satellites near space station, 217; space program, 60–61; US agreements with, 238; and world power balance, 224–25
China Academy of Space . . . (CA-SIC), 275
China Satellite Network Group, 275
Civilian-military space divide, 39–41, 80–81, 120
Civilian space policy, 9–10, 11, 51–52, 201–20;
civilian-military balance, 39; and competitiveness, 66–67
Clarke, Arthur C., 16, 257
Clementine lunar probe, 53
Clinton, Bill, 75, 80, 82, 136
Coats, Daniel, 155–56
Cobb, Wendy Whitman, 112
Cold War: Apollo-Soyuz mission, 35–37; competition transitioning to cooperation, 136; immediate post–World War II period, 26–28; legacy of, 48–49, 226, 285; Limited Test Ban Treaty, 42; NASA's role, 127; Outer Space Treaty, 41–44; prestige as motivation, 28–35, 223; security as goal, 37–41, 223; Soviet and US space programs, contrasted, 23–25, 44–46; space race milestones, 25, 34–35; technology, 196–97
Collisions, danger of, 17–18, 274
Columbia shuttle accident (2003), 36, 37, 54, 56–57, 71, 124, 130, 250
Commerce Control List (CCL), 165
Commercial Orbital Transportation Services (COTS), 57, 59, 124–25, 187, 250–51
Commercial policy, 11–12, 179–80, 246–69
Commercial Space Act (1998), 55
Commercial Space Launch Act (1984), 249, 250
Commercial Space Launch Amendments Act (1988), 249
Commercial Space Launch Amendments Act (2004), 251
Commercial Space Launch Competitiveness Act (2015), 261
Commercial space programs: background and history, 54–60, 77, 247–54; commercialization contrasted with privatization, 262–63; commercialization of space, 275–78; communication satellites, 256–58; human spaceflight, 254–56; launch services, 256; overview, 246–47, 268–69; policy challenges, 259–62; power of, 264–65; privatization of space, 262–64; regulations, 260–62; remote imaging, 258–59; security concerns, 265–67
Commons, space as, 215–16
Communication satellites, 248, 256–58
Communications Satellite Corporation (Comsat), 185, 248, 257
Competition over space, 226–30
Conestoga I rocket, 249
Congestion in space, 2, 16–17, 62

Congressional Budget Office (CBO), 173
Constellation Project, 57–59, 90, 252
Constructivism, 223, 225, 233–35, *236*, 241, 243
Cooper, Jim, 65, 96
Copernicus Earth observation, 242
Corona program, 31, 40, 153
COSMIC satellites, 207
Counterterrorism, 239
Cox, Chris, 63
Cox Report, 63
Crew Dragon (SpaceX), 66–67
Cuban Missile Crisis (1962), 34

Deep Impact (film), 203
Deep Space Climate Observatory (DSCOVR), 167
Defense Advanced Research Projects Agency (DARPA), 119
Defense contractors, 188
Defense Intelligence Agency (DIA), 144, 153, 155, 156, 282
Defense Space Strategy (2020), 83–84
Defense Support Program (DSP), 40
Delta launch vehicles/rockets, 249, 256
Department of Commerce, 164–67, 201, 217–18, 262, 275
Department of Defense (DOD): budget, 39; and goals of space program, 32; overview and functions, 143–45; space norms, 280; space shuttle program, 124
Department of Energy (DOE), 167–68
Department of State (DOS), 162–64
Department of Transportation (DOT), 168–72
Diplomacy, space, 162–63, 232–33
Directorate of Defense Trade Controls (DDTC), 163
DirecTV, 257
Discoverer 14 mission, 31
Discovery shuttle, 52
Dish satellite television, 257
Diversity and inclusion, 209–11
DNI (Director of national intelligence), 142, 152, 154
Dolman, Everett, 229, 238
Domestic policy, 7–9
Double Asteroid Redirection Test (DART), 203
Douglas Aircraft Company, 185
Douhet, Emilio, 27
Doves (Planet cubesats), 253

Downs, Anthony, 118
Dragon module, 187, 252, 254
Dream Chaser (spaceplane), 255, 256
Dryden, Hugh, 119
Dwight, Ed, 34
Dyna-Soar space plane, 40
Dynetics, 132–33

Early Bird (satellite), 248
Earth Observation Satellite Company (EO-SAT), 205, 259
Earth Resources Technology Satellite-A (ERTS-A), 204
Earthrise photo, 204
Easton, Richard D., 50
Economic concerns: balanced with security concerns, 268–69; impact of space activity, 187–88, 191–93; science vs. exploration funding, 211–12; in US, 164–66, *165*
Edwards, George, 80
Eisenhower, Dwight, 30, 31–32, 82, 87, 153, 184, 212; National Advisory Committee for Aeronautics (NACA), 119–20; New Look policy, 38
Electromagnetic pulse (EMP), 41
Environmental Science Services Administration (ESSA), 206–7
Eros asteroid, 53
European Space Agency (ESA), 214, 231, 234, 249, 286
European Union (EU), 231, 234; Galileo, 163
Eutelsat, 231
Executive orders (EOs), US presidential, 81–82
Explorer 1, 30–31, 120
Export Administration Regulations (EAR), 165

Falcon 1 rocket, 58, 187, 252
Falcon 9 rocket, 59, 179, 187, 252, 254, 257
FBC (faster, better, cheaper) approach, 53–54, 136
Federal Aviation Administration (FAA), 100, 169–70; launch regulations, 260; Office of Commercial Space Transportation, 251
Federal Communications Commission (FCC), 171–72, 261–62
Firefly Aerospace, 256
Five Eyes (intelligence alliance), 232
Foreign policy, 7–9

Fractional orbital bombardment system (FOBS), 40–41
France, 234
Frazier, Eric F., 50, 284
Funding. *see* Economic concerns

Gagarin, Yuri, 32
Galileo global navigation system, 242
Garver, Lori, 86, 125, 210, 211
Gates, Bill, 55, 186
Gemini spacecraft, 123
Gender gap: in space programs, 208, 209–11, 293n3; women in STEM, 209, 211
General Social Survey (GSS), 181, *182*
GEO (geostationary orbit), 16, 257
Geophysical Year (1957), 29–30, 120, 166
Geospatial intelligence (GEOINT), 155
Geostationary Operational Environmental Satellites (GOES), 207
Glenn, John, 34
Global commons, 215–16
Globalization, 198, 264
Global Positioning System (GPS), 43, 49–50, 65–66, 224
Globalstar, 55
Glushko, Valentin, 28
GNSS-RO (global navigation satellite system radio occultation), 207
Goddard, Robert, 28
Goddard Space Flight Center, 120
Goldin, Dan, 52, 53, 135–36
Gore, Al, 88–89
Government Accountability Office (GAO), 104–5, 133, 172, 174–76
Gravity (film), 17–18
Griffin, Mike, 57, 124–25, 250
Grumman Corporation, 185. *See also* Northrop Grumman (aerospace company)
Gulf War (1990/91), 44, 50, 62, 63, 223, 237, 286

Hamilton, Margaret, 34
Hidden Figures (film), 34
The High Frontier (O'Neill), 58
Hitchens, Theresa, 232
Honeywell Corporation, 185
Hubble Space Telescope, 52–53, 212
Human spaceflight: commercial space programs, 254–56; contrasted with robotic exploration, 212–13; funding and expense, 211–12; landing systems, 132, 175; privatization of, 263–64
Hussein, Saddam, 50
Huygens probe, 214

IBM (technology corporation), 185
ICBMs (intercontinental ballistic missiles), 30, 38
IC Commercial Space Council, 156
Incrementalism, 6–7
India: ASAT tests, 273, 280; space program, 61
Indian Space Research Organization (ISRO), 61
Individualism, 25
Inspiration 4 mission, 67, 179, 253
Intelligence community, 152–57
Intelligence Reform and Terrorism Prevention Act (2004), 152
Intermediate range ballistic missiles (IRBMs), 38
International Code of Conduct for Outer Space Activities, 235
International cooperation, 213–15
International Geophysical Year (1957), 29–30, 120, 166
International law, 264–65
International relations: overview, 222–26; in post–Cold War period, 237–40; shared meanings of space, 233–37; space as cooperative environment, 230–33; space as zero-sum environment, 226–30; theoretical frameworks for future work, *236*, 240–43
International Space Station (ISS), 36, 52; ISS National Laboratory, 54; SpaceX, 66–67
International Telecommunications Satellite Consortium (Intelsat), 248
International Telecommunications Union (ITU), 17
International Traffic in Arms Regulations (ITAR), 163–64, 188, 261
Internet services, 257
Iran, 234
Iraq, 239
Iridium, 55, 186
Iron Curtain, 32
Iron triangle concept, 106
Isaacman, Jared, 254
ISAM (in-space servicing, assembly, and manufacturing), 171

Israel, 234
ISR (intelligence, surveillance, and reconnaissance) functions, 49

Jackson, Mary, 34
James Webb Space Telescope, 54, 123, 174, 214
Japan Aerospace Exploration Agency (JAXA), 241, 273
Jet Propulsion Laboratory (JPL), 120
Johnson, Katherine, 34
Johnson, Lyndon B., 87, 96
Johnson-Freese, Joan, 232
Johnson Space Center, 120
Joint Force, 151–52
Joint Polar Satellite System (JPSS), 167
Jones, Bryan, 7, 135
Jupiter missile, 30, 39, 120

Kant, Immanuel, 232
Kennedy, John F.: Apollo program, 75, 77; Cold War, 31; as Democrat, 112; impact of space policies, 89–90; inaugural address, 25; May 1961 speech, 33, 70, 135, 285; weather observation system, support for, 206
Kennedy Space Center, 120, 192, 260
Kessler effect/syndrome, 17–18, 273–74
Khrushchev, Nikita, 29, 38–39
Klein, John, 238
Korolev, Sergei, 28
Kuiper satellite constellation, 258

Lambright, Henry, 135
Land Remote Sensing Policy Act (1992), 205, 250
Landsat satellites, 204–6, 248, 250, 258–59
Launch services: launch and landing sites, 260–61; launch vehicles/launch costs, 55–56; private investments in, 186–87; privatization of, 256; regulatory framework, 186; reusable systems, 58
Launius, Roger, 181
Leadership: contrasted with power, 286; indicators of, 284–85
Leadership in Space Act (2021), 287
Legislation, role of, 99–100
Lenin, Vladimir, 23, 26
LEO (low Earth orbit), 16, 62, 125, 252, 257–58; space debris collisions, 273

Liability and Registration Conventions, 19–20, 231
Liberalism, 223, 225 230–32, *236*
Lobbying, 54, 105, 189–90
Lockheed Martin (aerospace company), 55, 170, 185, 186, 188, 189, 190, 249
Luna 15, 34

MacDonald, Jason, 101
Machay, Martin, 98, 112, 192
Magnetosphere, 41
Maher, Neil, 210
Manned Spaceflight Center, Texas, 120
Manual on International Law Applicable to Military Activities in Outer Space (MILAMOS), 236
Mars Climate Orbiter, 53
Mars *Curiosity* mission, 54
Marshall Space Flight Center, 121–22
Mars *Pathfinder* mission, 53, 136
Mars *Perseverance* mission, 54
Mars Polar Lander, 53
Mars rover *Opportunity,* 212–13
Mars Surveyor, 53
Maxar (company), 247, 259
Mayhew, David, 97
McCubbins, Matthew, 93
McCurdy, Howard, 122, 136
McDougall, Walter, 26
McNamara, Robert, 33
MEO (medium Earth orbit), 16, 257
Mercury spacecraft, 123; Mercury 13 pilots, 34
Military Orbital Development System (MODS), 40
Military Orbital Laboratory (MOL), 40
Military space policy, 9–10, 11, 49–51, 141–43; civilian-military balance, 39
Mineral mining, 3, 252, 276
Minuteman missile, 39
Missile Defense Alarm System (MIDAS), 31, 40
Missile gap, US and Soviet Union, 24, 34, 153, 228
Mississippi Test Facility, 120
MIT Instrumentation Laboratory, 185
Moltz, James Clay, 45, 196, 229, 235
Monroe, James, 33, 82
Moon exploration: China's program, 281; Moon landings, 33, 42; Soviet attempts, 34–35. *See also* Apollo program

318 · Index

Moon Treaty, 20, 277
Morgenthau, Hans, 227
Motorola (telecommunications company), 55
Munitions List, 188
Musk, Elon, 58, 188, 189, 191, 251, 261, 266

Nadeau, François, 183
Nanoracks (in-space services company), 256
NASA (National Aeronautics and Space Administration): administrator's role, 135–36; authorization bills, 100; budget and funding, 6–7, *9*, 57, 174, *175,* 183, 193, 209; and bureaucracy, 116–19, 129, 130–31, 136–37; Commercial Crew Program, 254, 255; contracting process/disputes, 132–33; cultural impact of, 116; engineering and research, 247–48; FBC approach, 53–54, 136; formation of, 31–32, 39, 99, 119–22; goals and agenda, 32–33; mission failures, 53–54; and new technology, 192; organizational structure and tension, 122–25; partnership with USGS, 204–6; policy tools, 125–33; role of, 219; significance of in policymaking, 134–37; Space Exploration Initiative (SEI), 51–52, 53; Space Launch System (SLS), 66
National Defense Authorization Act (NDAA), 143
National Advisory Committee for Aeronautics (NACA), 119–20, 145, 219
National Aeronautics and Space Act (1958), 86–87, 99
National Aeronautics and Space Council (NASC), 87
National Cislunar Science and Technology Strategy (2022), 142
National Defense Authorization Act, 100
National Defense Strategy (NDS), 83, 143–44
National Institute of Standards and Technology (NIST), 166
National Nuclear Security Administration (NNSA), 168
National Oceanographic and Aeronautical Administration (NOAA), 166–67, 201
National Reconnaissance Office (NRO), 31, 64, 153–54
National Security Council (NSC), 38, 87, 142
National Security Space Strategy (2011), 83–84, 241
National Security Strategy (NSS), 83

National Space Council (NSpC), 86–89, 142, 287
National Space Policy (2010), 241
National Spaceport Intergovernmental Working Group, 170, 171
National Telecommunications and Information Administration (NTIA), 166
Navigation System with Timing and Ranging (NAVSTAR), 43
Navy Research Lab, 30
Near Earth Asteroid Rendezvous (NEAR), 53
Nelson, Bill, 86, 179, 283
Neoliberalism, 231
NEO (near-Earth objects) Observations Program, 202–4
New Glenn rocket, 60, 255, 256
New Look policy, Eisenhower administration, 38
New Shepard system (Blue Origin), 15, 60, 255
Nigeria, 286
9/11 terror attack, 64–65
Nixon, Richard, 36, 76; Space Task Group, 87, 88
Non-state actors. *see* Commercial space programs
North American Aviation, 185, 248
Northrop Grumman (aerospace company), 185, 188, 189, 190, 256
Nuclear fuel, 168
Nuclear Posture Reviews (NPR), 83
Nuclear weapons/arms race, 24, 30, 37–38, 41

Obama, Barack: China, negotiations with, 64, 65; commercialization, shift to, 77, 125, 179; Constellation program, 90, 108–9; and executive power, 81, 90; global leadership, shift to, 240–41; human spaceflight, approaches to, 59, 66, 110; NASA appointments, 86, 210; SIG-Space, 87
Office of Commercial Space Transportation (AST), 169–70
Office of Information and Regulatory Affairs (OIRA), 173
Office of Management and Budget (OMB), 84
Office of Science and Technology Policy (OSTP), 64, 87
Office of Secretary of Defense (OSD), 144
Office of Space Commerce, 165–66, 218
Office of Technology Assessment (OTA), 105

O'Keefe, Sean, 85
OMB (Office of Management and Budget), 172–74
O'Neill, Gerard, 58
OneWeb (satellite Internet service company), 60, 247, 257–58
Operation Paperclip, 27
Opportunity, Mars rover, 212–13
Orbital Reef space station, 60, 256
Orbital Sciences Corporation, 124
Orbiting requirements and distances, 16–17
Orion vehicle, 57
Outer Space Treaty (OST), 12, 17, 19–20, 41–44, 231, 252, 264–65
Overflight, principle of, 31, 38, 40, 153

Parker Solar Probe, 54
PE (punctuated equilibrium), 7
Pegasus rocket, 124
Peloponnesian War, 227
Pence, Mike, 66
Perseverance, Mars rover, 213
Persian Gulf War (1990/91), 44, 50, 62, 63, 223, 237, 286
Planet (satellite provider), 60, 188, 199, 247, 253, 259
Planetary defense, 202–4. *See also* Security concerns
Polaris program, 254
Polar Operational Environmental Satellites (POES), 207
Policymaking: and bureaucracy, 116–19; civilian space policy, 51–52; civilian space policy contrasted with military, 80–81; commercial space programs, 259–62; incrementalism, 6–7; military space policy, 49–51; multiple agency involvement, 160–61; punctuated equilibrium (PE), 7; shift to commercial providers, 179–80. *See also* Civilian space policy; Commercial policy; Domestic policy; Military space policy; Public policy
Popular culture and depictions of space, 140
Position, navigation, and timing (PNT) systems, 166
Precision-guided munitions (PGMs), 49–50
Presidential role, US, 71–74, 76–81, 287
Prestige as motivation during Cold War, 28–30, 35, 45, 228–29, 235

Privatization of space, 12, 262–64; contrasted with commercialization, 262–63
Project Artemis, 66, 77, 88, 132–33
Project Constellation, 57–59
Project Mercury, 121
Public opinion, 180–184
The Public Papers of the President, 78–79, *79*
Public policy: definitions and terminology, 10–14; policy change, 6–7; policymaking, stages of, 4–6; Presidential role in setting, 71–74; scientific aspects, 13–14; types of policy, 710
Punctuated equilibrium (PE), 135
Putin, Vladimir, 61, 239

Quayle, Dan, 53

R-7 missile, 39
Racial issues, 208–9
Radio crowding, 17
Reagan, Ronald, 43–44, 48, 77, 80, 82, 112; Cold War, 127; space shuttle, 249
Realism, 223, 227–28, 229, *236,* 237–38
Reconnaissance, 40, 49, 153, 154
Redstone missile, 120
Registration Convention (1974), 20
Relativity Space (aerospace company), 256
Remote imaging, 258–259, 267
Rescue Agreement (1968), 19
Resource exploitation, 276–78. *See also* mineral mining
Ride, Sally, 36, 209, 293n3
Robots, 122–23, 212–13
Rocket Lab, 256
Rockets: and private industry, 185–86; von Braun's work, 120
Rogers, Mike, 65, 96
Roosevelt, Franklin D., 89
Rumsfeld, Donald, 63, 85
Rumsfeld report, 63, 64
Russia: ASAT tests, 273, 280; commercial interests post-Soviet era, 55; cooperation with US, 179, 238; GLONASS, 207; ISS cooperation, 214; post-Cold War space program, 52, 61; Roscosmos, 238; Ukraine invasion, 258, 265–66
Rutan, Burt, 59

Salyut space station, 35–36, 48
Samos satellite program, 40

Sanctuary, space as, 2, 32, 48, 62–66, 223, 237–38
Sanders, Bernie, 193
Satellites: anti-satellite (ASAT) capabilities, 43; commercial uses, 253; communication systems, 248, 256–58; and data collection, 31; geostationary (GOES network), 166–67; and GPS (Global Positioning System), 49–50; intelligence gathering, 38, 40, 258; launch costs, 205; LEO constellations, 187; microsatellites, 275; military uses, 49–50; Navigation System with Timing and Ranging (NAVSTAR), 43; orbiting requirements and distances, 16–17; public vs. private purposes, 185, 188–89; rocket technology, 30; satellite providers, 60; Teledesic constellation, 186; weather observation, 163, 185, 206–8
Satellite Sentinel Project, 211
Saturn V rocket, 66, 123, 131, 185
Scaled Composites, 251
Schriever, Bernard, 39
Schwartz, Thomas, 93
Scientific exploration/research, 13–14, 211–12
Security concerns: after Outer Space Treaty (OST), 42–44; balanced with economic concerns, 268–69; *Challenges to Security in Space,* 282; commercial space programs, 265–67; and data collection, 31; as goal during Cold War, 37–41; politics of space security, 45; security, collective, 231–32; security, need for, 226–27; and space debris, 274
Senior Interagency Group on Space (SIG-Space), 87
September 11 attack, 64–65
Sexual harassment, 210
Shepard, Alan, 32
Shotwell, Gwynne, 210
Shuttle systems, 36–37, 124; *Atlantis,* 179; Columbia shuttle accident (2003), 36, 37, 54, 56–57; launching services, 250; private actors, 249; SST (single-stage-to-orbit) rockets, 195
Sierra Nevada Corporation, 255
SiriusXM, 186, 257
Skylab, 36
Social interactionism, 235
Sojourner rover, 53
Soviet Union: missile gap with US, 228; Russian Revolution (1917), 23; space program, contrasted with US, 23–25. *See also* Cold War; Russia
Space: as competitive environment, 226–30; as cooperative environment, 230–33; dangers, protection from, 155, 202–4; as global commons, 215–16; importance of, 2–4; privatization of, 262–64; as sanctuary, 2, 32, 48, 62–66, 223, 237–38; shared meanings of, 233–37; space norms, 278–80; space science, 52–54; as zero-sum environment, 226–30
Space and Missile Defense Command (SMDC), 146–47
Space-Based Infrared System (SBIRS), 40
Space Command, 65
Space debris, 17–18, 48, 216, 217, 272–74; mitigation measures, 18
Space Exploration Initiative (SEI), 51–52, 53, 70–71, 77, 127–28
Space industry, 184–91; infrastructure in southern US, 33, 75–76, 170–71
Space Launch System (SLS), 66
Space National Guard, 173
Space planes, 40
Space policy: domestic vs. foreign, 98–99; military space, 141–43
Space Policy Directives (SPDs), Trump administration, 82-83, 166
Space programs: economic growth from, 75–76; expenses, 209; human spaceflight, focus on, 121, 122–23; physical and technological requirements, 14–18; ubiquity of, 175–77
Space Resource Exploration and Utilization Act (2015), 276
Space Services Incorporated, 248–49
SpaceShipOne, 251–52, 254, 255
Space shuttle. *See* Shuttle systems
Space situational awareness (SSA), 215–18, 262
Space Station Freedom, 52
Space Station Intergovernmental Agreement (1998), 238
Space stations. *See* International Space Station (ISS)
Space Task Group, 87
Space tourism, 169–70, 255, 261
Space traffic management (STM), 166, 215–18, 275–77
Space Transportation System (STS), 36
SpaceX: Artemis program, 175; COTS awards, 251–52; Crew Dragon launch, 66–67;

Dragon cargo capsule, 254; Falcon 1, 187; Falcon 9 recovery, 15; ISS launches, 179–80; launch systems/sites, 254–55, 260–61; lobbying efforts, 189; NASA, contracts with, 132–33; political donations and public campaigns, 190–91; power of, 198–99; and profit motive, 188; and reusability, 58–59, 125; small satellites in LEO, 275; Starliner, 255; Starlink, 16, 217, 266; Starship system, 253; technology development, 246
SPOT (French satellite), 259
Sputnik, 30
SpyMeSat, 259
Stalin, Joseph, 23, 29
Starfish Prime test, 41
Starliner, 255, 256
Starlink system, 257, 258
Starship system, 253
Star Wars (SDI), 77. *See also* Strategic Defense Initiative (SDI)
Steer, Cassandra, 235–36
Steinberg, Alan, 97–98, 112, 183, 192–93
Stennis Space Center, 120
Stewart Ingersoll, Robert, 284
Strategic arms limitations (SALT), 42
Strategic Defense Initiative (SDI), 43–44, 49, 77–78

Technology: and space policy, 193–98; technological determinism, 196–98
Teledesic (satellite internet constellation), 55, 186
Television Infrared Observation Satellite (TIROS-1), 206
Telstar satellites, 41, 185
Tereshkova, Valentina, 34, 293n3
Thor-Delta, 185
Thor missile, 39
Thucydides, 227
Titan missile, 39
Treaty on the Prevention of the Placement of Weapons . . . (PPWT), 235
Truly, Richard, 86
Truman, Harry, 89, 135
Truman Doctrine (1947), 30
Trump, Donald, 65; Artemis program, 75, 77; and human spaceflight, 66; National Space Council (NSpC), 88, 287; Space Force, 78–79, 149–50; SPD-3, 166, 217; SPDs (Space Policy Directives), 82
Tsiolkovsky, Konstantin, 26, 28
Twitter, 189

U-2 (high-altitude plane), 38
UNCOPUOS, 20, 41, 225, 279, 285
United Arab Emirates (UAE), 282
United Kingdom, 234
United Launch Alliance (ULA), 56, 170, 186, 189, 252, 256
United Nations: Committee on the Peaceful Uses of Outer Space (UNCOPUOS), 20, 41, 225, 279, 285; Disarmament Commission, 20; General Assembly, 287; General Assembly and Outer Space Treaty, 42
United States: budget process, 6–7, 84–85, 95, 101–2, 111–12; civilian component of space exploration, 31–32, 37; competition vs. cooperation, 240–42; Congress, House contrasted with Senate, 110–11; Congress, motivations of, 96–99; Congress, policy tools, 99–106; Congress, role in policymaking, 71–73, 93–96, 113–14; Congress, significance of, 106–8; Congress and authorization bills, 100, 111–12; Congressional hearings, 102–5, *103, 104;* Congressional vs. Presidential roles, 108–9; leadership of in space, 283–88; National Space Council, 86–89; polarization of government, 73–74, 80, 100, 111–13; policymaking and policy tools, 76–86; Presidential motivations in policymaking, 74–76; Presidential role in policymaking, 71–74; as superpower, 223–24, 237–38, 239; vice presidential role, 88–89
United States Space Priorities Framework (2021), 142
US Air Force (USAF), 39–41, 64–65, 148–49, 252; Air Force Space Command, 43
US Army, 146–47
US Commercial Space Launch Competitiveness Act (2015), 252
US Geological Survey (USGS), 204
US Marine Corps Forces Space Command, 147
US Munitions List (USML), 163–64
US Navy, 147–48
US Northern Command, 65
USSPACECOM, 43, 64–65, 147, 150, 152, 174
US Space Command, 65

US Space Force (USSF), 65, 78–79, 96; and commercial companies, 253; functions and mission, 149–51; public opinion, 181–83
US Space Priorities Framework, 287
US Strategic Command, 151

Van Allen, James, 212–13
Van Allen radiation belt, 41
Vandenberg Space Force Base, 260
Vanguard program, 30, 120
Vaughan, Dorothy, 34
Vergeltungswaffe 2 (V-2) rockets, 26–27, 120
Virgin Galactic, 59–60, 247, 251; SpaceShipOne, 254; space tourism, 255
Vision for Space Exploration (VSE), 57, 70–71, 77, 84, 124–25, 187
von Braun, Werner, 27–28, 120
Voyager 1 and 2, 285
Vulcan rocket, 256

Wallops facility, Virginia, 260
Waltz, Kenneth, 227
Ward, Dan, 136

War on Terror, 239
The Washington Post, 189
Weaponization of space, accusations of, 11
Weapons of mass destruction, 233
Weather Research and Innovation Act (2017), 207–8
Weather satellites, 206–8
Webb, James, 33, 135
Weber, Max, 117
Wendt, Alex, 233
Whitman Cobb, Wendy, 183
Wilson, Woodrow, 23
Wolf, Frank, 101–2
Wolf Amendment, 64, 102, 281
WorldVu, 257–58
WS-117L satellite program, 40

XM radio, 186

Yutu 2 rover, 283

Zero-sum frameworks, 228–30, 239–40, 283

Wendy N. Whitman Cobb, PhD, is professor of strategy and security studies at the School of Advanced Air and Space Studies. Her work has been published in journals including *Space Policy, Astropolitics, Congress and the Presidency,* and *Strategic Studies Quarterly.* Her most recent book is *Privatizing Peace: How Commerce Can Reduce Conflict in Space.*

Derrick V. Frazier, PhD, is professor of strategy and security studies and deputy commandant at the School of Advanced Air and Space Studies. He has published in a variety of peer-reviewed journals including *Defense Studies, International Politics,* the *European Journal of International Relations, International Interactions, International Negotiation,* and *Foreign Policy Analysis.* His most recent book is *Regional Powers and Security Orders,* coauthored with Robert Stewart Ingersoll.

Printed in the USA
CPSIA information can be obtained
at www.ICGtesting.com
LVHW042119300324
775949LV00002B/194